# Statistical Mechanics

**Bipin K. Agarwal**
University of Allahabad
Allahabad, India

**Melvin Eisner**
University of Houston
Houston, Texas, U.S.A.

JOHN WILEY & SONS
New York Chichester Brisbane Toronto Singapore

First Published in 1988
Reprinted in 1989
WILEY EASTERN LIMITED
4835/24 Ansari Road, Daryaganj
New Delhi 110 002, India

**Distributors:**

*Australia and New Zealand*
JACARANDA WILEY LTD.
GPO Box 859, Brisbane, Queensland 4001, Australia

*Canada:*
JOHN WILEY & SONS CANADA LIMITED
22 Worcester Road, Rexdale, Ontario, Canada

*Europe and Africa:*
JOHN WILEY & SONS LIMITED
Baffins Lane, Chichester, West Sussex, England

*South East Asia:*
JOHN WILEY & SONS, INC.
05-04, Block B, Union Industrial Building
37 Jalan Pemimpin, Singapore 2057

*Africa and South Asia:*
WILEY EASTERN LIMITED
4835/24 Ansari Road, Daryaganj
New Delhi 110 002, India

*North and South America and rest of the world:*
JOHN WILEY & SONS, INC.
605 Third Avenue, New York, NY 10158 USA



**Library of Congress Cataloging-in-Publication Data**

Agarwal, B.K. (Bipin Kumar), 1931 –
Statistical mechanics.

Bibliography: p.
1. Statistical mechanics. I. Eisner, Melvin.

II. Title.

QC175.A33 1988 530.1'3 87-10859

**ISBN 0-470-20866-X John Wiley & Sons, Inc.**
**ISBN 81-224-0022-1 Wiley Eastern Limited**

Printed in India at Rajbandhu Industrial Co., Mayapuri, New Delhi-110 064.

# PREFACE

This book introduces the beginners to the basic philosophy and technique of Statistical Mechanics as first formulated by J W Gibbs. It is a powerful method that calculates and relates different kinds of things in almost all branches of physics.

Statistical Mechanics is an extremely fascinating field of study. However, it presents difficulties to both the reader and the teacher. We felt that a simple and straightforward book on it is very much needed. Our colleagues in other universities confirmed this. We then decided to give a final shape to the original working draft of this book.

Recently there has been an explosion of new ideas and techniques to solve problems that were intractable so far. The relationship between mechanics and statistical mechanics is also now understood a little better. Some idea of the thrust of modern concerns is given by our treatment of critical phenomena. It exposes the reader to the concepts of scaling, universality and renormalization.

We are grateful to Dr. Ranjana Prakash and Dr. V.P. Verma for the help given in preparing the corrections for reprinting the book.

We welcome any suggestions for improvement in the text.

B.K.A.
M.E.

March 1989

# CONTENTS

# 1
# BASIS OF CLASSICAL STATISTICAL MECHANICS

## 1.1 INTRODUCTION

In statistical mechanics we study the physical systems consisting of very large number of particles ($N \sim 10^{23}$). The simplest physical system of interest is a perfect gas in thermal equilibrium. From the macroscopic point of view it appears to be a continuum. A complete set of thermodynamic variables, characterizing its equilibrium state, is the energy $E$, volume $V$ and the number of molecules $N$. The $N$, although referred to molecules for convenience, is a macroscopic variable because it is directly related to the mass of the gas.

From the microscopic point of view the gas (matter) consists of discrete particles, like atoms or molecules. In classical mechanics, the microscopic description will specify, at a given time, the positions and velocities (or momenta) of all the particles in the gas. It is impossible to measure them instantaneously as $N \sim 10^{23}$. With lapse of time, the description of the behaviour of the gas will require solving an enormous number of equations of motion involving collisions. To get out of this impasse, we can try to relate the macroscopic description based on a few variables with the microscopic description based on a large number of variables, by using the method of (i) kinetic theory, or (ii) statistical mechanics.

The drastic reduction in the number of variables occurs because a measurement of the macroscopic property, like pressure $P$, gives an average of the values over a finite time interval ($\sim$ 1 s). During this period the molecules undergo a very large number of collisions as the time interval between two successive collisions is of the order of $10^{-10}$s. The mathematical process of averaging over a coordinate obviously eliminates it, resulting in simplicity. For example, consider the kinetic theory calculation of pressure of a gas in a cubical box of side length $L$.

Let $c_{xi}$ be the velocity component along the edge parallel to the $x$ axis for the $i$th molecule of mass $m$. The rate of momentum transfer to a wall normal to the $x$ axis is given by (number of collisions per s) $\times$ (momentum imparted per collision) $= (c_{xi}/2L) \times (2mc_{xi})$. Therefore, the pressure $P$ exerted by the molecules on that wall is

$$P = \frac{\text{total force on the wall}}{\text{area of the wall}} = \frac{m(c_{x1}^2 + \ldots + c_{xN}^2)}{L^3}$$

$$= \frac{1}{L^3} mN\overline{c_x^2}, \qquad \overline{c_x^2} = \frac{1}{N}\sum_{i=1}^{N} c_{xi}^2,$$

where $\overline{c_x^2}$ is the average value of $c_x^2$ for all the molecules. Thus $N(\sim 10^{23})$ velocity coordinates are reduced to a single suitably averaged coordinate.

In kinetic theory certain basic assumptions are made regarding the nature of molecules and their mutual interactions. Statistical mechanics does not concern itself with such details as it deals mainly with the energy aspects of the molecules. It makes assumptions of a more general nature, uses the theory of probability and is mathematically simpler. We shall discuss the method of statistical mechanics as first formulated by Josiah Willard Gibbs.

## 1.2 PHASE SPACE

First consider a very simple case. A bead of mass $m$ moves freely and arbitrarily on a string stretched along the $x$ axis. It has one degree of freedom. The position of the bead at time $t$ is $x(t)$ and its velocity $v_x = \dot{x}$ (or momentum $p_x = m\dot{x}$) at that instant. The state of the bead at any instant can be represented by a point $P$ in a hypothetical two-dimensional space, called the *phase space*, whose coordinates are $x$ and $p_x$. As the bead moves on the string, the value of $x$ changes. Under accelerating forces, $p_x$ also changes. As a result the point $P$ traces a trajectory in the phase space with the passage of time (Fig. 1.1).

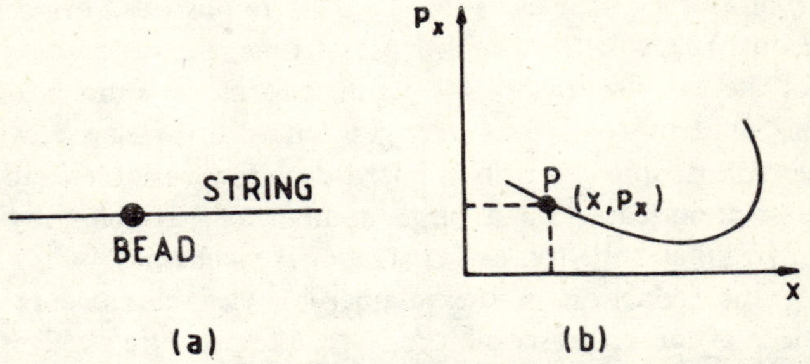


**Fig. 1.1 (a) A bead sliding on a ring. (b) Phase space and phase line for the bead.**

A molecule of an ideal gas can be represented as a structureless particle. Such a molecule has three translational degrees of freedom. Its phase space has six dimensions whose Cartesian coordinates are $x_1, x_2, x_3, p_1, p_2, p_3$. It is called the *$\mu$-space*, where $\mu$ stands for molecule. The instantaneous translational state of the molecule is given by the representative point in this hypothetical space. For a *system* of $N$ molecules (gas) the instantaneous state (Fig. 1.2a) is represented by a set of $N$ points in the $\mu$-space, one for each molecule (Fig. 1.2b). It is a symbolic picture of the space because it is not possible to display a six-dimensional space. The total number of translational degrees of freedom is $3 \times N = 3N$. Following Ehrenfest, we can construct a *phase space*, for all the molecules, which has $6N$ dimensions. It is called the $\Gamma$ *space*, where $\Gamma$ stands for gas. It is spanned by $3N$ coordinate axes and $3N$ momentum axes. The $6N$ coordinates

$$(x_{11}, x_{21}, x_{31}, \ldots, x_{1N}, x_{2N}, x_{3N}, p_{11}, p_{21}, p_{31}, \ldots, p_{1N}, p_{2N}, p_{3N})$$

represent the positions and momenta of all the molecules (state of the system) at a given time. In the $\Gamma$ space, the instantaneous state of the whole system (gas of $N$ molecules) is given by a single *representative point* (or *phase point*), Fig. 1.2c. The notation $[x]$, $[p]$ stands for the $3N$ coordinate axes and $3N$ momentum axes.

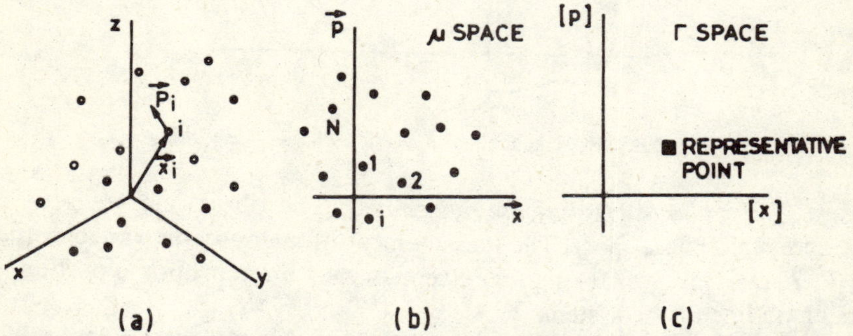


**Fig. 1.2** **(a) A gas containing $N$ molecules (system). (b) $\mu$ space for the system. (c) $\Gamma$ space and the representative point for the entire system.**

In general, if $f$ independent position coordinates and $f$ momentum coordinates are required to fully specify the state of a system, then the system is said to possess *$f$ degrees of freedom*. Any set of $f$ generalized coordinates $q_1, q_2, \ldots, q_f$ (Cartesian, polar or some other convenient set) can be used to uniquely determine the configuration of the system. The corresponding generalized momenta are $p_1, p_2, \ldots, p_f$. The $\Gamma$ space is then a conceptual Euclidean space having $2f$ rectangular axes $[q]$, $[p]$. The microscopic state of the whole system is specified by a representative point in this space. With the lapse of time, some or all of the $2f$ coordinates take on different values (Fig. 1.3). As a result, the representative point traces a *phase line* (or *phase trajectory*) in the accessible phase space (Fig. 1.4). Each point on the phase line represents one such possible microscopic state. A point in the phase space is accessible if it corresponds to the physical specification of the system under observation. For example, the states of the crystalline form of sodium are inaccessible at very high temperature. The system is likely to pass through all the accessible states. In this sense, the $2f$ coordinates take on *all* possible

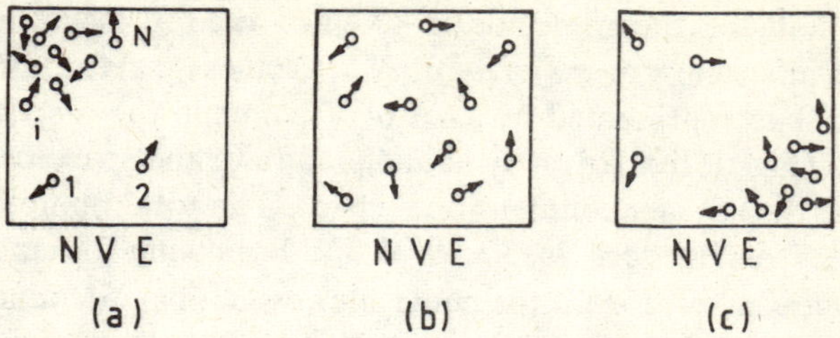


**Fig. 1.3** A few possible states of the system (gas containing $N$ molecules). For convenience only few molecules are shown.

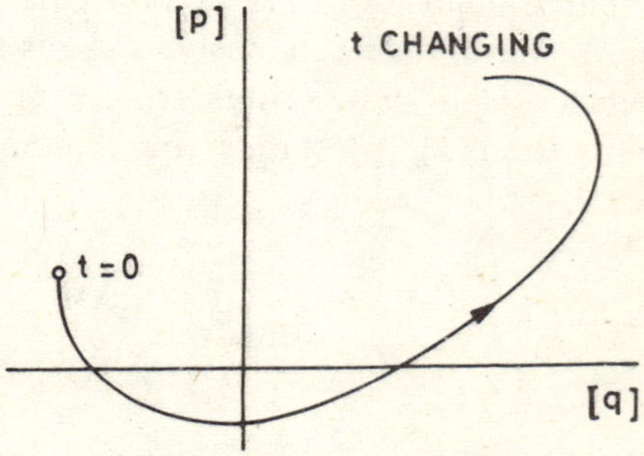


**Fig. 1.4** Phase space and a portion of the phase line.

values. We can say that they are *randomized.* The phase line tends to fill the accessible phase space. The measurement of macroscopic variables (like $P$, $V$, $T$, etc.) involves taking time averages over an appropriate portion of the phase line of the system.

So far we have not introduced any concept of statistical mechanics. The problem of solving an enormous number of equations of motion and of calculating the time averages of interest is still with us. Around 1900 Gibbs suggested that a way out is to introduce the idea of an *ensemble* of systems.

## 1.3 ENSEMBLE

Each phase point on the phase line of a single system develops out of the previous point in time, according to the laws of mechanics. Gibbs replaced this time dependent picture by a static picture in which the *entire phase line exists at one time* (Fig. 1.5 a). Then each phase point represents a separate system with the same macroscopic properties ($N$, $V$, $E$) as the system of interest but a different microscopic state. In other words, we 'imagine' a large number $M$ ($M \to \infty$) of systems, similar in structure to the system of interest but suitably *randomized* in the accessible, unobservable, microscopic states. Instead of taking the time average, we take an average over this artificially constructed group existing simultaneously at one time. Such a group of replicas or *collection of similar, noninteracting, independent, imagined systems* is called an *ensemble* by Gibbs (Fig. 1.5b). We have assumed that *the*

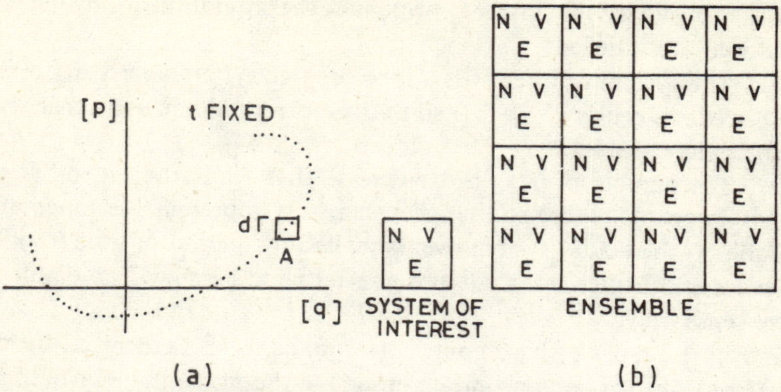


**Fig. 1.5** **(a) The ensemble (a small portion) at one time. Any region like A will appear to contain a swarm of phase points. (b) Schematic 'lattice' representation of an ensemble of $M$ ($M \to \infty$) imaginary systems, at one time, each with same $N$, $V$, $E$.**

*time average of some property of a system in equilibrium is same as the instantaneous ensemble average.* This is known as the *ergodic hypothesis.**

All the members of an ensemble, which are identical in features like $N$, $V$, $E$, are referred to as *elements*. These elements, though identical in structure (same macroscopic state), are randomized in the sense that they differ from one another in the coordinates and momenta of the individual molecules, that is, the elements differ in their unobservable microscopic *states*. The various elements, being imaginary, do not interact with each other. Each element behaves independently in accordance with the laws of mechanics (classical or quantum).

A clear difference exists between the actual system of interest and an element of the ensemble. The system is the physical object about which we intend to make predictions. The elements of the ensemble are mental copies of it to enable us to use the probability theory.

Thus an ensemble of systems consists of randomized 'mental' pictures of the system of interest that exist simultaneously. It is to be viewed as an intellectual exercise to imitate and represent at one time the states of the actual system as developed in the course of time. It is easier to compute the statistical behaviour of such a suitably chosen ensemble than to study the behaviour of any particular complex system. Results so obtained enable us to predict the probable behaviour of the system of interest.

An *ensemble average* is the average at a fixed time over all the elements in an ensemble. It is difficult to prove the exact equivalence of the ensemble average and the time average over a single system. However, one can hope

*The erogodic hypothesis is not rigorously true. One of the elements may behave in a strange fashion and upset it. However, experiments show that if an element is chosen at random, the odds are overwhelming that it has the same time average properties as predicted by the ensemble average.

that the former would closely approximate the latter if the following essential conditions are satisfied:

1. The system of interest is a macroscopic system consisting of a large number of molecules ($N \to \infty$) so that we can randomize in a true sense the microscopic variables.

2. The number of imagined elements that form the ensemble at one time is large ($M \to \infty$) so that they can truly represent the range of states available to the actual system over a really long period of time ($t \to \infty$). In statistical mechanics we shall use the terms system and ensemble in this above sense only.

In Fig. 1.5a, each phase point corresponds to an element in the ensemble (Fig. 1.5b). In an appropriate ensemble the phase points would be distributed continuously.

## 1.4 ENSEMBLE AVERAGE

### (A) Average Values

Consider the simple case of a set of $N$ points distributed arbitrarily along a line. If $x(i)$ is the distance of the $i$th point from the origin, then the average distance $\bar{x}$ from the origin is given by

$$\bar{x} = \frac{1}{N} \sum_{i=1}^{N} x(i).$$

If the line is divided into cells, and $N_i$ is the number of points in the $i$th cell located at $x(i)$, then we can write

$$\bar{x} = \frac{1}{N} \sum_{i} N_i\, x(i). \tag{1.1}$$

If the distribution is known in the form of a continuous function $N(x)$,

$$\bar{x} = \frac{1}{N} \int_{-\infty}^{+\infty} x\, N(x)\, dx, \qquad N = \int_{-\infty}^{+\infty} N(x)\, dx. \tag{1.2}$$

In general, if $R(x)$ is any arbitrary property of the points,

$$\bar{R} = \frac{\int_{-\infty}^{+\infty} R(x)\, N(x)\, dx}{\int_{-\infty}^{+\infty} N(x)\, dx}. \tag{1.3}$$

Generalization to higher dimensions is straightforward.

### (B) Density of Distribution in the Phase Space

The use of ensembles in statistical mechanics is guided by the following factors:

1. The aim is to know only the number of systems or elements that would be found in different states, that is, in different regions of the $\Gamma$ space,

at any time. All the elements being similar in structure, we need not distinguish between them.

2. The number of elements in the ensemble is so large ($M \to \infty$) that there is a continuous change in their number in passing from one region of the phase space to another.

We can, therefore, describe the condition of an ensemble by a density $D$ with which the phase points are distributed in the $\Gamma$ space. It is called the *distribution function* (or *density of distribution* or *probability density*).

In an ensemble of systems of $f$ degrees of freedom, $D$ is a function of $2f$ position and momentum coordinates $q_1, q_2, \ldots, q_f, p_1, p_2, \ldots, p_f$ which correspond to the $2f$ axes in the phase space. It can also depend on time $t$ explicitly. The reason is that although we are free to fix the distribution at any given time $t_0$, we have as yet no assurance for the distribution to remain same. If it remains same, the particular distribution would be one of equilibrium. We shall discuss this later on. Thus, in general,

$$D = D(q_1, \ldots, q_f, p_1, \ldots, p_f, t) \equiv D(q, p, t). \tag{1.4}$$

Consider a small region $A$ of the $\Gamma$ space such that the position coordinates lie between $q_1$ and $q_1 + dq_1, \ldots, q_f$ and $q_f + dq_f$, and the momenta lie between $p_1$ and $p_1 + dp_1, \ldots, p_f$ and $p_f + dp_f$ (Fig. 1.5a). The hypervolume of this region is

$$d\Gamma = dq_1 \ldots dq_f \, dp_1 \ldots dp_f \equiv dq dp. \tag{1.5}$$

By the definition of density, the number of systems or elements $dM$ lying in the specified infinitesimal region situated at the phase point $q_1, \ldots, p_f$ at the instant $t$ is

$$dM = D(q, p, t) \, d\Gamma. \tag{1.6}$$

If $M$ is the total number of elements in the phase space, then at every instant $t$,

$$M = \int D \, d\Gamma, \tag{1.7}$$

where the integration is over the whole phase space.

As suggested by (1.3), the *ensemble average* of a quantity $R(q, p)$ is defined by

$$\bar{R} = \frac{\int R(q, p) \, D(q, p, t) \, d\Gamma}{\int D(q, p, t) \, d\Gamma} = \frac{1}{M} \int R(q, p) \, D(q, p, t) \, d\Gamma. \tag{1.8}$$

If a system is selected at random from the ensemble, the probability of selecting one whose phase point lies in the small region at the point $q_1, \ldots, p_f$ is simply $\rho d\Gamma$, where

$$\rho = \frac{D}{\int D \, d\Gamma} = \frac{D}{M}, \quad \int \rho \, d\Gamma = 1. \tag{1.9}$$

In terms of $\rho\,(q, p, t)$, called the *normalized density of distribution*,

$$\bar{R} = \frac{\int R(q, p)\,\rho(q, p, t)\,dq\,dp}{\int \rho(q, p, t)\,dq\,dp} = \int R\,\rho\,d\Gamma. \tag{1.10}$$

The ensemble average (1.10) gives the average value of the physical quantity $R$ for the actual system of interest.

The macroscopic average properties (like $N, V, E$) of a system in thermodynamic equilibrium do not change with time. Therefore, our ensemble representing it must be such that the ensemble averages are time independent. This is a reasonable requirement. It follows that to construct a suitable ensemble we should study the behaviour of $\rho$ (or $D$) with time.

## 1.5 LIOUVILLE THEOREM

In 1838, Liouville showed, in another connection, how to use the classical Hamilton equations of motion,

$$\dot{q}_i = \frac{\partial H}{\partial p_i}, \qquad \dot{p}_i = -\frac{\partial H}{\partial q_i}, \qquad (i = 1, 2, ..., f), \tag{1.11}$$

where $H$ is the Hamiltonian (total energy expressed as a function of $q$'s and $p$'s), to obtain a statement about $dD/dt$. A knowledge of $H$ at every point in the phase space yields the element of the trajectory passing through the point, (1.11). The uniqueness theorem for systems of ordinary, nonlinear differential equations* implies that through every phase point in $\Gamma$ space there passes one and only one trajectory uniquely determined by (1.11). Consequently *no two trajectories can ever cross in* $\Gamma$ *space.*

Consider at any point $q_1, \ldots, q_f, p_1, \ldots, p_f$ situated in the $\Gamma$ space a small region of hypervolume $d\Gamma = dq_1 \ldots dq_f\,dp_1 \ldots dp_f$ (Fig. 1.5a). At any instant, the number $dM$ of phase points in this region is given by (1.6),

$$dM = D(q, p, t)\,dq_1 \ldots dp_f. \tag{1.12}$$

This number will, in general, change with time due to the flow of phase points. The change will occur when the number of phase points entering the hypervolume through any one face is different from the number leaving the opposite face.

Let us consider two faces normal to the $q_1$ axis with coordinates $q_1$ and $q_1 + dq_1$ (Fig. 1.6). The number of phase points entering the first face ($q_1$ = constant) in time $dt$ is given by

$$D\cdot(\dot{q}_1\,dt)\cdot(dq_2 \ldots dq_f\,dp_1 \ldots dp_f),$$

*R. Courant, *Differential and Integral Calculus*, (E.J. McShane, transl.), Vol II, Blackie, London, 1937, p. 454.

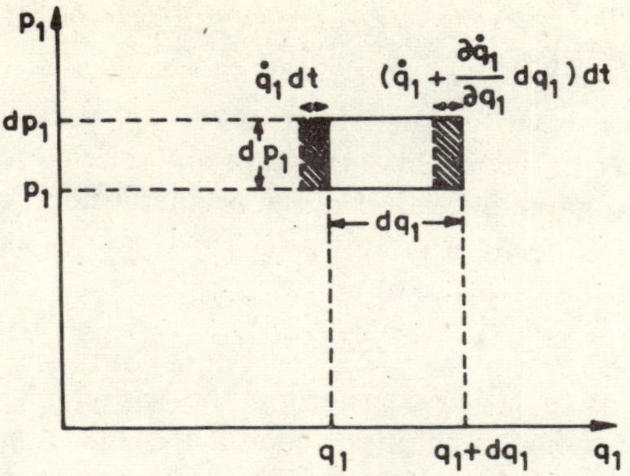


**Fig. 1.6 A fixed volume of (two-dimensional) phase space.**

where $\dot{q}_1$ is the component of velocity, in the direction of $q_1$ axis, of representative points at $q_1, \ldots, p_f$. The number of phase points leaving the opposite face ($q_1 + dq_1 =$ constant) in time $dt$ is

$$\left(D + \frac{\partial D}{\partial q_1} dq_1\right) \cdot \left(\dot{q}_1 + \frac{\partial \dot{q}_1}{\partial q_1} dq_1\right) dt . (dq_2 \ldots dq_f \, dp_1 \ldots dp_f),$$

after neglecting the higher differentials. Subtracting the latter expression from the former and again neglecting the second order differential, we get the net number of phase points entering the hypervolume $d\Gamma$ in time $dt$ as

$$-\left(D \frac{\partial \dot{q}_1}{\partial q_1} + \frac{\partial D}{\partial q_1} \dot{q}_1\right) dt \, d\Gamma.$$

A similar expression exists for the $p_1$ coordinate,

$$-\left(D \frac{\partial \dot{p}_1}{\partial p_1} + \frac{\partial D}{\partial p_1} \dot{p}_1\right) dt \, d\Gamma.$$

The total rate of change with time in the number of phase points $\partial (dM)/\partial t$ in region $d\Gamma$ is obtained by summing the net numbers of phase points entering the hypervolume through all the faces labelled by $q_1, \ldots, p_f$,

$$\frac{\partial (dM)}{\partial t} dt = -\sum_{i=1}^{f} \left[ D \left(\frac{\partial \dot{q}_i}{\partial q_i} + \frac{\partial \dot{p}_i}{\partial p_i}\right) + \left(\frac{\partial D}{\partial q_i} \dot{q}_i + \frac{\partial D}{\partial p_i} \dot{p}_i\right)\right] dt \, d\Gamma. \quad (1.13)$$

From equations of motion (1.11),

$$\frac{\partial \dot{q}_i}{\partial q_i} + \frac{\partial \dot{p}_i}{\partial p_i} = \frac{\partial^2 H}{\partial q_i \, \partial p_i} - \frac{\partial^2 H}{\partial p_i \, \partial q_i} = 0. \quad (1.14)$$

From (1.12–14), we get for the rate of change of density $\partial D/\partial t$ at the fixed phase point $(q, p)$ under consideration,

$$\left(\frac{\partial D}{\partial t}\right)_{q, p} = -\sum_{i=1}^{f} \left(\frac{\partial D}{\partial q_i} \dot{q}_i + \frac{\partial D}{\partial p_i} \dot{p}_i\right). \quad (1.15)$$

If we take into consideration the full dependence of the density $D(q, p, t)$ on the coordinates, momenta and time, (1.15) can be expressed as

$$\frac{dD}{dt} \equiv \left(\frac{\partial D}{\partial t}\right)_{q, p} + \sum_{i=1}^{f} \frac{\partial D}{\partial q_i} \frac{dq_i}{dt} + \sum_{i=1}^{f} \frac{\partial D}{\partial p_i} \frac{dp_i}{dt} = 0. \tag{1.16}$$

Here $dD/dt$ is the total time derivative of $D(q_1, \ldots, p_f, t)$. It gives the rate of change of $D$ in the neighbourhood of any selected moving phase point ($q$'s and $p$'s changing), instead of in the neighbourhood of a fixed point in the $\Gamma$ space. The relation (1.15) or (1.16) is known as *Liouville theorem.* Using (1.9), we can write (1.16) as

$$d\rho/dt = 0. \tag{1.17}$$

The simplicity of (1.15) depends on the use of (1.14), which in turn depends on the choice of conjugate coordinates and momenta for constructing the $\Gamma$ space. Use of velocities in place of momenta would have led to a more complex result.

From (1.16) we note that $D$ can vary with time under two separate conditions. (1) There is an explicit dependence on time, $(\partial D/\partial t)_{q, p}$. The density can vary with time at a given point in $\Gamma$ space. (2) There is an implicit dependence as some or all of the coordinates of the system vary with time, and the phase point wanders in the $\Gamma$ space. This implicit dependence of the density in the vicinity of a selected moving phase point is described by the two terms under the summation signs in (1.16).

The time rate of change of $D$ due to changes in $q_i$ alone is given by $\partial D/\partial q_i$ multiplied by the component of velocity in $q_i$ direction, that is, by $(\partial D/\partial q_i)(dq_i/dt)$. If we consider changes in all $q$'s,

$$\sum_{i=1}^{f} \frac{\partial D}{\partial q_i} \frac{dq_i}{dt} = \left(\frac{\partial D}{\partial t}\right)_{p, t},$$

where the suffixes suggest that the variation is being considered with respect to $q$. Similarly,

$$\sum_{i=1}^{f} \frac{\partial D}{\partial p_i} \frac{dp_i}{dt} = \left(\frac{\partial D}{\partial t}\right)_{q, t}$$

and (1.16) can be written as

$$\frac{dD}{dt} = \left(\frac{\partial D}{\partial t}\right)_{q, p} + \left(\frac{\partial D}{\partial t}\right)_{p, t} + \left(\frac{\partial D}{\partial t}\right)_{q, t} = 0.$$

This clearly shows that *the total rate of change of density $d\rho/dt$ in the vicinity of any selected phase point of a system, as it moves through the $\Gamma$ space, is zero.* Following Gibbs, this is known as the *principle of the conservation of density in phase.* It implies that the density of a group of phase points remains constant along their trajectories in the $\Gamma$ space, that is, it does not disperse. The distribution of representative points moves in phase space like an incompressible fluid.

We can use Liouville theorem to construct distribution functions that are independent of time. For example, if at any time the phase points are distributed uniformly in the $\Gamma$ space, they will have uniform density for ever.

As a group of phase points move away from a given region of Γ space, a different group would move in, and because they keep the same density, the density in the given region would not change. Thus there is no crowding together of the phase points into some favoured regions of Γ space. Later on we shall construct more useful and realistic stationary distributions.

## 1.6 CONSERVATION OF EXTENSION IN PHASE

We can use (1.16) to obtain one more important principle of statistical mechanics. Consider a small region of hypervolume $\Delta\Gamma$ in the phase space which is small enough for the density $D$ (or $\rho$) to be taken as uniform throughout its extension. From (1.12, 16),

$$\Delta M = D\Delta\Gamma,$$

$$\frac{d(\Delta M)}{dt} = \frac{dD}{dt}\Delta\Gamma + D\frac{d(\Delta\Gamma)}{dt} = D\frac{d(\Delta\Gamma)}{dt}. \tag{1.18}$$

Here $\Delta\Gamma$ is the hypervolume of a closed region bounded by a $(2f-1)$-dimensional hypersurface in a $2f$-dimensional phase space (Fig. 1.7). The phase points lying on this hypersurface form a movable boundary which changes its shape and moves about in the Γ space due to the 'flow' of the phase points. The phase points can neither enter nor leave through this boundary because with every point on each phase line a definite phase velocity is associated.

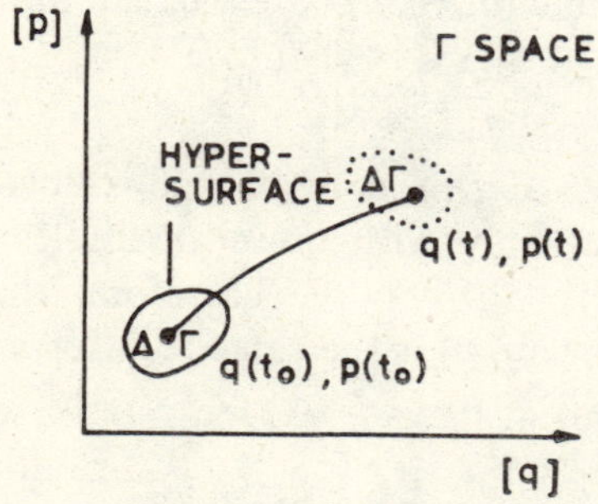


Fig. 1.7 Motion of phase volume in Γ space.

The phase points on the boundary form a kind of continuous skin which permanently encloses the phase points of the hypervolume $\Delta\Gamma$ under consideration. Also, these points can neither be created nor destroyed, because each point represents a definite element in the ensemble. Consequently, the number $\Delta M$ of phase points enclosed in the region $\Delta\Gamma$ must remain constant, $d(\Delta M)/dt = 0$, and (1.18) becomes

$$\frac{d(\Delta\Gamma)}{dt} = 0. \tag{1.19}$$

This means that the volume $\Delta\Gamma$, or extension-in-phase in $\Gamma$ space, bounded by a moving surface and containing a definite number of phase points, does not change with time in spite of the displacements and distortions. Every finite arbitrary extension-in-phase can be regarded as composed of infinitesimal parts, and so this result can be generalized. Following Gibbs, the result expressed in (1.19) is called the *principle of conservation of extension-in-phase.*

## 1.7 EQUATION OF MOTION AND LIOUVILLE THEOREM

The *Poisson bracket* $\{a, b\}$ is defined by

$$\{a, b\} = \sum_i \left( \frac{\partial a}{\partial q_i} \frac{\partial b}{\partial p_i} - \frac{\partial a}{\partial p_i} \frac{\partial b}{\partial q_i} \right). \tag{1.20}$$

The canonical equations of motion (1.11) and total time derivative $dD/dt$ can be written as

$$\{q_i, H\} = \frac{\partial H}{\partial p_i} = \dot{q}_i, \ \{p_i, H\} = -\frac{\partial H}{\partial q_i} = \dot{p}_i, \tag{1.21}$$

$$\begin{aligned} \frac{dD}{dt} &= \frac{\partial D}{\partial t} + \sum_i \frac{\partial D}{\partial q_i}\frac{dq_i}{dt} + \sum_i \frac{\partial D}{\partial p_i}\frac{dp_i}{dt} \\ &= \frac{\partial D}{\partial t} + \sum_i \left( \frac{\partial D}{\partial q_i}\frac{\partial H}{\partial p_i} - \frac{\partial D}{\partial p_i}\frac{\partial H}{\partial q_i} \right) \\ &= \frac{\partial D}{\partial t} + \{D, H\}. \end{aligned} \tag{1.22}$$

The last result is the equation of motion in terms of Poisson bracket.

The Liouville theorem (1.16), $dD/dt = 0$, can now be expressed as

$$\frac{\partial D}{\partial t} = -\{D, H\}. \tag{1.23}$$

The implicit dependence of $D(\dot{q}, p, t)$ on time is thus given by the Poisson bracket. We shall deal only with those ensembles for which $D$ does not depend on time explicitly, $\partial D/\partial t = 0$. Therefore, the *stationary ensemble* described by $D(q, p)$ or $\rho(q, p)$ is the same for all times,

$$\partial\rho/\partial t = 0, \ \{\rho(q, p), H\} = 0. \tag{1.24}$$

## 1.8 EQUAL A PRIORI PROBABILITY

In the ensemble we have large number of elements (replicas of the original system) distributed in different possible (accessible) microscopic states but characterized by the same macroscopic variables $N$, $V$, $E$. This is all the knowledge we can claim about them. To this extent we have no reason to prefer one microscopic state over the other.

A fundamental postulate of statistical mechanics is that *a macroscopic system in equilibrium is equally likely to be in any of its accessible microscopic states satisfying the macroscopic conditions of the system.* It is called *the postulate of equal a priori probability.* We have no direct proof for it. It is reasonable and does not contradict any known laws of mechanics. It leads to results that agree with the observations.

In classical mechanics every point in the phase space represents a possible microscopic state of the system. Therefore, it is reasonable to say that the number of states in a given region of phase space is proportional to the hypervolume $d\Gamma = dq_1 \dots dq_f\, dp_1 \dots dp_f$ of that region. As classically the possible states form a continuum, the constant of proportionality cannot be fixed. (In the quantum theory we can think of the phase space as subdivided into cells of volume $h^f$ each, where $h$ is Planck's constant. The constant of proportionality involves $h$ as a result of quantal picture of discrete states).

## 1.9 STATISTICAL EQUILIBRIUM

A statistical ensemble is defined by the distribution function $\rho$ which characterizes it. In general, there are seven constant independent additive integrals of motion in mechanics: the energy, the three components of the momentum vector, and the three components of the angular momentum vector. Usually energy is the only constant known. For a large system the total momentum and angular momentum have zero value or can be reduced to zero by a suitable choice of the coordinate system. We shall therefore consider only those ensembles that are functions of the energy and so useful in thermodynamics.

The energy $E$ is a constant of the motion for a conservative system. Let us take $\rho$ as a function of energy which in turn can be expressed as a function of $q$ and $p$,

$$\rho = \rho(E), \tag{1.25}$$

$$\frac{\partial \rho}{\partial q_i} = \frac{d\rho}{dE}\frac{\partial E}{\partial q_i}, \qquad \frac{\partial \rho}{\partial p_i} = \frac{d\rho}{dE}\frac{\partial E}{\partial p_i}. \tag{1.26}$$

The Liouville theorem (1.15) becomes

$$\left(\frac{\partial \rho}{\partial t}\right)_{q,p} = -\frac{\partial \rho}{\partial E}\sum_i \left(\frac{\partial E}{\partial q_i}\dot{q}_i + \frac{\partial E}{\partial p_i}\dot{p}_i\right). \tag{1.27}$$

By hypothesis, $E = E(q, p)$, and $dE/dt = 0$, so that

$$\frac{dE}{dt} = \sum_i \left(\frac{\partial E}{\partial q_i}\dot{q}_i + \frac{\partial E}{\partial p_i}\dot{p}_i\right) = 0. \tag{1.28}$$

From (1.27, 28),

$$\left(\frac{\partial \rho}{\partial t}\right)_{q,p} = 0, \tag{1.29}$$

in agreement with the condition (1.24) for the stationary ensemble. Obviously, for such an ensemble

$$\{\rho\,(E), H\} = 0. \tag{1.30}$$

Thus, an ensemble characterized by (1.25) is in *statistical equilibrium.* Such ensembles enable us to apply statistical mechanics to thermodynamics, where we are interested in a system for which the total energy $H(q, p) = E$ is conserved,

$$E\,(q_1, \ldots, p_f) = \text{constant}. \tag{1.31}$$

Locus of phase points corresponding to (1.31) forms a $(2f-1)$-dimensional hypersurface, called an *energy surface* or *ergodic surface*, in the $\Gamma$ space (Fig. 1.8a). We can imagine a family of such energy surfaces constructed in the $\Gamma$ space (Fig. 1.8b). Each energy surface divides the phase space in two parts, one of lower and the other of higher energy. Clearly, two surfaces of constant energy cannot intersect.

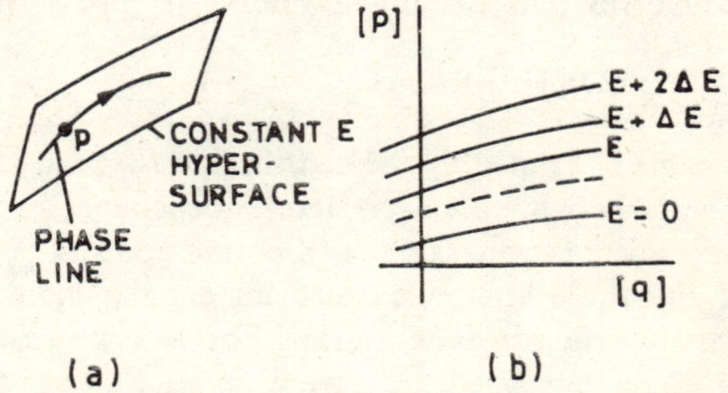


**Fig. 1.8 (a) Ergodic surface. (b) Ergodic surfaces in the $2f$-dimensional phase space.**

The representative point of a conservative system remains always on the same ergodic surface. The ensemble for a conservative system at one time will populate one such ergodic surface. Values of $q$, $p$ have not been specified in (1.29). Therefore, (1.29) will hold good for any point in the $\Gamma$ space.

## 1.10 MICROCANONICAL ENSEMBLE

An ensemble defined on the ergodic surface (1.31) satisfies the condition (1.29) of statistical equilibrium for any $\rho$ which is a function of energy alone. A simple choice is

$$\begin{aligned} \rho &= \text{constant} \qquad \text{for } E = E_0 \\ &= 0 \qquad\qquad\quad \text{otherwise.} \end{aligned} \tag{1.32}$$

The ensemble characterized by this density distribution is called the *microcanonical ensemble*. It can also be expressed as (Fig. 1.9a)

$$\rho(E) = \text{constant} \times \delta\,(E - E_0). \tag{1.33}$$

The $\delta$ is the Dirac delta function with the property

$$f(x') = \int_{-\infty}^{+\infty} \delta\,(x - x')\, f(x)\, dx, \tag{1.34}$$

where $f(x)$ is an arbitrary well-behaved function. In particular,

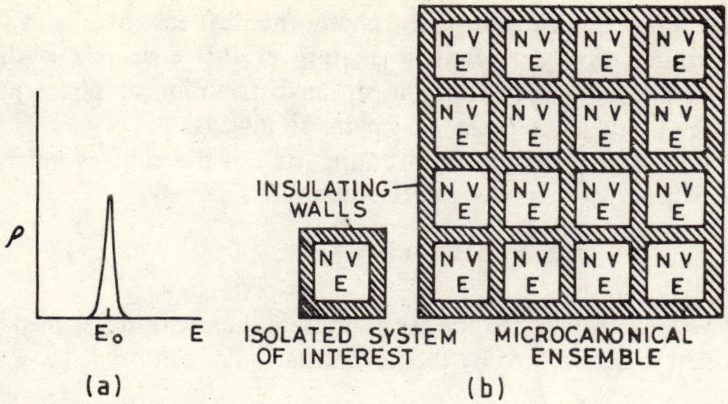


**Fig. 1.9** **(a) Density distribution for the microcanonical ensemble. (b) Schematic lattice representation of the microcanonical ensemble. The shaded walls are nonconducting for heat and impermeable to all molecules.**

$$\int_{-\infty}^{+\infty} \delta\,(x-x')\,dx = 1.$$

Effectively, $\delta\,(x - x') = 0$ except at $x = x'$.

The microcanonical ensemble is appropriate for an *isolated* system $(N, V, E)$, because the energy of an isolated system is constant. Schematically it is shown in Fig. 1.9b.

We cannot specify exactly the energy of a system. However, we can certainly specify the energy within some narrow range, say, between $E$ and $E + \Delta E$. We can then select two neighbouring ergodic surfaces, one at $E$ and the other at $E + \Delta E$ (Fig. 1.10). The phase point of a conservative system remains always on the same ergodic surface. Consequently, phase points lying within the shell will always remain in it. The microcanonical ensemble can now be defined as

$$\begin{aligned} \rho &= \text{constant} \qquad \text{in the range } E \text{ to } E + \Delta E \\ &= 0 \qquad \text{outside this range.} \end{aligned} \tag{1.35}$$

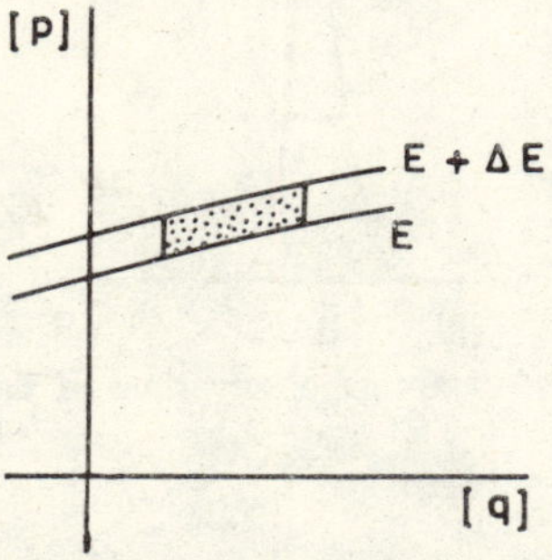


**Fig. 1.10** **The microcanonical ensemble representing an isolated system (gas contained in a finite volume) with energy between $E$ and $E + \Delta E$. The shaded region represents the accessible portion of the phase space. It corresponds to a uniform distribution of representative points bounded by the ergodic surfaces of constant energy $E$ and $E + \Delta E$ and the surfaces corresponding to the physical boundary of the container.**

As $\rho$ is a function of energy, the microcanonical ensemble is in statistical equilibrium. The characteristic feature of this ensemble is that $\rho$ is constant within the energy shell and so the distribution of phase points is uniform; by Liouville theorem, it remains so always.

For an isolated system in equilibrium, each of the allowed microscopic states is realized with equal a priori probability, namely,

$$\rho(P) = \text{constant} = \left[ \int_{E<E(P)<E+\Delta E} d\Gamma \right]^{-1}, \qquad (1.36)$$

where $P = (q, p)$ belongs to the set of phase points within the shell (1.35). The constant is determined by the normalization condition

$$\int \rho \, d\Gamma = 1, \qquad \rho \, d\Gamma = \text{probability}. \qquad (1.37)$$

The microcanonical ensemble, defined by the distribution (1.36), forms the basis for statistical mechanics.

If some physical quantity $R$ is measured relating to an isolated system, the distribution of our observations will follow a curve of the type shown in Fig. 1.11. We claim that the sharp maximum of the curve will fall at the ensemble average $\bar{R}$, that is,

$$R_{\text{obs}} \cong \bar{R} = \int R(P) \, \rho(P) \, d\Gamma, \qquad (1.38)$$

where $\rho(P)$ is given by (1.36). The system of interest spends most of its time in the region of $\Gamma$ space for which $R \cong \bar{R}$. This illustrates an important feature of statistical mechanics. It is assumed that the time interval between two successive observations is longer than the time required for a fluctuation in the properties of the system to die out (relaxation time).

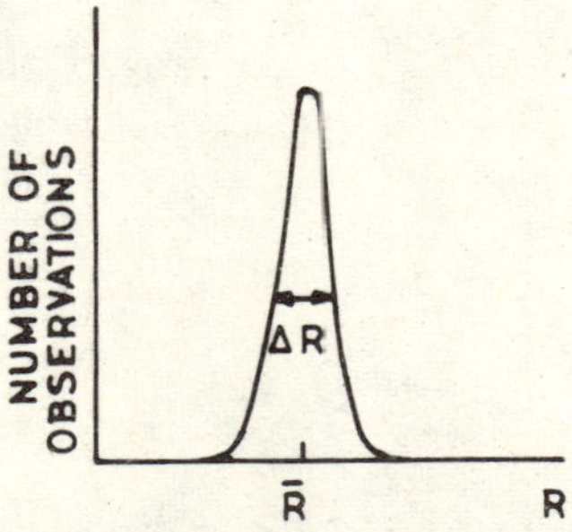


**Fig. 1.11** Expected distribution of observations of the physical quantity $R$.

## 1.11 IDEAL GAS

To illustrate the use of microcanonical ensemble let us consider an ideal gas. It is a system of $N$ noninteracting point particles of equal mass enclosed in a cubical box of side length $L$ with perfectly elastic walls. Let the edges form the rectangular coordinate axes $x_1$, $x_2$, $x_3$.

Suppose the $i$th particle has $p_{1i}$, $p_{2i}$, $p_{3i}$ as the three components of momentum. The pressure $P_1$ exerted by the molecules on the wall normal to the $x_1$ axis is $\sum_i p_{1i}^2/mL^3$.

We construct a microcanonical ensemble and calculate the ensemble average $\overline{P}_1$,

$$\overline{P}_1 = \frac{\frac{1}{mL^3}\int \sum_{i=1}^{N} p_{1i}^2 \rho d\Gamma}{\int \rho d\Gamma}, \tag{1.39}$$

where $d\Gamma = dq\, dp$ and $\rho$ = constant in the region of phase space specified by $E < \sum_i p_i^2/2m < E + \Delta E$. We can put the average of $\sum_i p_{1i}^2$, $\sum_i p_{2i}^2$, $\sum_i p_{3i}^2$ as equal to each other. Then $\overline{P}_1 = \overline{P}_2 = \overline{P}_3 = P$ and (1.39) gives

$$P = \left[\frac{1}{3mV}\int \sum_{i=1}^{N} p_i^2\, dq\, dp\right] \Big/ \int dq\, dp, \tag{1.40}$$

where $V = L^3$.

In the limit $\Delta E \rightarrow 0$, we can replace $\sum_i p_i^2$ by the constant factor $2mE$ to get

$$PV = \frac{2}{3}E. \tag{1.41}$$

The ideal gas law is $PV = nRT = NkT$ where $n$ is the number of moles of gas, $R$ the gas constant for one mole, $N$ the number of molecules, $k$ the Boltzmann constant, and $T$ the absolute temperature. Comparing it with (1.41), we obtain

$$E = \frac{3}{2}NkT. \tag{1.42}$$

This result is expected from heat capacity determinations.

## *PROBLEMS*

1.1 A system can with equal probability be in any of its $N$ states. What is the probability of the system being in one of its states?

1.2 A two-dimensional vector **A** of given length $A = |\mathbf{A}|$ is equally likely to point in any direction specified by the angle $\theta$ from the $x$-axis. Show that the probability that the $x$ component of this vector lies between $A_x$ and $A_x + dA_x$ is $(dA_x/\pi)(A^2 - A_x^2)^{-1/2}$ for $-A \leqslant A_x \leqslant A$.

1.3 A simple pendulum oscillates according to $\theta = \theta_0 \cos(2\pi/T)t$, $T = 2\pi(1/g)^{1/2}$. Show that the probability that in a random measurement the angle of deviation is between $\theta$ and $\theta + d\theta$ is $(d\theta/\pi)(\theta_0^2 - \theta^2)^{-1/2}$.

1.4 Show that for a random variable $x$, the probability of an event in which $x$ becomes greater than a given value $A$ satisfies the inequality $p(x > A) \leqslant \overline{x^2}/A^2$.

1.5 A particle initially at origin jumps by one unit to the right or left, with probability $\frac{1}{2}$. In $\frac{1}{2}(e^{i\phi} + e^{-i\phi})$, $-\pi \leqslant \phi \leqslant \pi$, we can interpret the coefficient of $e^{i\phi}$ as the probability for moving one step to the right, and of $e^{-i\phi}$ for the left. Find the probability $P_n(l)$ that after $n$ steps the particle will be at the point $l$ of the one-dimensional grating.

1.6 By analogy with Prob. 1.5, find $P_n(l)$ of a similar random walk on a two-dimensional square and a three-dimensional cubic grating.

1.7 Convert $I = \iint f(x, y)\, dxdy$, for a given region of integration, to an integration in the $u$-$v$ space, assuming a one-to-one relationship between the points $(x, y)$ and the points $(u, v)$ given by $x = x(u, y)$ and $y = y(u, v)$.

1.8 In a one-dimensional box of length $2a$, a particle with constant velocity is mirror-reflected at the ends. Draw its phase trajectory.

1.9 A particle of mass $m$ moves vertically upward with an initial velocity $v_0$ in a constant gravitation field from point $z_0$. Find the phase trajectory.

1.10 Find the phase trajectory for an oscillator, with a small friction,

$$\ddot{x} + \gamma \dot{x} + \omega_0^2 x = 0,\ \omega_0 = (k/m)^{1/2} \gg \gamma.$$

What is the change in the phase volume with time?

1.11 For a simple pendulum $E = (p_\theta^2/2ml^2) - mgl \cos\theta$. Draw the phase space orbits when (i) $|\theta| < \cos^{-1}(-E/mgl)$, (ii) $E = mgl$, and (iii) $E > mgl$.

1.12 Verify Liouville's theorem for:
(a) an elastic impact between two spheres moving along a line; (b) the motion of three particles in a constant gravitational field with initial phase points $A(z_0, p_0)$, $B(z_0 + a, p_0)$, $C(z_0, p_0 + b)$.

1.13 Apply Liouville's theorem to a fully nonelastic collision between two spheres.

1.14 In general, $\rho$ = constant for all points in phase space which correspond to given constant values of the energy ($E_0$), momentum ($\mathbf{P}_0$) and angular momentum ($\mathbf{M}_0$) of the system, and $\rho = 0$ at all other points, defines a microcanonical distribution. Show that the points defined by

$$E(q, p) = E_0,\ \mathbf{P}(q, p) = \mathbf{P}_0,\ \mathbf{M}(q, p) = \mathbf{M}_0$$

form a manifold of only $2f - 7$ dimensions and not $2f$ dimensions like the phase volume. Does this create any problem? If yes, how can you resolve it?

1.15 How can we exclude the momentum and angular momentum from the definition of $\rho$? What is the remaining integral of the motion?

1.16 Show that the ergodic surfaces are ellipses in the phase space of a linear oscillator of frequency $\nu$. Find the phase space volume enclosed by an ergodic surface in terms of energy and frequency.

1.17 Use the canonical transformation $m^{1/2} q = Q$, $m^{-1/2} p = P$, $p\delta q - P\delta Q = 0$, to express the Hamiltonian for the linear oscillator in a form which does not contain the mass $m$. Discuss the effect of this transformation on the ergodic surface and on the phase volume.

1.18 A particle of mass $m$ moves in a box with walls at $x = 0$ and $x = L$. Plot the trajectory of the particle in the phase space. Calculate the phase space volume $\Gamma(E)$ with energy less than $E$. Show that $\Gamma(E)$ does not change as the wall at $x = L$ is slowly moved.

# 2
# QUANTUM PICTURE

## 2.1 MICROCANONICAL ENSEMBLE

The correct description of any system, small or large, is provided by quantum mechanics. The classical statistical mechanics is useful only as a certain approximation to quantum statistical mechanics. We had to use the method of classical statistical mechanics due to the complexity of the problem ($N \sim 10^{23}$). We now have another reason to use the method of statistical mechanics arising from the uncertainty principle of Heisenberg. According to it we cannot simultaneously measure exactly both the position and the conjugate momentum coordinates of a particle, needed for the classical description. Therefore, no single microscopic state of any system can be found from measurement. We must then use the method based on probabilities and averages.

Due to the uncertainty principle, the notion of the classical phase space cannot be used as such in the quantum statistical mechanics.

In quantum statistical mechanics, a microscopic state is defined in a quantum mechanical sense. In particular, a stationary system of $N$ particles in volume $V$ can be in any one of the quantum states determined by the Schrödinger eigenvalue equation

$$\hat{H}_N \psi_i ([q]) = E_i \psi_i ([q]), \tag{2.1}$$

where $\hat{H}_N$ is the Hamiltonian operator of the $N$-particle system, $\psi_i ([q])$ is the wave function for the entire system in the *quantum state i*, and $E_i$ the *energy of the quantum state i.* The set of microscopic states in quantum statistical mechanics is thus a *discrete* denumerable set $\{i\}$ of quantum states denoted by the quantum number $i$. If we know $N$ and $V$, we can always, in principle, solve (2.1) and know the allowed quantum states (accessible microscopic states) $i$. The system must be in one or another of these states.

We can construct mentally an ensemble to represent what we know about the physical system of interest. Each element of the ensemble can be in one of the discrete quantum states allowed by the system. The probability $P_i$ of finding an element in the state $i$ is determined in such a way as to reflect the initial information (like energy) on the system of interest.

The number of different quantum states which have a given energy is called the *degeneracy* $g$ of the energy level. The word 'level' will always connote the value of the energy for one or more states. Thus, energy levels can have degeneracies while quantum states do not.

A basic assumption of equilibrium statistical mechanics is that the probability $P_i$ of the $i$th $N$-particle quantum state being occupied is a function of $E_i$ only, (1.25),

$$P_i = P_i(E_i). \tag{2.2}$$

Further, all quantum states with the same energy (say $E_i = E_j = E_k = E_l = E$) have the same probability ($P_i = P_j = P_k = P_l = P(E)$). The probabilities of all degenerate quantum states in one level are equal. We can say that *in a state of macroscopic equilibrium, all stationary quantum states of equal energy have equal a priori probability.* This replaces the familiar hypothesis of equal a priori probability of microscopic state in the classical phase space.

For an isolated system the energy $E_{\text{system}}$ is constant. The $P_i$ should then depend on $i$ such that zero probability is assigned to all states $i$ unless $E_i = E_{\text{system}}$. If $\Omega$ is the degeneracy of the energy level $E_{\text{system}}$ then these states are the only ones which are represented in the ensemble. Each such state has equal probability,

$$P_i = \text{constant} \equiv a, \qquad (E_i = E_{\text{system}}) \tag{2.3}$$

$$\sum_{i=1}^{g} P_i = \sum_{i=1}^{g} a = ga = 1, \qquad g = \Omega(E_i, V, N). \tag{2.4}$$

This defines the *microcanonical ensemble* with probability distribution

$$P_i = \begin{cases} 1/\Omega & E_i = E_{\text{system}} \\ 0 & E_i \neq E_{\text{system}} \end{cases} \tag{2.5}$$

where $\Omega = \Omega(E_{\text{system}}, V, N)$ and $E_{\text{system}}$ is well determined so as to lie between $E$ and $E + \Delta E$. One should regard $\Delta E$ as large compared with the Heisenberg uncertainty $\delta E \sim h/\delta t$, where $\delta t$ is the time available for observation, and $h$ is Planck constant.

Note that as the energy of a macroscopic system increases, the degeneracies of the different energy levels increase rapidly. Thus, the larger the system energy, the more quantum states are available to the system and the smaller the probability of any one state being occupied, as it should be (Fig. 2.1).

## 2.2 QUANTIZATION OF PHASE SPACE

Due to uncertainty principle,

$$\delta x \delta p_x \sim h, \tag{2.6}$$

the classical states in a cell of size $h$ per degree of freedom, or $h^f$ per $f$ degrees of freedom, in the phase space, merge into a single quantum

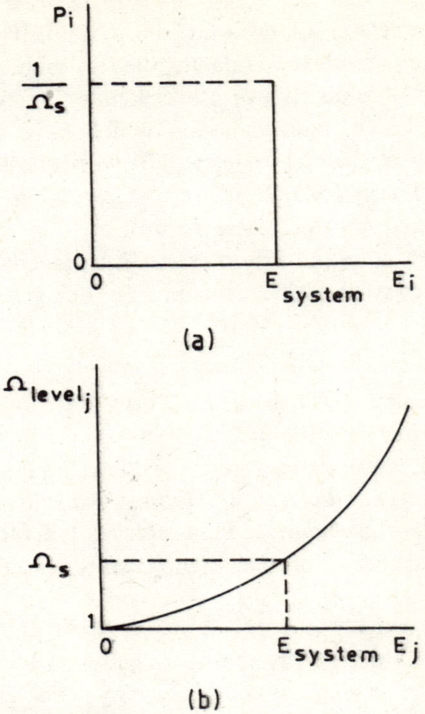


**Fig. 2.1** (a) The microcanonical distribution showing $P_i$ as a function of energy $E_i$ of state $i$. Here $\Omega_s = \Omega$ system. (b) A typical variation of degeneracy $\Omega_j$ of the energy level $j$ as a function of energy $E_j$.

state. Within such a cell the states cannot be further differentiated. If we think of the phase space as divided into cells, each of size $h^f$, then the set of microscopic states contained in a volume element $\Delta\Gamma$ corresponds to a set of

$$\Delta\Gamma/h^f \tag{2.7}$$

quantum states. As $h$ has a small value, ($h = 6.626\ 10^{-34}$ Js), this correspondence is a good approximation. In this sense, a discrete quantum state occupies a volume equal to $h^f$ in the 'cellular' quantum mechanical phase space. This result of converting continuous classical phase space into a 'cellular' one was proposed by Planck in connection with his work on the blackbody radiation.

As a simple illustration, consider a particle enclosed in a three-dimensional cubical box of size $L$ ($V = L^3$). We have

$$\hat{H}_{N-1}\,\psi_i(q) = \epsilon_i\,\psi_i(q),\ \hat{H}_{N-1} = \hat{p}^2/2m = -(h^2/8\pi^2 m)\,\nabla^2,$$

$$\psi_i(q) = \psi_{n_1, n_2, n_3} = \psi_{n_1}\psi_{n_2}\psi_{n_3},$$

$$= A \sin\frac{n_1\pi x_1}{L} \sin\frac{n_2\pi x_2}{L} \sin\frac{n_3\pi x_3}{L},$$

and the energy levels are

$$\epsilon_n = \epsilon_{n_1} + \epsilon_{n_2} + \epsilon_{n_3} = \frac{h^2}{8mL^2}\,n^2,\ n^2 = n_1^2 + n_2^2 + n_3^2,$$

$$n_1, n_2, n_3 = 1, 2, 3, \ldots \tag{2.8}$$

In the $(n_1, n_2, n_3)$-space each lattice point, whose coordinates are all positive integers, corresponds to a quantum state (eigenstate). Therefore, for large $n$ (Bohr's correspondence principle) we can write (Fig. 2.2)

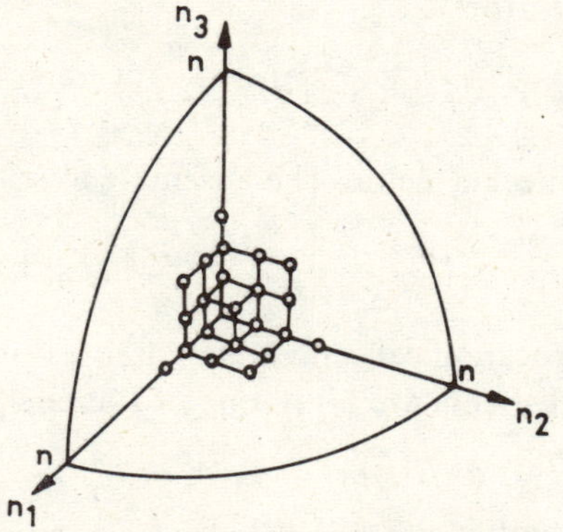


**Fig. 2.2 Lattice points for the combinations of integers $n_1$, $n_2$, $n_3$.**

$$\Omega(\epsilon_n) = \frac{1}{8}\cdot\frac{4}{3}\pi n^3 = \frac{4}{3}\pi\frac{L^3}{h^3}(2m\epsilon_n)^{3/2} = \frac{\Omega_\Gamma(\epsilon_n)}{h^3} \tag{2.9}$$

because classically the volume in the $\Gamma$ space is given by

$$\Omega_\Gamma(\epsilon) = \int_{p^2 \leq 2m\epsilon} d^3x\, d^3p = L^3 \int d^3p = L^3\,\frac{4}{3}\pi(2m\epsilon)^{3/2}. \tag{2.10}$$

The $\Omega(\epsilon_n)$ is the number of microstates or $\Gamma$-cells which are accessible to the system.

Because any combination of three integers $(n_1, n_2, n_3)$ gives only one number $n^2 = n_1^2 + n_2^2 + n_3^2$, for a given $\epsilon_n$ (or $n$) there correspond several states $\psi_{n_1, n_2, n_3}$. Thus the six states $\psi_{1,2,3}$, $\psi_{2,1,3}$, $\psi_{3,2,1}$, $\psi_{3,1,2}$, $\psi_{2,3,1}$, $\psi_{1,3,2}$ belong to the same energy level (*degeneracy*). The degree of degeneracy grows with $n$, that is, with energy. If we consider a thin energy shell of average energy $\epsilon_n$ in the accessible $\Gamma$ space, then we imagine that all quantum cells corresponding to the degeneracy of the energy level $\epsilon_n$ have been included.

The difference between (2.1) and (2.8) is important. The wave function $\psi_i$ in (2.8) describes a single particle in the $i$th particle quantum state. The $\psi_i$ is a function of the coordinates of one particle. The energy eigenvalue is the energy of a single particle when it is in the particle quantum state $i$.

An *ideal gas* is one in which there is no interaction between the particles. Each particle is described quantum mechanically as if it is alone in the box. Each particle then is in one of the accessible single particle quantum states given by (2.8).

For a macroscopic system we can apply the classical equations of motion in view of Bohr's correspondence principle. Thus, Liouville's theorem will again hold. In the classical limit, we replace $P_i$ in (2.3) by a probability density $\rho$.

## 2.3 BASIC POSTULATES

Let $\Psi_n$ be the state function of the $n$th element in an arbitrary ensemble $A$. The expectation value of the energy is the quantum-mechanical average of the Hamiltonian operator

$$\langle H\rangle_n = \int \Psi_n^* \hat{H} \Psi_n \, dq \equiv \langle \Psi_n \mid \hat{H} \mid \Psi_n \rangle, \quad \langle \Psi_m \mid \Psi_n \rangle = \delta_{mn}, \tag{2.11}$$

where $\delta_{mn}$ is the Kronecker delta. The ensemble average is

$$\overline{\langle H\rangle_A} = \frac{1}{M} \sum_{n=1}^{M} \langle H\rangle_n = \frac{1}{M} \sum_{n=1}^{M} \langle \Psi_n \mid \hat{H} \mid \Psi_n \rangle. \tag{2.12}$$

In the ideal (microcanonical) ensemble $I$, representing an isolated system in equilibrium, the elements are all in one of the energy eigenstates $\psi_i$,

$$\hat{H}\, \psi_l = E_l\, \psi_l, \qquad \psi_n = \psi_{l_n}. \tag{2.13}$$

Therefore, in this ensemble

$$\begin{aligned} \overline{\langle H\rangle_I} &= \frac{1}{M} \sum_{n=1}^{M} \langle \psi_{l_n} \mid \hat{H} \mid \psi_{l_n} \rangle \\ &= \frac{1}{M} \sum_{n=1}^{M} E_{l_n} = \sum_l P_l\, E_l, \end{aligned} \tag{2.14}$$

where $P_l$ is the probability that an element in ensemble $I$ is in the state $\psi_l$. Note that (2.14) is also the definition of quantum mechanical average. Thus, for an isolated system in equilibrium (pure case) the ensemble average gives nothing new. This situation has no classical analogue because in spite of the maximal knowledge given by the wave function we must use an averaging process prescribed by quantum mechanics itself.

We can expand $\Psi_n$ for the arbitrary ensemble as

$$\Psi_n = \sum_j a_{nj}\, \psi_j, \qquad \langle \psi_j \mid \psi_k \rangle = \delta_{jk}. \tag{2.15}$$

This would be so if the system of interest interacts, however weakly, with the surroundings. The system would not be in a stationary state (mixed case). Its state is a superposition of pure states, (2.13). Then we do not possess such complete knowledge, as in the pure case, about the quantum mechanical description of the system. There may exist many wave functions compatible with the incomplete information about the system. The effect of these must be suitably averaged (ensemble average). In this case

$$\begin{aligned} \overline{\langle H\rangle_A} &= \frac{1}{M} \sum_{n=1}^{M} \langle \sum_j a_{nj}\, \psi_j \mid \hat{H} \mid \sum_k a_{nk}\, \psi_k \rangle \\ &= \frac{1}{M} \sum_{n=1}^{M} \sum_j a_{nj}^* \, a_{nj}\, E_j. \end{aligned} \tag{2.16}$$

If we choose

$$P_l = \frac{1}{M} \sum_{n=1}^{M} a_{nl}^* \, a_{nl} \equiv \overline{a_{nl}^* \, a_{nl}}, \tag{2.17}$$

we get $\overline{\langle H \rangle_I} = \overline{\langle H \rangle_A}$.

Let us find the expectation value of some observable $R$ such that

$$\hat{R}\,\phi_i = R_i\,\phi_i, \qquad \langle \phi_j \mid \phi_i \rangle = \delta_{ji}. \tag{2.18}$$

We can write

$$\phi_i = \sum_j b_{ij}\,\psi_j, \qquad \psi_j = \sum_i c_{ji}\,\phi_i. \tag{2.19}$$

The quantity $|\,c_{ji}\,|^2 \equiv c_{ji}^*\,c_{ji}$ gives the probability that if the element is in the state $\psi_j$, a measurement of $R$ will give the value $R_i$ corresponding to the eigenfunction $\phi_i$. Now

$$\begin{aligned}
\overline{\langle R \rangle_I} &= \frac{1}{M}\sum_{n=1}^{M} \langle \Psi_n \mid \hat{R} \mid \Psi_n \rangle = \frac{1}{M}\sum_{n=1}^{M} \langle \psi_{i_n} \mid \hat{R} \mid \psi_{i_n} \rangle \\
&= \frac{1}{M}\sum_{n=1}^{M} \sum_j \sum_k c_{i_n j}^*\, c_{i_n k}\, \langle \phi_j \mid \hat{R} \mid \phi_k \rangle \\
&= \frac{1}{M}\sum_{n=1}^{M} c_{i_n j}^*\, c_{i_n j}\, R_j = \sum_i P_i \sum_j (c_{ij}^*\, c_{ij})\, R_j,
\end{aligned} \tag{2.20}$$

$$\begin{aligned}
\overline{\langle R \rangle_A} &= \frac{1}{M}\sum_{n=1}^{M} \sum_j \sum_k a_{nj}^*\, a_{nk}\, \langle \psi_j \mid \hat{R} \mid \psi_k \rangle \\
&= \frac{1}{M}\sum_{n=1}^{M} \sum_j \sum_k \sum_l \sum_m a_{nj}^*\, a_{nk}\, c_{jl}^*\, c_{km}\, \langle \phi_l \mid \hat{R} \mid \phi_m \rangle \\
&= \frac{1}{M}\sum_{n=1}^{M} \sum_j \sum_k \sum_l a_{nj}^*\, a_{nk}\, c_{jl}^*\, c_{kl}\, R_l\,(1-\delta_{jk}) \\
&\qquad + \frac{1}{M}\sum_{n=1}^{M} \sum_j \sum_k \sum_l a_{nj}^*\, a_{nk}\, c_{jl}^*\, c_{kl}\, R_l\, \delta_{jk}.
\end{aligned} \tag{2.21}$$

The last term in (2.21) is just $\overline{\langle R \rangle_I}$. Therefore, the condition for $\overline{\langle R \rangle_A} = \overline{\langle R \rangle_I}$ is that the first term on the right side in the last step of (2.21) vanishes. A sufficient condition for this is

$$\frac{1}{M}\sum_{n=1}^{M} a_{nj}^*\, a_{nk}\,(1-\delta_{jk}) = 0 \tag{2.22}$$

or, with $a_{nj} = r_{nj}\exp(i\theta_{nj})$,

$$\frac{1}{M}\sum_{n=1}^{M} r_{nj}\, r_{nk} \exp[i\,(\theta_{nk}-\theta_{nj})]\,(1-\delta_{jk}) = 0. \tag{2.23}$$

We have no information on the phases $\theta_{ni}$ because the measurable quantity in quantum mechanics is $\Psi_n^*\,\Psi_n$ and not $\Psi_n$. The condition (2.23) will hold if the phase angles $\theta_{ni}$ are completely random over the ensemble $A$. Thus the arbitrary ensemble $A$ will show the statistical behaviour of the ideal ensemble $I$ if we postulate *random phases*. This hypothesis is known as the *postulate of random phases.* It is needed for real physical systems that interact, however weakly, with the surroundings

We can now state the postulates of statistical mechanics as follows:

**Postulate of Equal a priori Probabilities**

$$P_l \equiv \overline{a_{nl}^* a_{nl}} = \begin{cases} 1/\Omega & E < E_l < E + \Delta E \\ 0 & \text{otherwise} \end{cases} \tag{2.24}$$

**Postulate of Random Phases**

$$\overline{a_{nj}^* a_{nk}} = 0, \qquad j \neq k, \tag{2.25}$$

where bar denotes the ensemble average. The constant $1/\Omega$ follows from the normalization $\sum_l \overline{a_{nl}^* a_{nl}} = 1$. For the mixed case we need both the quantum mechanical and the ensemble average. For the microcanonical ensemble only (2.24) is required.

We need (2.25) for systems interacting with surroundings (say, the canonical ensemble to be discussed later). It ensures that the relevant probability amplitudes do not interfere. We have an incoherent superposition of states. In the absence of interference, we can assume that the elements are in definite energy eigenstates.

In the classical limits, under Bohr's correspondence principle, it is found that these allowed eigenstates divide the phase space into cells of equal size, (2.7). We can then replace the eigenstates by equal regions in phase space. These equal regions in phase space are assigned equal a priori probabilities. In the classical case there is no analogue to the postulate of random phases because unknown phases arise solely in connection with the quantum mechanical description in terms of the wave functions that are determined only within a phase factor.

For the purpose of calculations we now formulate the statistical hypothesis in terms of the following postulates:

(1) *Postulate of Ensemble Average*: The average behaviour of a macroscopic system in equilibrium is given by the average taken over a suitable ensemble consisting of an infinite number of randomized mental copies of the system of interest.

(2) *Postulate of Equal a priori Probabilities*: In a state of macroscopic equilibrium, all stationary quantum states of equal energy have equal a priori probability. (This statement is devoted to the microcanonical ensemble).

(3) *Postulate of Equilibrium State*: Equilibrium state is the one which occupies the maximum volume in $\Gamma$ space (classical or 'cellular' quantal).

Some implications of these postulates are:

(a) The method of calculation is *statistical* in nature.

(b) The predictions are to be regarded as true on the average rather than precisely expected for any particular system, $R_{\text{obs}} \cong \overline{R}$.

(c) The probability of finding a system in a given state being proportional to the phase space volume associated with it, the most probable state would be one which occupies the maximum volume in phase space. It follows that *the equilibrium state is the state of maximum probability.*

Note that, in (2.24), $\Omega = \Omega$ ($N$, $V$, and $E$ given) is the number of accessible quantum states (degeneracy associated with the energy level $E$) for an isolated system in equilibrium and $P_i$ is a function of quantum numbers. The value of $E$ is one of the energy levels of the quantum mechanical system defined by $N$ and $V$. Because $V$ and $N$ are very large, the energy levels for such a macroscopic system will be so close together as to form almost a continuum of states, and moreover, each of these levels will have an extremely high degeneracy. This degeneracy increases with energy, (2.9).

## 2.4 CLASSICAL LIMIT

In classical mechanics we can specify simultaneously both $q_i$ and $p_i$ for a particle. In quantum mechanics the uncertainty principle prevents this. A classical description is a reasonable approximation only when the effect of $h$ is negligible, that is,

$$\delta q\, \delta p \gg h. \tag{2.26}$$

Consider the motion of a molecule in a gas. If $p_{av}$ denotes its mean momentum and $r_{av}$ its mean separation from other identical molecules, then a classical description is valid when

$$r_{av}\, p_{av} \gg h, \tag{2.27}$$

or, using the de Broglie wavelength, $p = h/\lambda$, when

$$r_{av} \gg \lambda_{av}, \qquad (\textit{classical limit}). \tag{2.28}$$

Since $\lambda_{av}$ is a measure of the spread of molecule in space, it means that when (2.28) holds the molecular wave functions do not overlap and therefore they are *distinguished* by their position.

Suppose the gas contains $N$ identical particles in a volume $V$. Let the $N$ one-particle wave functions for the $N$ particles be

$$\psi_a(1),\ \psi_b(2),\ \ldots,\ \psi_z(N), \tag{2.29}$$

where $\psi_a(1)$ means particle 1 is in state $a$. Their product gives the wave function for the total system

$$\Psi(1, 2, \ldots, N) = \psi_a(1)\, \psi_b(2) \ldots \psi_z(N). \tag{2.30}$$

When (2.28) holds, the individual wave functions $\psi_a(1)$, $\psi_b(2)$, ... do not overlap appreciably, and an exchange of particles (say, 1 and 2) in (2.30) produces a new state,

$$\Psi'(1, 2, \ldots, N) = \psi_a(2)\, \psi_b(1) \ldots \psi_z(N). \tag{2.31}$$

In the familiar classical language a new microscopic state results and the phase point is shifted from one cell in the $\Gamma$ space to another. In all we get $N!$ product type wave functions for the $N!$ possible permutations of the particles among the one-particle wave functions. Thus, the particles are *distinguishable* and classical statistics is valid.

If

$$r_{av} \ll \lambda_{av}, \qquad (\textit{quantum limit}), \tag{2.32}$$

then the one-particle wave functions overlap. A given particle in the gas cannot be localized. The state of the whole gas is described by a single wave function $\Psi(1, 2, \ldots, N)$ which cannot be decomposed in any meaningful simple way. Thus, the particles are *indistinguishable* and quantum statistics must be employed.

To give a physical content to (2.28), we imagine that each particle occupies a tiny cube of side $r_{av}$ and these cubes fill the volume $V$,

$$r_{av}^3 N = V, \quad r_{av} = (V/N)^{1/3}. \tag{2.33}$$

If we anticipate and relate the temperature $T$ with the average energy $\bar{\epsilon}$ by

$$p_{av}^2/2m \simeq \bar{\epsilon} = \frac{3}{2}kT, \quad p_{av} \simeq (3mkT)^{1/2},$$

where $k$ is the Boltzmann constant, then

$$\lambda_{av} \simeq h/(3mkT)^{1/2}. \tag{2.34}$$

Therefore, the condition (2.28) becomes

$$(V/N)^{1/3} \gg h/(3mkT)^{1/2}, \qquad (\textit{classical limit}). \tag{2.35}$$

This means that the classical description is valid when

(i) $N$ is small (dilute gas),
(ii) $T$ is large, and
(iii) $m$ is not too small. (2.36)

As an example of classical particles, consider the molecules in a gas at *NTP*. The molecular density is $10^{19}$ mol/cm$^3$, and so the volume available to each molecule is $10^{-19}$ cm$^3$. If the molecular radius is taken to be of the order of $10^{-8}$ cm, the molecular volume is about $10^{-24}$ cm$^3$. The molecule being much smaller than the volume available to it, we can, in principle, identify each molecule in the gas. Therefore, the molecules are localized and distinguishable.

As an example of quantum particles, consider the conduction electrons in a metal. The density of electrons is of the order of $10^{22}$ per cm$^3$. The volume available to each electron is $10^{-22}$ cm$^3$. For a 1 eV electron the momentum is $p_x = (2mE)^{1/2} = 0.5 \times 10^{-19}$ erg. s. cm$^{-1}$. The corresponding de Broglie wavelength is $h/p_x = 13 \times 10^{-8}$ cm. So the volume of conduction electron is about $2 \times 10^{-21}$ cm$^3$, which is larger than the volume available to the electron ($10^{-22}$ cm$^3$). Hence the electron wave functions overlap considerably. We cannot localize the electrons, they are indistinguishable, and quantum statistics must be applied.

## 2.5 SYMMETRY OF WAVE FUNCTIONS

For simplicity, first consider a two-particle system described by the wave function $\Psi(1, 2)$. We have seen that in the quantum region, unlike the classical region, it is not possible to distinguish between identical particles. To state this in a formal way we introduce the *permutation operator* $\hat{P}_{12}$, which, acting on a state, interchanges *all* coordinates of particles 1 and 2,

$$\hat{P}_{12}\,\Psi(1, 2) = \Psi(2, 1). \tag{2.37}$$

If identical particles are indistinguishable, the interchange (2.37) should not produce any observable effect. This would be so if the wave function changes at most by a phase factor $\eta$ which leaves $\Psi^*\Psi$ unchanged,

$$\hat{P}_{12}\,\Psi(1, 2) = \Psi(2, 1) = e^{i\eta}\,\Psi(1, 2). \tag{2.38}$$

If the interchange is repeated, the original wave function must be obtained,

$$(\hat{P}_{12})^2\,\Psi(1, 2) = \hat{P}_{12}\,\Psi(2, 1) = \Psi(1, 2) = e^{2i\eta}\,\Psi(1, 2). \tag{2.39}$$

Thus

$$e^{2i\eta} = 1, \text{ or } e^{i\eta} = \pm 1, \tag{2.40}$$

$$\Psi(1, 2) = \Psi(2, 1), \qquad (\textit{symmetric}), \tag{2.41}$$

$$\Psi(1, 2) = -\,\Psi(2, 1), \qquad (\textit{antisymmetric}). \tag{2.42}$$

It is a *law of nature* that the symmetry or antisymmetry under the interchange of two particles is a property of the particles themselves. Pauli first stated this law as follows:

1. Systems consisting of identical particles of integral spin, 0, $1\hbar$, $2\hbar$, ..., are described by symmetric wave functions, $\Psi^{(S)}$.
2. Systems consisting of identical particles of half-odd-integral spin, $\frac{1}{2}\hbar$, $\frac{3}{2}\hbar$, $\frac{5}{2}\hbar$, ..., are described by antisymmetric wave functions, $\Psi^{(A)}$.

Particles of type (1) are called *bosons* and obey *Bose-Einstein statistics.* Particles of type (2) are called *fermions* and obey *Fermi-Dirac statistics.* Thus there is a deep lying connection between spin and quantum statistics. Examples of bosons are photon, $\pi$ meson, $^4$He atom, etc. Examples of fermions are neutrino, electron, proton, $^3$He atom, etc.

The simple product type wave function (2.30), sufficient for classical statistics, must now be properly symmetrized in quantum statistics. For the two-particle system,

$$\Psi(1, 2) = \psi_a(1)\,\psi_b(2), \qquad (\textit{classical}), \tag{2.43}$$

$$\Psi^{(S)}(1, 2) = 2^{-1/2}\,[\psi_a(1)\,\psi_b(2) + \psi_a(2)\,\psi_b(1)] = \Psi^{(S)}(2, 1), (\textit{Bose-Einstein}), \tag{2.44}$$

$$\Psi^{(A)}(1, 2) = 2^{-1/2}\,[\psi_a(1)\,\psi_b(2) - \psi_a(2)\,\psi_b(1)]$$
$$= \frac{1}{\sqrt{2}}\begin{vmatrix} \psi_a(1) & \psi_a(2) \\ \psi_b(1) & \psi_b(2) \end{vmatrix} = -\,\Psi^{(A)}(2, 1), (\textit{Fermi-Dirac}). \tag{2.45}$$

Two fermions (like, electrons) cannot be in the same state, $a = b$, because then $\Psi^{(A)}(1, 2)$ vanishes, (2.45). This is true for a system of $N$ fermions as well. Then (2.45) has the form of the *Slater determinant*,

$$\Psi^{(A)}(1, 2, \ldots, N) = \frac{1}{(N!)^{1/2}}\begin{vmatrix} \psi_a(1) & \psi_a(2) & \cdots & \psi_a(N) \\ \psi_b(1) & \psi_b(2) & \cdots & \psi_b(N) \\ \vdots & \vdots & \vdots & \vdots \\ \psi_z(1) & \psi_z(2) & \cdots & \psi_z(N) \end{vmatrix}, \tag{2.46}$$

where $1/(N!)^{1/2}$ is the normalization factor. The interchange of two particles involves the interchange of two columns in the determinant. This changes the sign of the determinant. If we attempt to put two fermions in the same state, then the determinant vanishes. Thus two otherwise noninteracting fermions appear to stay away from one another because of the requirement of antisymmetry.

The statement that *no two fermions can be in the same quantum state* is called the *Pauli exclusion principle*. If we have discrete states labelled by $\psi_a, \ldots, \psi_r, \ldots$ then for fermions the occupation number $n_r$ of electrons in any state is

$$n_r = 0, 1, \quad \text{(all } r\text{, for fermions)}. \tag{2.47}$$

For bosons there is no restriction on the occupation numbers,

$$n_r = 0, 1, 2, 3, \ldots, \quad \text{(all } r\text{, for bosons)}. \tag{2.48}$$

## 2.6 EFFECT OF SYMMETRY ON COUNTING

We can associate a definite number $g_i$ of elementary cells, or single-particle wave functions, with the same energy eigenvalue $\epsilon_i$, where $g_i$ is the degeneracy of the energy state $\epsilon_i$. For the non-degenerate case $g_i=1$. By hypothesis, every quantum state has an equal a priori probability. Therefore, $g_i$ is also called the *statistical weight* of the concerned energy state or level.

### Maxwell-Boltzmann (MB) Statistics

In the classical limit (2.28) the $N$ particles are distinguishable, there are no symmetry restrictions on $\Psi$, and *any* combination of the $\psi$'s such as

$$\Psi(1, 2, \ldots, N) = \psi_a(q_1)\, \psi_b(q_2) \cdots \psi_z(q_N) \tag{2.49}$$

is a possible state. The distinguishability of particles means that any interchange of particles among the occupied states $\psi_a, \psi_b, \ldots$, leads to a new state for the system, without a change in total system energy.

To include degeneracy, we divide the $\psi$'s into the groups $1, 2, \ldots, i, \ldots, k$, with respective energies $\epsilon_1, \epsilon_2, \ldots, \epsilon_i, \ldots \epsilon_k$, so that the energy eigenvalues for all the $\psi$'s in the $i$th group lie between $\epsilon_i$ and $\epsilon_i + d\epsilon_i$. We assume that the number of such $\psi$'s is $g_i$. Each $g_i$ is assumed to be large, but its exact value is unimportant.

For the particular macroscopic state or macrostate $\{n_i\}$ ($n_i$ particles in the energy region corresponding to the group $i$) the number of microscopic states or microstates is

$$\Omega\{n_i\} = \frac{N!}{\prod_{i=1}^{k} n_i!}. \tag{2.50}$$

This number is obtained by counting the different possible ways of arranging $N$ distinguishable objects, so that there are $n_1$ in the first group, $n_2$ in the second group, and so on. We can arrange $N$ objects in $N!$ different ways.

However, the permutation of $n_1$ objects among themselves in group 1, etc., will not alter the groupings. We now wish to calculate the number of Ψ's corresponding to this distribution. We denote this number by $\Omega\{g_i, n_i\}$. The number of ψ's or states is $g_i$ in the $i$th group (level). Among the $n_i$ particles in this group, the first can occupy these states $g_i$ ways. The second and also the subsequent ones can occupy these states $g_i$ ways, because there is no restriction on the number of particles that can occupy a given ψ. Therefore, there are $(g_i)^{n_i}$ different ways, each corresponding to a new Ψ, in which $n_i$ particles can occupy the $g_i$ states. This gives for the $k$ possible groups of eigenstates,

$$\Omega\{g_i, n_i\} = N! \prod_{i=1}^{k} \left(\frac{(g_i)^{n_i}}{n_i!}\right), \tag{2.51}$$

where $n_1 + n_2 + \ldots + n_k = N$. Note that $N$ is a constant number.

**Bose-Einstein (BE) Statistics**

Consider a system containing $N$ indistinguishable bosons. The total wave function $\Psi(1, 2, \ldots, N)$ must be symmetrical with respect to the interchange of two particles. A symmetric wave function can be represented by the linear combination

$$\Psi(s) = \sum_{\nu} P_\nu [\psi_a(q_1)\, \psi_b(q_2) \ldots \psi_z(q_N)], \tag{2.52}$$

where $\nu$ is the number of binary permutations in the permutation $P$.

There is no restriction on the number of particles which can coexist in a given state ψ, (2.48). The number of ways of distributing $n_i$ bosons among the $g_i$ states of energy $\epsilon_i$ can be obtained as follows (Appendix I). The problem is equivalent to that of distributing $n_i$ white balls among $g_i$ labelled cells. A particular distribution (2 particles in cell 1, 1 in cell 2, 0 in cell 3, 4 in cell 4, 1 in cell 5, ...) can be represented as shown in Fig. 2.3. A cell may be empty or may even contain all the particles. As is obvious from the

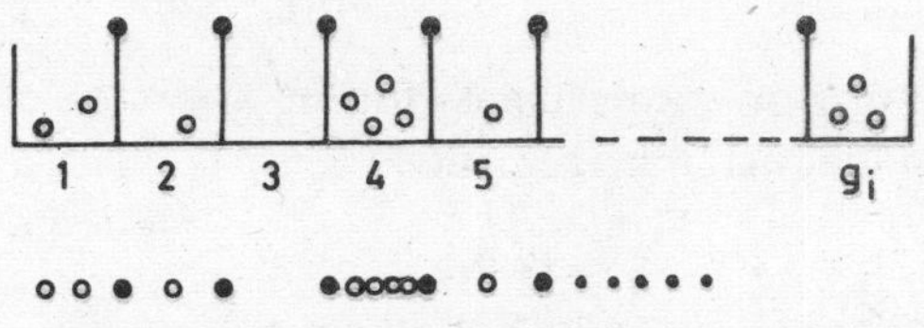


**Fig. 2.3 Counting for the Bose-Einstein statistics. ○ particles, ● dividing walls.**

figure, we can get the desired number by finding the number of permutations of arranging in a row the $n_i$ white balls (representing the particles) together with $g_i - 1$ black balls (representing the dividing walls). If we label all the balls irrespective of their colour with the numbers 1, 2, ..., $n_i + g_i - 1$, the number of permutations is $(n_i + g_i - 1)!$ For each distinct distribution there are $n_i!$ permutations of the white balls among themselves, and $(g_i-1)!$ permutations of the black balls among themselves, which do not change the distribution (2 particles in cell 1, 1 in cell 2, 0 in cell 3, 4 in cell 4, 1 in cell 5, ...). Therefore, the number of distinct distributions of the bosons for state of energy $\epsilon_i$ is

$$\Omega_i^{(S)} = \frac{(n_i + g_i - 1)!}{n_i!\,(g_i-1)!} \simeq \frac{(n_i + g_i)!}{n_i!\,g_i!}, \tag{2.53}$$

remembering that $g_i \gg 1$.

The bosons being indistinguishable, the total number of distributions of the $N$ particles in the system among the $k$ possible groups of eigenstates is

$$\Omega^{(S)}\{g_i, n_i\} = \prod_{i=1}^{k} \frac{(n_i + g_i - 1)!}{n_i!\,(g_i-1)!}, \qquad \text{(BE)}. \tag{2.54}$$

**Fermi-Dirac (FD) Statistics**

The antisymmetric wave function (2.46) can be expressed as

$$\Psi^{(A)} = \sum_{\nu} (-1)^{\nu} P_{\nu} [\psi_a(q_1)\,\psi_b(q_2) \ldots \psi_z(q_N)]. \tag{2.55}$$

The occupation number for fermions is 0 or 1, (2.47). Therefore, we must have $n_i \leqslant g_i$, in the $i$th group of states with energy $\epsilon_i$. The number of distributions in this case is bound to be different from (2.54) because we must exclude in our counting all those states where the occupation number exceeds 1. We, therefore, expect $\Omega^{(A)} \ll \Omega^{(S)}$ for any situation in which $n_i \leqslant g_i$.

If we place $n_i$ particles among $g_i$ states, the first particle has a choice of $g_i$ states, the second has a choice of $(g_i-1)$ states, because of the Pauli exclusion principle, and so on. The total number of arrangements for the $i$th group of states, taking into account the indistinguishability of $n_i$ fermions, is

$$\frac{g_i\,(g_i-1) \ldots (g_i - n_i + 1)}{n_i!} = \frac{g_i!}{n_i!(g_i - n_i)!}.$$

The total number of arrangements for the $k$ groups of energy states is the product of individual arrangements,

$$\Omega^{(A)}\{g_i, n_i\} = \prod_{i=1}^{k} \frac{g_i!}{n_i!(g_i-n_i)!}, \qquad \text{(FD)}. \tag{2.56}$$

It is useful to compare $\Omega$, $\Omega^{(S)}$ and $\Omega^{(A)}$. For simplicity we take $N = 3$, $k = 2$, $n_1 = 2$, in $\epsilon_1$, $n_2 = 1$ in $\epsilon_2$, $g_1 = 2$, and $g_2 = 1$. Then

$$\Omega^{(A)} = \prod_i \frac{g_i!}{n_i!(g_i-n_i)!} = \frac{2!}{2!0!}\cdot\frac{1!}{1!0!} = 1, \qquad \text{(FD)},$$

$$\Omega^{(S)} = \prod_i \frac{(n_i + g_i - 1)!}{n_i!\,(g_i-1)!} = \frac{3!}{2!1!}\cdot\frac{1!}{1!0!} = 3, \qquad \text{(BE)},$$

$$\Omega = N! \prod_i \frac{g_i^{n_i}}{n_i!} = 3! \frac{2^2.1^1}{2!1!} = 12, \qquad \text{(MB)}, \tag{2.57}$$

where the particles are distinguishable in the last case (MB).

## 2.7 VARIOUS DISTRIBUTIONS USING MICROCANONICAL ENSEMBLE (IDEAL GASES)

The simplest system is of $N$ identical noninteracting particles occupying a volume $V$. We have three cases: the *ideal Boltzmann gas* (*classical gas*), the *ideal Bose gas*, and the *ideal Fermi gas*. We can represent such a system by a microcanonical ensemble.

The postulate of equal a priori probabilities, applicable to the microcanonical ensemble, enables us to express the probability $W$ of a particular macrostate $\{n_i\}$ to occur as proportional to $\Omega\ \{g_i, n_i\}$,

$$W = \begin{cases} C_{MB}^{-1} \prod_i (g_i^{n_i}/n_i!), & \text{(MB)}, \\ C_{BE}^{-1} \prod_i (n_i + g_i - 1)!/n_i!\,(g_i-1)!, & \text{(BE)}, \\ C_{FD}^{-1} \prod_i g_i!/n_i!\,(g_i-n_i)!, & \text{(FD)}, \end{cases} \tag{2.58}$$

where $C^{-1}$'s are the constants of proportionality. Instead of maximizing $W$, it is easier to maximize $\ln W$. Neglecting 1 compared to $g_i$, taking the logarithm and using the Stirling approximation $\ln n! \simeq n \ln n - n$,

$$\ln W = \begin{cases} \Sigma\, n_i \ln g_i - \Sigma \ln n_i! - \ln C_{MB}, \\ \Sigma \ln (n_i + g_i - 1) - \Sigma \ln n_i! - \ln (g_i - 1)! - \ln C_{BE}, \\ \Sigma \ln g_i! - \Sigma \ln n_i! - \Sigma \ln (g_i - n_i)! - \ln C_{FD}, \end{cases}$$

$$= \begin{cases} \Sigma\, n_i (\ln g_i - \ln n_i) - \text{constant}, & \text{(MB)}, \\ \Sigma\, [(n_i + g_i) \ln (n_i + g_i) - n_i \ln n_i - g_i \ln g_i - \text{constant}, & \text{(BE)}, \\ \Sigma\, [g_i \ln g_i - n_i \ln n_i - (g_i - n_i) \ln (g_i - n_i) - \text{constant}. & \text{(FD)}. \end{cases} \tag{2.59}$$

The function $\ln W$ can be approximately taken to be continuous, so that

$$-\delta \ln W = \begin{cases} -\Sigma\, (\ln g_i - \ln n_i - 1)\, \delta n_i, & \text{(MB)}, \\ \Sigma\, [-\ln (n_i + g_i) + \ln n_i]\, \delta n_i, & \text{(BE)}, \\ \Sigma\, [\ln n_i - \ln (g_i - n_i)]\, \delta n_i, & \text{(FD)}. \end{cases} \tag{2.60}$$

The macroscopic conditions for the system $(N, V, E)$ are the same in all cases,

$$N = \sum_{l=1}^{k} n_l, \quad \delta N = \sum_l \delta n_l = 0, \tag{2.61}$$

$$E = \sum_{l=1}^{k} n_l \, \epsilon_l, \ \delta E = \sum_l \epsilon_l \, \delta n_l = 0. \tag{2.62}$$

The variation in $\ln W$ for the maximum value is taken subject to these two conditions,

$$0 = \begin{cases} \Sigma (\ln n_l - \ln g_l + \alpha + \beta\epsilon_l)\, \delta n_l, & \text{(MB)}, \\ \Sigma [\ln n_l - \ln (n_l + g_l) + \alpha + \beta\epsilon_l]\, \delta n_l, & \text{(BE)}, \\ \Sigma [\ln n_l - \ln (g_l - n_l) + \alpha + \beta\epsilon_l],\, \delta n_l, & \text{(FD)}, \end{cases} \tag{2.63}$$

where $\alpha$, $\beta$ are the Lagrange multipliers which make the variations $\delta n_l$ independent. Each term in the brackets must be zero. They give the most probable distributions as

$$\bar{n}_l = g_l \, e^{-(\alpha+\beta\epsilon_l)}, \qquad \text{(MB)}, \tag{2.64}$$

$$\bar{n}_l = \frac{g_l}{e^{\alpha+\beta\epsilon_l} - 1}, \qquad \text{(BE)}, \tag{2.65}$$

$$\bar{n}_l = \frac{g_l}{e^{\alpha+\beta\epsilon_l} + 1}, \qquad \text{(FD)}. \tag{2.66}$$

The $\alpha$ is determined by the normalization condition (2.61), $N = \sum_l \bar{n}_l$, or from (2.64) for an ideal Boltzmann gas,

$$Ne^{\alpha} = \sum_l g_l e^{-\beta\epsilon_l} = \sum_{p_l} \exp(-\beta p_l^2/2m) \tag{2.67}$$

$$= \frac{V}{h^3}\int_0^{\infty} \exp(-\beta p^2/2m)\, 4\pi p^2 \, dp = \frac{V}{\lambda^3}, \ \lambda = \left(\frac{\beta h^2}{2\pi m}\right)^{1/2}$$

The $\beta$ should be determined from the relation (2.62), $E = \sum_l \bar{n}_l \, \epsilon_l$, which cannot be solved explicitly. In the next chapter we shall show that $\beta = 1/kT$.

Note that $n_l/g_l$ is the fraction of the $g_l$ states that are occupied. Therefore, the quantity

$$f(\epsilon_l) = \bar{n}_l/g_l \tag{2.68}$$

is called the *occupation index* for the states of energy $\epsilon_l$.

In the limit $g_l \gg n_l$,

$$\frac{g_l!}{n_l!\,(g_l - n_l)!} = \frac{g_l\,(g_l - 1)\ldots(g_l - n_l + 1)}{n_l!} \ \text{slightly} < \frac{g_l^{n_l}}{n_l!},$$

$$\frac{(n_l + g_l - 1)!}{n_l!\,(g_l - 1)!} = \frac{(n_l + g_l - 1)\ldots(g_l + 1)\,(g_l)}{n_l!} \ \text{slightly} > \frac{g_l^{n_l}}{n_l!}.$$

Therefore, (2.57) gives

$$\Omega_{FD} \simeq \Omega_{BE} \simeq \frac{\Omega_{MB}}{N!}, \qquad (g_l \gg n_l), \tag{2.69}$$

or for the distributions

$$\bar{n}_{i\,FD},\ \bar{n}_{i\,BE} \to \bar{n}_{i\,MB}$$

$$\frac{g_i}{e^\alpha e^{\beta\epsilon_i}+1},\ \frac{g}{e^\alpha e^{\beta\epsilon_i}-1} \to \frac{g_i}{e^\alpha e^{\beta\epsilon_i}},\quad (g_i \gg \bar{n}_i). \tag{2.70}$$

This implies that when the number of quantum states in a level is much greater than the number of indistinguishable particles (bosons or fermions) in that level, $g_i \gg n_i$, we can safely use the MB distribution. Thus the classical limit is reached when

$$g_i \gg n_i, \quad \text{for all } i, \tag{2.71}$$

$$e^\alpha e^{\beta\epsilon_i} \gg 1,\ \text{or}\ e^\alpha = \frac{V/N}{\lambda^3} \gg 1. \tag{2.72}$$

For the MB distribution $\bar{n}_i/\bar{n}_0 = e^{-(\epsilon_i-\epsilon_0)/kT}$, where the suffix 0 refers to the lowest level and we have taken $g_i/g_0 = 1$. For $kT = 4 \times 10^{-21}$ J, corresponding to room temperature $T \approx 300$ K, we get

| $\epsilon_i - \epsilon_0$(J) | 0 | $10^{-21}$ | $10^{-20}$ | $10^{-19}$ |
|---|---|---|---|---|
| $\bar{n}_i/\bar{n}_0$ | 1 | 0.8 | 0.1 | $4 \times 10^{-6}$ |

It means that energy levels lying higher than about $10\,kT$ above the ground state are occupied to a negligible degree.

We have derived the quantum distributions (2.65, 66) by using the method of the most probable distribution. This is not entirely satisfactory because of the use of Stirling approximation which requires both $g_i$ and $g_i - n_i$ to be large quantities in going from (2.59) to (2.60). The main objection is that for the FD case when $e^\alpha e^{\beta\epsilon_i} \ll 1$, we have $\bar{n}_i \simeq g_i$ in (2.66). This contradicts the assumption that $(g_i - n_i)$ is large. It is desirable, therefore, to derive these results in a rigorous way by using the method based on the canonical and grand canonical partition functions (to be discussed later). The method of Darwin and Fowler also avoids the use of the Stirling approximation but will not be discussed here.

## PROBLEMS

2.1 Calculate the de Broglie wavelength of (i) an electron with kinetic energy $2.5 \times 10^{-10}$ ergs $\simeq 160$ eV, (ii) for neutrons with kinetic energy $1.3 \times 10^{-13}$ ergs $\simeq 0.08$ eV, and (iii) a droplet, 0.1 mm in size, moving with a velocity of 10 cm/s. Discuss your results.

2.2 Show that for an electron $mc^2 \simeq 0.51$ MeV, $\hbar/mc \simeq 3.9 \times 10^{-11}$ cm, and $\hbar/mc^2 \simeq 1.3 \times 10^{-21}$ s, where $\hbar = h/2\pi$.

2.3 Calculate the velocity of a particle due to the uncertainty principle (i) when it is an electron with $\Delta x \sim 1$Å, and (ii) a piece of chalk of mass 1 gm with $\Delta x \sim 0.1$ mm.

2.4 An operator whose expectation value for all admissible wave functions is real is known as a *hermitian operator*. Show that $\hat{p} = -i\hbar\partial/\partial x$ is hermitian.

2.5 Show that $\langle p_x \rangle = 0$ for the particle in a box.

2.6 Show that $L\sqrt{\langle p_x^2 \rangle} > \hbar$ for a particle in a box.

2.7 Define a function by

$$\Psi(x) = \int_{-\infty}^{\infty} \phi(k)\, e^{ikx}\, dk, \quad k = p/\hbar.$$

Calculate $\Psi(x)$ and $|\Psi(x)|^2$ for the choice $\phi(k) = e^{-\alpha(k-k_0)^2}$. Plot $\Psi(x)$ and $\phi(k)$, and estimate $\Delta x\, \Delta k$.

2.8 Construct the energy level diagram for the particle in a cubic box and indicate the degree of degeneracy of each level. Find the number of quantum states for this particle and compare it with the classical phase volume. Find the density of states.

2.9 For a system of two non-interacting particles the Hamiltonian is $\hat{H} = (\hat{p}_1^2/2m_1) + (\hat{p}_2^2/2m_2)$. The particles being independent, the probability of finding one at $x_1$ and the other at $x_2$ is simply the product of two independent probabilities, $P(x_1, x_2) = P(x_1)\, P(x_2)$. Show that this implies that in the wave equation $\hat{H}\, \psi(x_1, x_2) = E\, \psi(x_1, x_2)$, the $\psi(x_1, x_2)$ is separable as $\psi(x_1, x_2) = \psi_1(x_1)\, \psi_2(x_2)$. Write down the equations satisfied by $\psi_1$ and $\psi_2$, and solve them. Express the solutions in the coordinates

$$x = x_1 - x_2, \quad X = \frac{m_1 x_1 + m_2 x_2}{m_1 + m_2} = \frac{\mu}{m_2}\, x_1 + \frac{\mu}{m_1}\, x_2,$$

where $\mu = m_1 m_2/(m_1 + m_2)$ is the *reduced mass*.

2.10 Solve the wave equation for a particle in a box with sides at $x = 0$ and $x = L$ with the boundary condition $\psi(0) = \psi(L)$.

2.11 Solve the Schrödinger equation for the one-dimensional harmonic oscillator problem.

2.12 The probability of finding any macrostate, $W$(macro), can be expressed, according to the principle of equal a priori probabilities, as

$$W(\text{macro}) = \frac{\text{number of microstates corresponding to the given macrostate}}{\text{total number of microstates}} = \frac{\Omega\{g_i, n_i\}}{\sum_{\{N\}} \Omega\{g_i, n_i\}},$$

where the $\sum_{\{N\}}$ indicates a sum over all possible sets of $n_i$'s consistent with the constraints (2.61, 62). Show that the total number of microstates is fixed in any system and so justify (2.58), that is, $W$ is proportional to $\Omega\{g_i, n_i\} = \prod_i \Omega_i$.

2.13 A total of 40 members of a club have to elect a guest speaker out of the two, *A* or *B.* If 16 vote for *A* and 24 for *B,* calculate $\Omega\ \{g_i, n_i\}$, $\sum_{\{N\}} \Omega\ \{g_i, n_i\}$, and $W$.

[Ans: $40!/(16!\ 24!) = 6.29\times 10^{10}$, $2^{40}$, 0.06.]

2.14 The quantity $\Omega\ (E, V, N) = \sum_{\{N\}} \Omega\ \{g_i, n_i\}$ is called the *microcanonical partition function.* Show that

$$\ln \Omega\ (E, V, N) = \ln (\Omega_1 + \Omega_2 + \ldots) \simeq \ln \Omega_{\max},$$

where $\Omega_{\max}$ is the largest $\Omega_i$ in the series.
[Hint: Assume that there are as many as $N$ systems with $\Omega_i$'s comparable to $\Omega_{\max}$, that is, as many such systems as there are particles in any one system. Then $\ln \Omega\ (E, V, N) \simeq \ln \Omega_{\max} + \ln N$.]

2.15 Use the method of steepest descent to derive the FD distribution. (see, for example, *An Introduction to Statistical Mechanics* by P. Dennery, John Wiley, 1972).

# 3
# STATISTICAL MECHANICS AND THERMODYNAMICS

## 3.1 ENTROPY

The connection between statistical mechanics and thermodynamics is provided by entropy.

Let $\Omega_\Gamma(E)$ denote the volume in $\Gamma$ space enclosed by the ergodic surface of energy $E$,

$$\Omega_\Gamma(E) = \int_{E(q,\,p)\,<\,E} dq_1 \ldots dq_f \, dp_1 \ldots dp_f. \tag{3.1}$$

The volume of the shell $\Delta\Gamma(E)$ between the ergodic surfaces $E$ and $E + \Delta$, which is occupied by the microcanoncial shell, is given by

$$\Delta\Gamma(E) = \Omega_\Gamma(E + \Delta) - \Omega_\Gamma(E) = \frac{\partial \Omega_\Gamma(E)}{\partial E} \Delta = g(E)\,\Delta$$

$$\equiv \int_{E\,<\,H\,(q,\,p)\,<\,E\,+\,\Delta} d^f q \; d^f p \; \rho\,(q, p), \tag{3.2}$$

where $\rho(q, p) = 1$ if $E < H(q, p) < E + \Delta$ and 0 otherwise. The quantity $g(E) = \partial \Omega_\Gamma(E)/dE$ is called the *density of states* of the system.

The *entropy* $\sigma$ of a system in statistical equilibrium is defined by

$$\sigma(V, E) \equiv \ln \Delta\Gamma(E). \tag{3.3}$$

This definition is the same in quantum statistical mechanics except that $\Delta\Gamma(E)$ is then calculated in quantum mechanics, (2.7). The dimension of $\Omega_\Gamma(E)$, or of $\Delta\Gamma$, is $(\text{length} \times \text{momentum})^f = (\text{action})^f$. If $\Delta\Gamma$ is imagined to be divided into elementary cells of volume $h^f$, where the Planck constant $h$ has the dimension of action, then $\Delta\Gamma/h^f$ is dimensionless. Defining

$$\sigma = \ln\,(\Delta\Gamma/h^f) = \ln \Delta\Gamma - f \ln h, \tag{3.4}$$

we find that the change in entropy, $\delta\sigma = \delta \ln \Delta\Gamma$, is independent of the unit, as it should be.

If (3.3) is to be acceptable, it should exhibit the well-known properties of the *thermodynamic entropy S*. To show this we consider the equilibrium between two systems $(N', V', E')$ and $(N'', V'', E'')$ in thermal contact.

**Composite System**

Consider two systems, $(N', V', E')$ and $(N'', V'', E'')$, that are isolated from each other (Fig. 3.1a). We can construct the microcanonical ensemble for each taken alone (Fig. 3.1b).

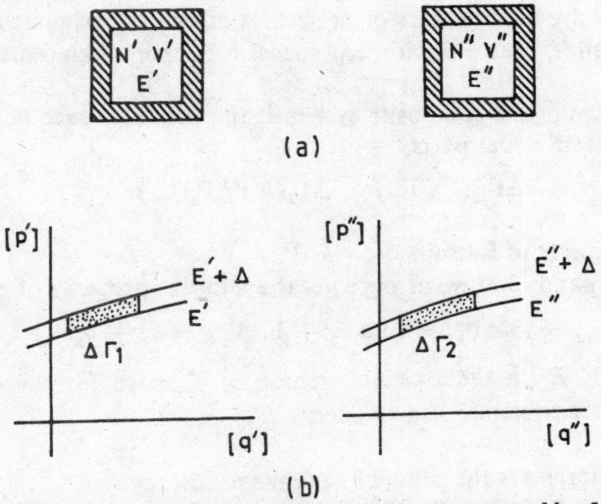


**Fig. 3.1** (a) Two isolated systems. (b) Microcanonical ensembles for the two systems. The coordinates and momenta of the molecules in the two systems are denoted by $(q', p')$ and $(q'', p'')$, respectively.

$$
\begin{aligned}
P_1 \text{ or } \rho_1 &= \text{constant} && \text{for energy between } E' \text{ and } E' + \Delta \\
&= 0 && \text{otherwise,} \\
P_2 \text{ or } \rho_2 &= \text{constant} && \text{for energy between } E'' \text{ and } E'' + \Delta \\
&= 0 && \text{otherwise.}
\end{aligned}
$$

Let $\Delta\Gamma_1(E')$ and $\Delta\Gamma_2(E'')$ be the volumes occupied by the two ensembles in their respective $\Gamma$ spaces.

Now imagine the two systems to be brought in thermal contact and thereby form a composite system in equilibrium (Fig. 3.2). The micro-

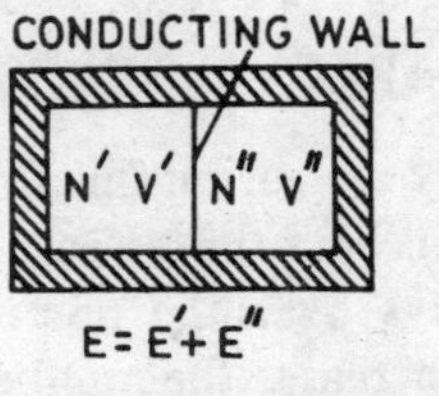


**Fig. 3.2** Composite system. The total energy is $E = E' + E''$.

canonical ensemble for the composite system is defined by

$$P_{12} \text{ or } \rho_{12} = \text{constant} \qquad \text{for energy between } (E' + E'') \text{ and } E' + E'' + 2\Delta$$
$$= 0 \qquad \text{otherwise.} \tag{3.5}$$

This ensemble will contain all mental copies of the composite system for which:

(a) $N'$ particles with coordinates and momenta $(q', p')$ are contained in the volume $V'$.

(b) $N''$ particles with coordinates and momenta $(q'', p'')$ are contained in the volume $V''$.

(c) $E = E' + E''$, the total energy, lies between $E$ and $E + 2\Delta$.

(d) $n_i', n_j''$ are the numbers of molecules of the two kinds that occupy the various cells $i, j$ into which the $\mu$ spaces for different kinds of molecules are divided.

The $\Gamma$ space of the composite system is the product space of the phase spaces of the individual parts,

$$\Delta\Gamma_{12}(E', E'') = \Delta\Gamma_1(E')\,\Delta\Gamma_2(E''). \tag{3.6}$$

**Extensive Property of Entropy**

From (3.6) and (3.3) we at once get the additive property of entropy,

$$\sigma_{12} = \ln \Delta\Gamma_{12} = \ln \Delta\Gamma_1 + \ln \Delta\Gamma_2 = \sigma_1 + \sigma_2. \tag{3.7}$$

In (3.6), $\Delta\Gamma_2(E, E')$ is the accessible region of $\Gamma$ space for the composite system when the system 1 has an energy $E'$.

**State of Equilibrium is the State of Maximum Entropy**

From the basic postulates of statistical mechanics, the equilibrium state is the one for which $\Delta\Gamma$ is maximum. Therefore, the entropy of a system in equilibrium should really be defined by

$$\sigma(V, E) = \ln (\Delta\Gamma)_{MP}, \tag{3.8}$$

where the suffix denotes $\Delta\Gamma$ associated with the most probable distribution. However, the distinction between the original definition (3.3) and (3.8) is unimportant for large systems. This is so because $\sigma$ depends on $\ln \Delta\Gamma$ which is a very slowly varying function of $\Delta\Gamma$. For example, for a system of $N \simeq 10^{20}$ particles $(\Delta\Gamma)_{MP} \approx N!$, $\sigma_{MP} \approx N \ln N - N \approx N \ln N \approx 10^{20} \ln 10^{20} \simeq 46 \times 10^{20}$. If in measuring $\Delta\Gamma$ we make an error of as much as $10^{10}$, the error in entropy is only $\delta\sigma = \ln 10^{10} \approx 23 \ll \sigma_{MP}$.

For the composite system at equilibrium, we can write (3.6) as

$$\Delta\Gamma_{12}(E, E') = \Delta\Gamma_1(E')\Delta\Gamma_2(E'') = \Delta\Gamma_1(E')\,\Delta\Gamma_2(E-E'). \tag{3.9}$$

The $\Omega_r$ and so $\Delta\Gamma$, is a rapidly increasing function of energy $E$ for a large system. The physical reason is that the energy $E$, as it increases, can be distributed in many more ways over the microscopic degrees of freedom of the system. This rapidly increases the number of accessible microstates.

For an estimate, let us consider the case of an ideal gas, $f = 3N$. Then (3.1) can be expressed as

$$\Omega_r(E) = V^N \int_0^E dp_1 \dots dp_{3N}, \tag{3.10}$$

where

$$V^N = \int dq_1 \dots dq_{3N} = \int dx_{11}dx_{21}dx_{31} \int dx_{12}dx_{22}dx_{32} \dots$$

$$\dots \int dx_{1N}dx_{2N}dx_{3N},$$

because the energy of an ideal gas does not depend on the positions of molecules. Since $E = \sum_{i=1}^{3N} p_i^2/2m$, this integral is just the volume contained in the $3N$-dimensional hypersphere of radius $(2mE)^{1/2}$.

The volume of an $n$-dimensional hypersphere of radius $R$ is (see Appendix V)

$$V_n(R) = a_n R^n, \quad a_n = \pi^{n/2}/\Gamma(\tfrac{1}{2}n + 1) = \pi^{n/2}/(\tfrac{1}{2}n)!. \tag{3.11}$$

Therefore, (3.10) can be written as

$$\Omega_r(E) = \text{const. } E^{3N/2}, \tag{3.12}$$

which, for $N \sim 10^{20}$, certainly varies very rapidly with $E$. In fact, it does not matter whether we write $\Omega_r \sim E^{f/2}$ or $\Omega_r \sim E^f$. Since $\Delta\Gamma = g(E)\Delta$, for a given $\Delta$ we have $g(E) \sim E^{f-1} \sim E^f$. More precisely we can write (3.2) as

$$\begin{aligned} \Delta\Gamma &= \Omega_r(E) - \Omega_r(E-\Delta) = \text{const.}\,[E^f - (E-\Delta)^f] \\ &= \text{const. } E^f \left[1 - \left(1 - \frac{\Delta}{E}\right)^f\right] \\ &\simeq \text{const. } E^f\,(1 - e^{-f\Delta/E}). \end{aligned} \tag{3.13}$$

For $f \sim 10^{20}$, $f\Delta/E \gg 1$ and

$$\Delta\Gamma \simeq \text{const. } E^f. \tag{3.14}$$

Thus $\Delta\Gamma$ is practically the volume of the whole hypersphere of radius $(2mE)^{1/2}$, $\Delta\Gamma \sim \Omega_r(E)$.

It follows that in (3.9) the term $\Delta\Gamma_1(E')$ increases rapidly with $E'$ and the term $\Delta\Gamma_2(E-E')$ decreases rapidly with $E'$ (Fig. 3.3a). As a result $\Delta\Gamma_{12}$ or the combined density of states has a sharp maximum at, say, $\bar{E}'$, $\bar{E}'' = E - \bar{E}'$, and (3.9) becomes

$$\Delta\Gamma_{12}(E, \bar{E}') = \Delta\Gamma_1(\bar{E}')\,\Delta\Gamma_2(\bar{E}'') = \text{maximum}. \tag{3.15}$$

This implies, Fig. 3.3b,

$$\sigma_1(\bar{E}') + \sigma_2(\bar{E}'') = \text{maximum}. \tag{3.16}$$

The value $\bar{E}'$ which corresponds to the maximum of $\ln \Delta\Gamma_{12}(E, E')$ is determined by the condition

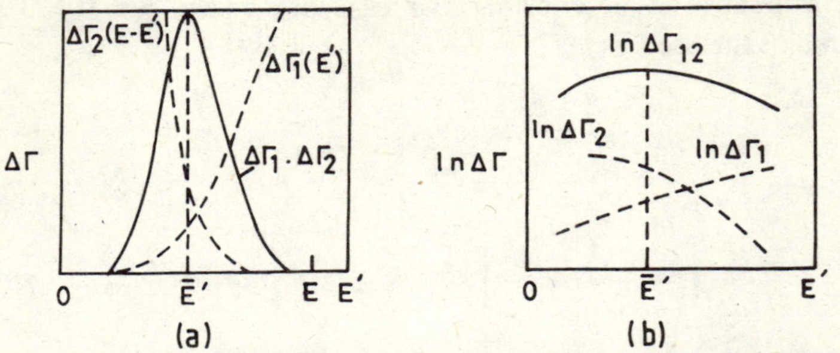


**Fig. 3.3** **(a) The combined density of states with a sharp maximum at $\bar{E}'$ (solid curve). (b) The dependence of ln $\Delta\Gamma_1(E')$ and ln $\Delta\Gamma_2(E-E')$ on $E'$ (dotted curves), showing a unique maximum (solid curve) at some value $\bar{E}'$ for $\Delta\Gamma_{12}(E, E')$.**

$$\frac{\partial \ln \Delta\Gamma_{12}(E, E')}{\partial E'} = \frac{1}{\Delta\Gamma_{12}(E, E')} \frac{\partial \Delta\Gamma_{12}(E, E')}{\partial E'} = 0, \tag{3.17}$$

or

$$\frac{\partial \ln \Delta\Gamma_1(E')}{\partial E'} + \frac{\partial \ln \Delta\Gamma_2(E'')}{\partial E''}(-1) = 0, (E = E' + E'' = \bar{E}' + \bar{E}''). \tag{3.18}$$

If we define a quantity $\theta$ by

$$\frac{\partial \sigma(V, E)}{\partial E} \equiv \frac{1}{\theta}, \tag{3.19}$$

then (3.18) can be written as

$$\theta_1(E') = \theta_2(E''), \qquad E'' = E - E'. \tag{3.20}$$

This is the fundamental condition which determines the particular value $E' = \bar{E}'$ (and so also the value $E'' = E - \bar{E}' = \bar{E}''$) for which the composite system occupies the maximum volume in $\Gamma$ space, that is, the system is in equilibrium (or most probable state). This is also the state of maximum entropy.

This discussion gives meaning to the quantity $\theta$ of an isolated system. Thus $\theta$ of an isolated system is the parameter that determines the equilibrium between one part of the system and another. Later on we shall identify it with temperature.

### Principle of Entropy Increase

Let us discuss the problem of the approach to thermal equilibrium. We have just seen that $\Delta\Gamma_{12}(E, E')$ has a sharp maximum (Fig. 3.3a) at $E' = \bar{E}'$. It means the system 1 almost always has an energy close to $\bar{E}'$ and the system 2 has an energy close to $\bar{E}'' = E - \bar{E}'$, when in contact.

Suppose 1 and 2 are initially separate. Their mean energies at equilibrium are $\bar{E}'_i$ and $\bar{E}''_i$, respectively. If they are brought in thermal contact, an exchange of energy takes place until the equilibrium of the composite system is reached. For this the final energies $\bar{E}'_f$ and $\bar{E}''_f$ of the system must be such that

$$\bar{E}'_f = \bar{E}', \quad \text{and} \quad \bar{E}''_f = E - \bar{E}' = \bar{E}'', \tag{3.21}$$

so that $\Delta\Gamma_{12}$ becomes maximum, or (3.16) is satisfied. Thus the exchange of energy goes on until the total entropy becomes maximum. It implies that the final probability can never be less than the initial probability,

$$\sigma_1(\bar{E}'_f) + \sigma_2(\bar{E}''_f) \geqslant \sigma_1(\bar{E}'_i) + \sigma_2(\bar{E}''_i). \tag{3.22}$$

The word 'entropy' is derived from the Greek word which roughly means 'evolution' (Gr. *en*, in + *trope*, turning). It indicates the 'turn', or direction, taken by the process.

We have shown that the definition (3.3) of entropy in statistical mechanics, based on microcanonical ensemble, possesses all the basic properties of the thermodynamic entropy $S$.

The microcanonical ensemble for the composite system is defined in the shell over a range of values $(E', E'')$. However, at equilibrium the most probable partition of energy $(\bar{E}', \bar{E}'')$ prevails for a macroscopic system. It implies that almost all the elements of the ensemble have the values $(\bar{E}', \bar{E}'')$.

The following definitions of $\sigma$ are equivalent up to additive constants:

$$\sigma = \ln \Delta\Gamma(E), \; \sigma = \ln g(E), \; \sigma = \ln \Omega_r(E),$$
$$\sigma = \ln \Omega(n_i), \; \sigma = \ln W. \tag{3.23}$$

The last relation is called *Boltzmann's definition of entropy.*

From (2.50) and (3.23),

$$\begin{aligned} \sigma &= \ln N! - \sum_i \ln n_i! \\ &= N \ln N - \sum_i n_i \ln n_i. \end{aligned}$$

At equilibrium $\Omega(n_i)$ has its maximum value and $n_i$ is given by the Maxwell-Boltzmann distribution (2.64). Therefore, apart from a constant,

$$\begin{aligned} \sigma &= N \ln N - \sum_i \bar{n}_i \, (-\alpha - \beta\epsilon_i) \\ &= N \ln N + \alpha N + \beta E. \end{aligned} \tag{3.24}$$

## 3.2 EQUILIBRIUM CONDITIONS

Consider three basic situations: thermal, mechanical and concentration equilibrium between two systems. They will define the basic intensive quantities: temperature, pressure and chemical potential.

For a system in equilibrium the entropy $\sigma$ will depend on the energy $E$ of the system, on some external variable $X$, such as volume, and on the number $N$ of molecules in the system,

$$\sigma = \sigma(E, X, N). \tag{3.25}$$

Consider an isolated composite system made up of two parts separated by a barrier (Fig. 3.4a). It can be represented by a microcanonical ensemble because the total energy is constant. Initially the barrier is insulating, rigid and nonpermeable.

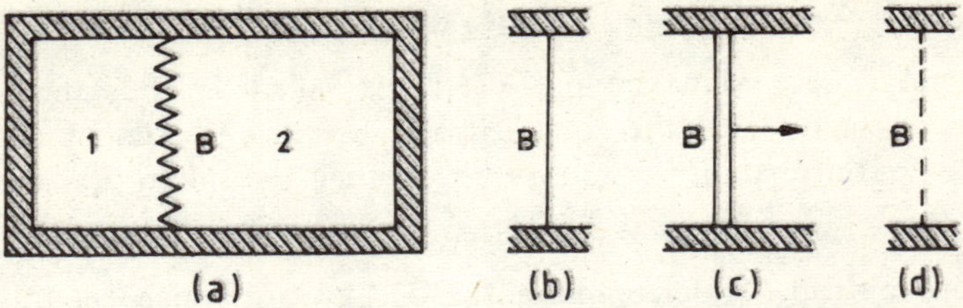


**Fig. 3.4** (a) Insulated system made up of two parts. Initially the barrier $B$ is insulating, rigid and nonpermeable. $B$ becomes (b) conducting, (c) conducting movable piston, and (d) porous.

### Thermal Equilibrium

Suppose the barrier becomes thermally conducting (Fig. 3.4b). At thermal equilibrium the entropy $\sigma = \sigma_1 + \sigma_2$ of the composite system is required to be maximum,

$$d\sigma = d\sigma_1 + d\sigma_2 = \left(\frac{\partial \sigma_1}{\partial E_1}\right) dE_1 + \left(\frac{\partial \sigma_2}{\partial E_2}\right) dE_2 = 0. \qquad (3.26)$$

Total energy $E = E_1 + E_2$ is constant, so that

$$dE = dE_1 + dE_2 = 0,$$

$$d\sigma = \left(\frac{\partial \sigma_1}{\partial E_1} - \frac{\partial \sigma_2}{\partial E_2}\right) dE_1 = 0. \qquad (3.27)$$

The variation $dE_1$ being arbitrary,

$$\partial\sigma_1/\partial E_1 = \partial\sigma_2/\partial E_2. \qquad (3.28)$$

The definition (3.19) of a quantity $\theta$,

$$\partial\sigma/\partial E = 1/\theta \qquad (3.29)$$

gives for thermal equilibrium

$$\theta_1 = \theta_2. \qquad (3.30)$$

Suppose initially the two subsystems are not in thermal equilibrium, but that $\theta_1 > \theta_2$. By the principle of entropy increase (3.22), after the thermal contact we must have $\sigma_{\text{final}} > \sigma_{\text{initial}}$, or $d\sigma > 0$. Then, by (3.27, 28),

$$\left(\frac{1}{\theta_1} - \frac{1}{\theta_2}\right) dE_1 > 0. \qquad (3.31)$$

Because $\theta_1 > \theta_2$, (3.31) is satisfied only if $dE_1 < 0$. This means the energy flows from the system of high $\theta$ to that of low $\theta$. Therefore, $\theta$ *behaves like temperature.*

### Mechanical Equilibrium

Let the barrier be replaced by a conducting piston which is free to move ($X = V$) (Fig. 3.4c). We have to maximize $\sigma$ while keeping $N_1$, $N_2$, $V = V_1 + V_2$, and $E = E_1 + E_2$ constant,

$$d\sigma = \left(\frac{\partial \sigma_1}{\partial V_1}\right) dV_1 + \left(\frac{\partial \sigma_2}{\partial V_2}\right) dV_2 + \left(\frac{\partial \sigma_1}{\partial E_1}\right) dE_1 + \left(\frac{\partial \sigma_2}{\partial E_2}\right) dE_2 = 0. \quad (3.32)$$

The last two terms add up to zero when thermal equilibrium is reached.

Using $dV = dV_1 + dV_2 = 0$, we get for arbitrary $dV_1$,

$$\partial\sigma_1/\partial V_1 = \partial\sigma_2/\partial V_2. \tag{3.33}$$

Defining a quantity by

$$(\partial\sigma/\partial V)_{E,N} = \Pi/\theta, \tag{3.34}$$

we obtain the condition for mechanical equilibrium,

$$\Pi_1/\theta_1 = \Pi_2/\theta_2, \qquad (\theta_1 = \theta_2). \tag{3.35}$$

Suppose initially the two subsystems in thermal equilibrium are not in mechanical equilibrium, but that $\Pi_1 > \Pi_2$. Then $d\sigma > 0$ implies

$$\frac{1}{\theta}(\Pi_1 - \Pi_2)\, dV_1 > 0. \tag{3.36}$$

For $\Pi_1 > \Pi_2$, this requires $dV_1 > 0$, that is, system at high $\Pi$ expands. Thus $\Pi$ *behaves like pressure P*.

**Concentration Equilibrium**

If the barrier is rigid but conducting and permeable, Fig. 3.4d, then particles and energy can be exchanged We have to maximize $\sigma$ while keeping $V_1$, $V_2$, $N_1 + N_2$, and $E_1 + E_2$ constant,

$$d\sigma = \left(\frac{\partial\sigma_1}{\partial N_1}\right) dN_1 + \left(\frac{\partial\sigma_2}{\partial N_2}\right) dN_2 + \left(\frac{\partial\sigma_1}{\partial E_1}\right) dE_1 + \left(\frac{\partial\sigma_2}{\partial E_2}\right) dE_2 = 0. \tag{3.37}$$

After thermal equilibrium has been reached, we get on using $dN = dN_1 + dN_2 = 0$, for the concentration equilibrium,

$$\partial\sigma_1/\partial N_1 = \partial\sigma_2/\partial N_2. \tag{3.38}$$

For convenience, define

$$\partial\sigma/\partial N = -\mu/\theta, \tag{3.39}$$

where $\mu$ is called the *chemical potential*. Then the condition for concentration equilibrium is

$$\mu_1/\theta_1 = \mu_2/\theta_2, \qquad (\theta_1 = \theta_2). \tag{3.40}$$

Suppose initially the two subsystems in thermal equilibrium are not in concentration equilibrium, but that $\mu_1 > \mu_2$. Then $d\sigma > 0$ implies

$$-\frac{1}{\theta}(\mu_1 - \mu_2)\, dN_1 > 0. \tag{3.41}$$

For $\mu_1 > \mu_2$, we must have $dN_1 < 0$, that is, the molecules move from a region of higher $\mu$ to that of lower $\mu$.

## 3.3 QUASISTATIC PROCESSES

Suppose the conditions of a system in equilibrium, $\sigma = \sigma(E, X, N)$, are changed slightly and reversibly so that the final state is also of equilibrium. Then

$$d\sigma = \left(\frac{\partial \sigma}{\partial E}\right) dE + \left(\frac{\partial \sigma}{\partial X}\right) dX + \left(\frac{\partial \sigma}{\partial N}\right) dN$$

$$= \frac{1}{\theta} dE + \frac{x}{\theta} dX - \frac{\mu}{\theta} dN \tag{3.42}$$

where $x = \Pi$ if $X = V$, (3.34). Rearranging

$$dE = \theta d\sigma - x dX + \mu dN. \tag{3.43}$$

First consider the simple case of fixed number of molecules,

$$dN = 0, \, x = \Pi, \, X = V, \tag{3.44}$$

$$dE = \theta d\sigma - \Pi dV. \tag{3.45}$$

The change in internal energy consists of two parts. The part $\theta d\sigma$ represents the change in internal energy when the external parameter $X$ is kept constant. This is the way we define *heat*. The quantity of heat $\hat{d}Q$ added to the system in a reversible process is therefore given by

$$\hat{d}Q = \theta d\sigma, \tag{3.46}$$

where the symbol $\hat{d}$ emphasizes that $\hat{d}Q$ is not an exact differential. The part $-\Pi dV$ represents the change in internal energy due to the change in the external parameter. This is the way we define *work*. Thus

$$W = -\Pi dV \tag{3.47}$$

and (3.45) becomes

$$dE = \hat{d}Q + \hat{d}W. \tag{3.48}$$

This is the *first law of thermodynamics*. In mechanics the work is given by $-PdV$. Therefore, $\Pi$ is the pressure $P$,

$$\Pi = P. \tag{3.49}$$

The effects of a reversible addition of heat and work on the distribution of systems among the energy levels given by the Schrödinger equation are shown in Fig. 3.5.

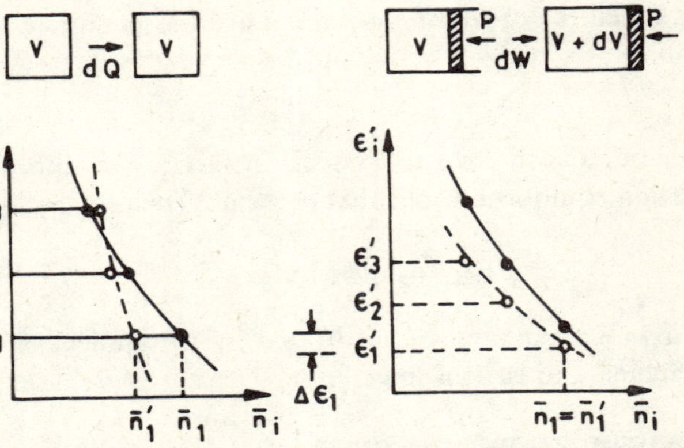

Fig. 3.5 (a) The change in the most probable distribution $\bar{n}_i$ on reversibly adding heat to system. Energy levels $\epsilon_i$ remain same but more particles occupy higher energy levels. (b) The change in $\bar{n}_i$ on reversibly performing work on the system. The energy levels are lowered leaving $\bar{n}_i$ unchanged. Note that $dE = (1/N) \sum_{i=1}^{k} \bar{n}_i d\epsilon_i.$

In thermodynamics we know that $1/T$ is the integrating factor of $\hat{d}Q$,

$$dS = \hat{d}Q/T, \tag{3.50}$$

and that all integrating factors for $\hat{d}Q$ differ only by a constant of proportionality. Comparing (3.46, 50),

$$1/T = k/\theta,$$

$$S = k\sigma, \tag{3.51}$$

where $k$ is the Boltzmann constant. This relates the statistical entropy $\sigma$ with the thermodynamic entropy $S$. We can take (3.51) to be an experimental fact, $\theta = kT$. From (3.24, 29, 51), $\partial\sigma/\partial E = \beta = 1/\theta$, or

$$\beta = 1/kT. \tag{3.52}$$

Once we know $E$ as a function of $\sigma$ and $V$, $E = E(\sigma, V)$, we can write

$$dE = \theta d\sigma - PdV = \left(\frac{\partial E}{\partial \sigma}\right)_V d\sigma + \left(\frac{\partial E}{\partial V}\right)_\sigma dV. \tag{3.53}$$

Therefore, we can define temperature and pressure as

$$\theta = (\partial E/\partial\sigma)_V, \quad -P = (\partial E/\partial V)_\sigma . \tag{3.54}$$

Instead of working with the parameters $(\sigma, V)$, it is sometimes more convenient to work with $(\theta, V)$, or with $(\theta, P)$. This requires the use of the thermodynamic functions $F(\theta, V)$ and $G(\theta, P)$.

*Helmholtz free energy* $F(\theta, V)$ is defined by

$$F = E - \theta\sigma = E - \sigma(\partial E/\partial\sigma)_V, \tag{3.55}$$

so that

$$\begin{aligned} dF &= dE - \theta d\sigma - \sigma d\theta = -\sigma d\theta - PdV \\ &= \left(\frac{\partial F}{\partial \theta}\right)_V d\theta + \left(\frac{\partial F}{\partial V}\right)_\theta dV. \end{aligned} \tag{3.56}$$

Comparing the coefficients,

$$-\sigma = (\partial F/\partial\theta)_V, \quad -P = (\partial F/\partial V)_\theta. \tag{3.57}$$

*Gibbs free energy* $G(\theta, P)$ is defined by

$$G = E - \theta\sigma + PV = E - \sigma(\partial E/\partial\sigma)_V - V(\partial E/\partial V)_\sigma , \tag{3.58}$$

$$dG = dE - \theta d\sigma - \sigma d\theta + PdV + VdP = -\sigma d\theta + VdP, \tag{3.59}$$

so that

$$-\sigma = (\partial G/\partial\theta)_P, \quad V = (\partial G/\partial P)_\theta. \tag{3.60}$$

From (3.43),

$$dE = \theta d\sigma - PdV + \mu dN. \tag{3.61}$$

Then

$$dG = -\sigma d\theta + VdP + \mu dN, \tag{3.62}$$

$$\left(\frac{\partial G}{\partial N}\right)_{\theta,\,P} = \mu, \tag{3.63}$$

or

$$G = N\mu. \tag{3.64}$$

## 3.4 ENTROPY OF AN IDEAL BOLTZMANN GAS USING THE MICROCANONICAL ENSEMBLE

Definitions (3.23) for entropy have been taken to be equivalent although $\Delta\Gamma(E)$ is the volume of a shell of thickness $\Delta E$ at $E$, while $\Omega_f(E)$ is the volume of the whole sphere from 0 to $E$ in the phase space. It works because in the microcanonical ensemble $\ln \Delta\Gamma$ is not sensitive to the value of $\Delta E$, (3.14). This is a great advantage because we do not know how to choose $\Delta E$ in each case of interest.

Consider an isolated ideal gas of $N$ molecules of mass $m$ in a volume $V$. The total energy is between $E$ and $E+\Delta E$. The microcanonical ensemble for this system occupies the phase space volume

$$\Delta\Gamma = \int_{E<H<E+\Delta E} d\Gamma$$

$$= \int dq_1 \ldots dq_{3N} \int dp_1 \ldots dp_{3N} = V^N \,\Delta\Gamma_p. \tag{3.65}$$

For $f(=3N)$ very large, $\Delta\Gamma_p$ is practically the volume of the whole sphere of radius $p = (2mE)^{1/2}$, $E = (1/2m) \sum_{i=1}^{N} p_i^2$, $\Delta\Gamma_p \simeq \Omega(E)$, as shown in (3.13). This means we can replace $\Delta E$ by the entire range from 0 to $E$ in the specification of the energy range of the microcanonical ensemble without causing any serious error. The constraint

$$E-\Delta E \leqslant \frac{1}{2m} \sum_{i=1}^{N} p_i^2 \leqslant E \tag{3.66}$$

is relaxed to read

$$0 \leqslant \frac{1}{2m} \sum_{i=1}^{N} p_i^2 \leqslant E. \tag{3.67}$$

Using (3.11), to sufficient accuracy,

$$\Delta\Gamma = V^N \,\Omega_f(E) = V^N \frac{\pi^{f/2}}{(\frac{1}{2}f)!} (2mE)^{f/2}, \tag{3.68}$$

$$\sigma = \ln \Delta\Gamma = N \ln [V\pi^{3/2}\,(2mE)^{3/2}\,(2e/3N)^{3/2}]$$
$$= N \ln [V(4\pi m/3)^{3/2}\,(E/N)^{3/2}] + 3N/2, \tag{3.69}$$

where we have used the Stirling approximation $n! \simeq (n/e)^n$ or $\ln n! \simeq n \ln n - n$ with $e = 2.71828$ as the base of the natural logarithm.

Note that the use of (3.68) makes the calculation of entropy to be quite an involved one even for an ideal gas in the method of microcanonical ensemble. This complication can be avoided by using other ensembles, as we shall see later on.

From (3.69) we can write the entropy of an ideal gas as

$$\sigma = N \ln [V(E/N)^{3/2}] + N\sigma_0, \tag{3.70}$$

$$\sigma_0 = \frac{3}{2}\left(\ln \frac{4\pi m}{3} + 1\right). \tag{3.71}$$

The result (3.70) for the entropy is not correct. It runs into two main difficulties:

(1) $\Delta\Gamma$ given by (3.65) has the dimension of (distance × momentum)$^{3N}$.

(2) $\sigma$ is not additive because the volume $V$ (and not $V/N$) occurs in the argument of the logarithm. This prevents us from dividing the system in two parts and writing $\sigma = \sigma_1 + \sigma_2$.

As shown in (3.4), the first difficulty is easily removed by measuring the phase space volume in terms of $h$. Thus $\Delta\Gamma$ is replaced by the dimensionless quantity $\Delta\Gamma/h^{3N}$ and (3.70, 71) read

$$\sigma = \ln (\Delta\Gamma/h^{3N}) = N \ln [V(E/N)^{3/2}] + N\sigma_0, \tag{3.72}$$

$$\sigma_0 = \frac{3}{2}\left(\ln \frac{4\pi m}{3h^2} + 1\right). \tag{3.73}$$

The second difficulty is not removed so easily. In fact, it leads to the famous Gibbs paradox.

## 3.5 GIBBS PARADOX

### (A) Mixing of Two Different Ideal Gases

The mixing of two different gases is an irreversible process. It is therefore attended by an increase of the entropy. Consider two different ideal gases $(N_1, V_1, T)$ and $(N_2, V_2, T)$, Fig. 3.6. They are allowed to mix by

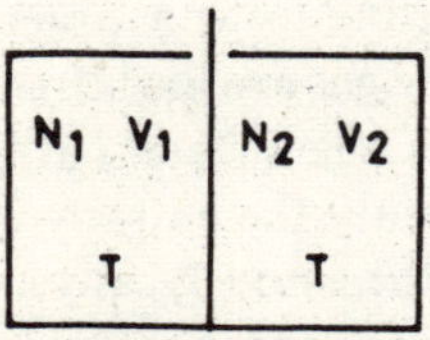


**Fig. 3.6 Mixing of two gases.**

removing the partition reversibly. It can be regarded as the expansion of each of the gases to the volume $V = V_1 + V_2$. The temperature, and therefore $E/N$, remains unchanged for each gas. From (3.72), the change in entropy is

$$\begin{aligned}\Delta\sigma &= \sigma_{12} - (\sigma_1 + \sigma_2)\\ &= (N_1 \ln V + N_2 \ln V) - (N_1 \ln V_1 + N_2 \ln V_2)\\ &= N_1 \ln (V/V_1) + N_2 \ln (V/V_2) > 0.\end{aligned} \tag{3.74}$$

This gives the entropy of mixing for two different ideal gases and is in agreement with experiments. For the case $N_1 = N_2 = N$ and $V_1 = V_2 = \frac{1}{2}V$, we get $\Delta\sigma = 2N \ln 2$.

### (B) Mixing of One Ideal Gas with the Same Ideal Gas

Suppose the two gases are the same. Then the removal of the partition should not affect the distribution of systems over the accessible states. The final entropy ought to be the same with, or without, the partition,

$$\Delta\sigma = \sigma_{12} - (\sigma_1 + \sigma_2) = 0. \tag{3.75}$$

This result is in agreement with the thermodynamics of a reversible process and also with experiments, but contradicts (3.72, 74). The derivation of (3.72, 74) does not depend on the identity of molecules and so would give the same increase in entropy (3.74) even in this case. In particular, for the case $N_1 = N_2 = N$, $V_1 = V_2 = \frac{1}{2}V$, we get an unobserved and therefore unaccountable increase of entropy by $2N \ln 2$ when a partition is simply removed from a box containing the same gas throughout. This is the *Gibbs paradox.*

This Gibbs paradox implies that the entropy of a given gas depends on the history of the gas. For example, if we imagine the present state of the gas to be achieved just by slowly removing one by one a large number of their partitions, then the final entropy can have any value one desires. This is certainly not tenable.

### Resolution of the Paradox

In the case (A), the removal of partition leads to the diffusion of the molecules throughout the whole volume $V$ (twice the volume if $V_1 = V_2 = \frac{1}{2}V$). There is a random mixing of the different molecules and so an increase of disorder. This is an irreversible process and the increase of entropy (3.74) makes sense.

We can imagine the mixing to be a process in which the positions of some of the molecules of one gas are interchanged with those of the other gas. Each such exchange creates a new state. Therefore, the number of accessible states increases or equivalently the entropy increases.

On the other hand, in the case (B), any such interchange of molecules is always an interchange between identical molecules. Therefore, no new state is created when the partition is removed. It follows that in this case the application of (3.72, 74) overestimates the number of accessible states because classically we have taken all the molecules, even of the same gas, as distinguishable.

The way out of the paradox is to regard all the identical molecules in the case (B) to be *indistinguishable*. If there are $N$ molecules, then $N!$ possible permutations among themselves do not lead to physically distinct situations. There is just one way of arranging them. Therefore, our estimation of number of accessible states, or equivalently of $\Delta\Gamma$, is too large by a factor of $N!$. We should replace $\Delta\Gamma$ by $\Delta\Gamma/N!$ (*Boltzmann counting*), so that (3.72) becomes

$$\sigma = \ln (\Delta\Gamma/N!\, h^{3N}) = \ln (\Delta\Gamma/h^{3N}) - \ln N!$$
$$= \ln (\Delta\Gamma/h^{3N}) - (N \ln N - N)$$
$$= N \ln [(V/N)\, (E/N)^{3/2}] + N\sigma_0, \tag{3.76}$$

$$\sigma_0 = \frac{3}{2} \ln \frac{4\pi m}{3h^2} + \frac{5}{2} \tag{3.77}$$

Use of (3.76) gives the correct result (3.75) for the case (B) and reproduces the result (3.74) for the case (A). Thus the Gibbs paradox is resolved because of the appearance of the extra term $-N \ln N$. It makes the entropy properly additive, as now $V/N$, rather than $V$, appears in the argument of the logarithm.

As indistinguishability of identical particles is assumed in quantum mechanics, the Gibbs paradox will not occur if $\Delta\Gamma(E)$ in (3.3) is calculated in quantum mechanics.

## 3.6 SACKUR-TETRODE EQUATION

The incorrect classical result (3.70) can be rescued, as shown above, by introducing the following two ideas:

(1) There exists an elementary cell of volume $h^f$ in the phase space in terms of which $\Delta\Gamma$ is to be measured.

(2) Molecules of a gas are indistinguishable.

Both these ideas occur naturally in quantum theory. In the classical picture we have to take them into account explicitly as ad hoc corrections. In this way, in 1911 Sackur obtained the correct equation for the entropy of an ideal gas which was verified experimentally by Tetrode in 1912.

The contact with the thermodynamic quantities is made by using $S = k\sigma$ and (3.54, 76),

$$\frac{1}{\theta} = \frac{1}{kT} = \left(\frac{\partial\sigma}{\partial E}\right)_{V,\,N} = \frac{\partial}{\partial E}\left(\frac{3N}{2} \ln E\right) = \frac{3N}{2E}, \tag{3.78}$$

or

$$E = 3N\theta/2 = 3NkT/2, \tag{3.79}$$

which is the familiar result for the internal energy of an ideal monatomic gas. Then (3.76) reads

$$S = Nk \ln [(V/N)\, (1/\lambda^3)] + 5Nk/2, \tag{3.80}$$

where

$$\lambda = h/(2\pi mkT)^{1/2} \tag{3.81}$$

is called the *thermal de Broglie wavelength* associated with the molecule of a gas at temperature $T$. The relation (3.80) is the famous *Sackur-Tetrode equation*. It is in complete agreement with experiments at higher temperatures.

Note that

$$k = \frac{\text{gas constant}}{\text{Avogadro number}} = \frac{R}{N_a} = \frac{8.314\ \text{JK}^{-1}}{6.02 \times 10^{23}} = 1.38 \times 10^{-23}\ \text{JK}^{-1},$$

$$Nk = (N/N_a)\, R = nR, \tag{3.82}$$

where $n$ is the number of moles in the system. From (3.80) we can write the entropy for one mole of an ideal gas as

$$S_{\text{mol}} = R \ln [(V/N_a)\,(2\pi mkT/h^2)^{3/2}\, e^{5/2}]. \tag{3.83}$$

For argon at 25°C and 1 atm, using

$$V = 22.414 \times 10^{-3}\ \text{m}^3,\ h = 6.62 \times 10^{-34}\ \text{Js},$$

$$m = 6.63 \times 10^{-26}\ \text{kg},\ T = 298.2\ \text{K},\ e = 2.718,$$

we get $S_{\text{mol}} = 154.7\ \text{JK}^{-1}\ \text{mol}^{-1}$.

From (3.39, 76),

$$\mu/kT = -(\partial\sigma/\partial N)_{E,\,V} = \ln (N\lambda^3/V). \tag{3.84}$$

The *equation of state* is obtained by using (3.34, 80),

$$P/T = (\partial S/\partial V)_{E,\,N} = kN/V, \quad \text{or} \quad PV = nRT. \tag{3.85}$$

## 3.7 ENTROPY AND PROBABILITY

The microcanonical ensemble is defined by (2.5), with the normalization

$$\sum_{i=1}^{\Omega} P_i = 1. \tag{3.86}$$

The quantity $\Omega$ is often called the *microcanonical partition function* and represents the number of states of the system between $E$ and $E + \Delta E$. $\Omega$ is proportional to $\Delta\Gamma$ or to $\Delta\Gamma/N!\,h^f$, and we can define entropy as

$$\begin{aligned}\sigma &= \ln \Omega \\ &= -\,\Omega\left(\frac{1}{\Omega}\right)\ln\left(\frac{1}{\Omega}\right) \\ &= -\sum_i P_i \ln P_i.\end{aligned} \tag{3.87}$$

Since $0 < P_i < 1$, we have $\ln P_i < 0$ and so the right hand side is properly positive.

We can interpret $\Delta\Gamma$ as a measure of the imprecision of our knowledge of the state of a system or as a measure of its randomness. In the macrostate of maximum $\Delta\Gamma$, the microstate of the system (in equilibrium) is least well defined. In fact, it is the macrostate of maximum disorder. The definitions (3.3, 8) imply that entropy is also a measure of the degree of disorder of a system in a given macrostate. It follows that *in the state of equilibrium the entropy attains its maximum value*. We can take (3.87) as the general definition of entropy. For an isolated system in equilibrium it reduces to Boltzmann's definition of entropy (3.23)

## 3.8 PROBABILITY DISTRIBUTION AND ENTROPY OF A TWO LEVEL SYSTEM

We can use the analogy between a cell in phase space and a quantum state to discuss at this stage an illustrative example of a simple two level system.

Consider a system of $N$ independent particles. Let each particle carry a magnetic moment $\mu$ that can be either parallel or antiparallel to an external magnetic field $H$. Let $+\epsilon$ be the energy associated with one orientation and $-\epsilon$ with the other, where $|\epsilon| = \mu H$ (Fig 3.7). The constant total energy $E$ of the isolated system is

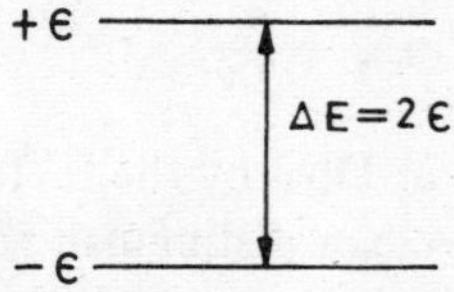


**Fig. 3.7 Two-level system.**

$$E = n\epsilon = -N_{-}\epsilon + N_{+}\epsilon, \quad n = N_{+} - N_{-}, \tag{3.88}$$

where $N_+(N_-)$ is the number of particles with energy $\epsilon(-\epsilon)$. Note that

$$N = N_{+} + N_{-}, \; N_{-} = \tfrac{1}{2}(N - n), \; N_{+} = \tfrac{1}{2}(N + n). \tag{3.89}$$

We would like to calculate the number of microstates accesssible to our isolated system with constant energy $E$.

We can arrange $N$ particles among themselves in $N!$ ways. Many of these ways do not give independent distinguishable arrangements into groups of $N_-$ and $N_+$ particles. The $N_-!$ interchanges of the $N_-$ particles just among themselves give no new arrangement. Similar situation holds for $N_+$ particles. Therefore, the total number $\Omega\{n\}$ of arrangements corresponding to a net magnetic moment $M = n\mu$ and energy $E = n\epsilon$ is

$$\Omega\{n\} = \frac{N!}{N_{-}!\,N_{+}!}. \tag{3.90}$$

For example, for the macrostate specified by $N = 4$, $E = 0$ (or $n = 0$, $N_+ = N_- = 2$), we have $4!/(2!\,2!) = 6$ possible microstates. Equal volumes of phase space are to be associated with each independent arrangement of the magnetic moments. Taking this volume to be unity for convenience, we have $\Omega\{n\} = W\{n\}$, where $W$ can be interpreted as the probability for a particular macrostate to occur. Then

$$\begin{aligned} \ln W &= \ln N! - \ln N_{-}! - \ln N_{+}! \\ &= N \ln N - N - N_{-} \ln N_{-} + N_{-} - N_{+} \ln N_{+} + N_{+} \\ &= -[N_{-} \ln(N_{-}/N) + N_{+} \ln(N_{+}/N)] \end{aligned}$$

$$= - \left[\frac{1}{2}N\left(1 - \frac{n}{N}\right)\ln\frac{1}{2}\left(1 - \frac{n}{N}\right) + \frac{1}{2}N\left(1 + \frac{n}{N}\right)\ln\frac{1}{2}\left(1 + \frac{n}{N}\right)\right], \tag{3.91}$$

where we have used the Stirling approximation. Using (3.84) and

$$\ln\left(1 \pm \frac{n}{N}\right) = \pm\frac{n}{N} - \frac{n^2}{2N^2} \pm \cdots, \tag{3.92}$$

we can write

$$\ln W \simeq - N_-\left[\ln\frac{1}{2} + \left(-\frac{n}{N} - \frac{n^2}{2N^2}\right)\right] - N_+\left[\ln\frac{1}{2} + \left(\frac{n}{N} - \frac{n^2}{2N^2}\right)\right]$$

$$= - N\ln\frac{1}{2} - \frac{n^2}{2N}. \tag{3.93}$$

We have not divided $\Delta\Gamma$ or $\Omega\{n\}$ by the factor $N!$, (3.76), because here the particles are *localized*: we can distinguish where the parallel and anti-parallel moments are located. The results (3.91, 93) can be used to obtain the following quantities.

**Probability $w(M)$ of a net Moment $M$**

In zero magnetic field the projection of each moment is equally likely to be $\pm\ \mu$. Consider a specific set of possible orientations. For each particle the probability is $\frac{1}{2}$ that it will take the orientation required by the specified assignment. Therefore, the probability that a given sequence of particles occurs is $(1/2)^N$.

The probability $w(M)$ of a net moment $M = n\mu$ is obtained on multiplying $W\{n\}$ by the probability $(1/2)^N$ of a specific sequence. Thus, using (3.93) for $\ln W$,

$$\begin{aligned} w(M) &= (\tfrac{1}{2})^N W\{n\} = (\tfrac{1}{2})^N \exp[\ln W] \\ &= (\tfrac{1}{2})^N[(\tfrac{1}{2})^{-N}\exp(-n^2/2N)] \\ &= \exp(-n^2/2N). \end{aligned} \tag{3.94}$$

This shows that the magnetization has a Gaussian distribution about the value $n = 0$ (Fig. 3.8). As expected, the average value of the magnetization

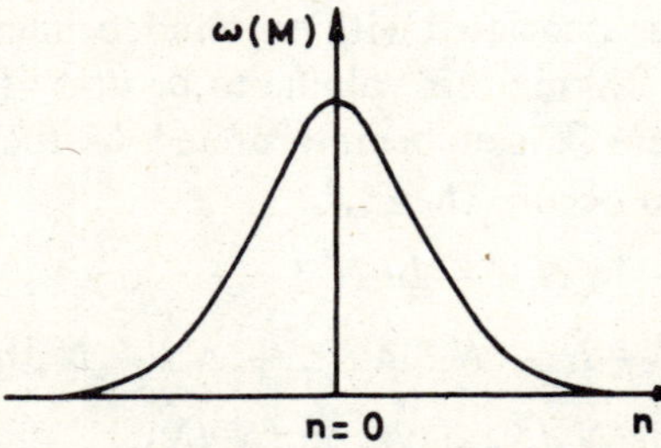


**Fig. 3.8 Gaussian distribution.**

in the absence of an external magnetic field is zero. Also, the most probable value is the same as the average value.

**Entropy**

The entropy is defined by (3.91),

$$S = -k\,[N_- \ln (N_-/N) + N_+ \ln (N_+/N)], \tag{3.95}$$

where $N_-$, $N_+$ depend on $E$ through $n$, (3.88, 89). In the approximation (3.93),

$$\begin{aligned} S \simeq k \ln W &= -k\left(N \ln \frac{1}{2} + \frac{n^2}{2N}\right) \\ &= -Nk\left[-\ln 2 + \frac{(E/\epsilon)^2}{2N^2}\right]. \end{aligned} \tag{3.96}$$

A plot of $S/Nk$ against $E/\epsilon N$ is shown in Fig. 3.9.

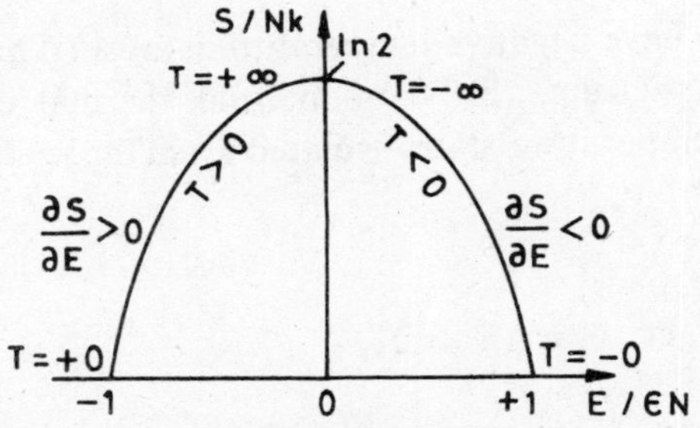


**Fig. 3.9** **Negative temperature in a two-level system. The slope $\partial S/\partial E$ gives the sign of $T$.**

The case of zero moment, $n = 0$, gives $S = kN \ln 2$ in the Stirling approximation ($N \rightarrow$ large). This is just the entropy of all possible arrangements ($2 \times 2 \times \ldots = 2^N$) because each particle can have two possibilities, parallel or antiparallel moment, regardless of how the other particles are. It shows that in calculating entropy no significant error is made if we assume that the entire accessible phase space has the properties of the most probable condition of the system with large $N$. In other words, for large $N$ the entropy is insensitive to the precise specification of the condition of the system in the neighbourhood of the most probable condition.

**Negative Temperature**

The temperature is defined by the slope of the $S$ versus $E$ curve (Fig. 3.9),

$$\frac{1}{T} = \left(\frac{\partial S}{\partial E}\right)_{V,N} = \frac{1}{\epsilon}\frac{\partial S}{\partial n} = \frac{1}{2}\frac{k}{\epsilon} \ln \frac{N-n}{N+n} = \frac{1}{2}\frac{k}{\epsilon} \ln \frac{N_-}{N_+}, \tag{3.97}$$

where we have used (3.89) for $N_+$, $N_-$ and (3.95) for $S$. At the conventional absolute zero all the particles are in the ground state ($\epsilon_-$), $n = N_+ - N_- = 0 - N = -N < 0$, giving $T = +0$ and $S = 0$ (complete order) at $E = -N\epsilon$. As the temperature is raised, or energy is given to the system, the population $N_+$ in the upper level ($\epsilon_+$) grows $T > 0$, till we have $n = 0$, giving $S = Nk \ln 2$ (maximum disorder) and $T = +\infty$. If more energy is given to the system, the upper level becomes more populated than the lower level, $N_+ > N_-$, $n \geqslant 0$, and we get a decrease in $S$ (more order) and $T < 0$. Thus the system is no longer normal in its behaviour. The *negative temperature* $T_-$ corresponds to higher energies than positive temperature $T_+$.

If $T_-$ and $T_+$ systems are brought into thermal contact, energy will flow from $T_-$ to $T_+$, that is, $T_-$ is hotter than $T_+$. A maser or a laser is a devise based on such a $T_-$ system.

As more and more energy is given to the system we get $n = N_+ - N_- = N - 0 = N > 0$, that is all the particles are in the upper states, $S=0$ (complete order), and $T = -0$ (Fig. 3.9). The difference between the two zeros $(+0, -0)$ is that we approach the first of them from the side of positive, and the second from the side of negative temperatures. The possible temperature ranges are from $+0$ through $+\infty$, $-\infty$ to $-0$, $+\infty$ and $-\infty$ coinciding with each other.

For a system to have negative temperature it must (i) have a finite upper limit to the energy spectrum, (ii) be in internal thermal equilibrium, and (iii) have negative temperature states isolated from those states that are at positive temperature.

**Specific Heat**

From (3.97), we get, with $\Delta E = 2\epsilon$,

$$\frac{N+n}{N-n} = \frac{N_+}{N_-} = \exp(-2\epsilon/kT) = \exp(-\Delta E/kT),$$

$$\frac{N_-}{N} = \frac{\exp(\epsilon/kT)}{\exp(\epsilon/kT) + \exp(-\epsilon/kT)} = \frac{1}{1 + \exp(-\Delta E/kT)},$$

$$\frac{N_+}{N} = \frac{\exp(-\epsilon/kT)}{\exp(\epsilon/kT) + \exp(-\epsilon/kT)} = \frac{1}{1 + \exp(\Delta E/kT)}. \tag{3.98}$$

These relations for $N_-/N$ and $N_+/N$ give the probabilities of finding any one particle in the states $-\epsilon$ and $\epsilon$, respectively. The energy $E$ and the specific heat $C$ are given by

$$E = n\epsilon = -(N_- - N_+)\epsilon = -N\epsilon \tanh(\epsilon/kT), \tag{3.99}$$

$$C = \frac{dE}{dT} = Nk \frac{(\epsilon/kT)^2}{\cosh^2(\epsilon/kT)}$$

$$= Nk\left(\frac{\Delta E}{kT}\right)^2 \frac{\exp(\Delta E/kT)}{[1 + \exp(\Delta E/kT)]^2}. \tag{3.100}$$

These quantities are plotted in Fig. 3.10a, b. The specific heat of the peaked form (Fig. 3.10b) is called the *Schottky anomaly*. It is observed when a body has a gap $\Delta E$ in its energy states.

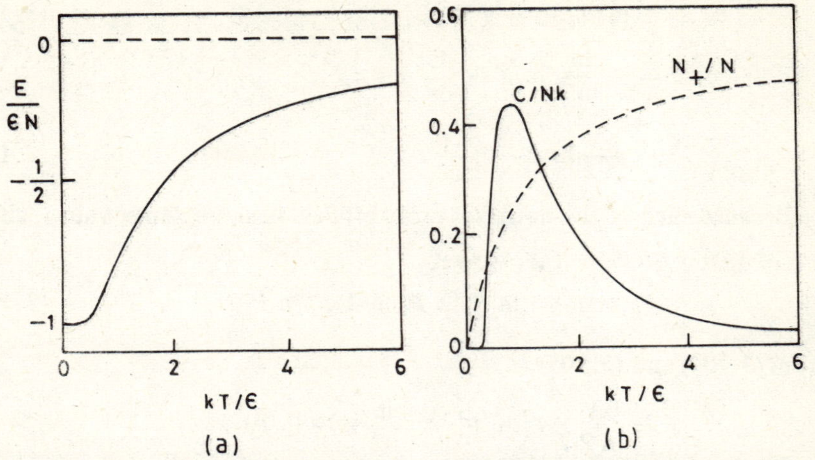


**Fig. 3.10 (a) Energy, and (b) specific heat and the fractional population $N_+/N$ in the upper state, as functions of $kT/\epsilon$ for a two-level system.**

## 3.9 ENTROPY AND INFORMATION THEORY

Comparison of (3.72) and (3.76) shows that the difference of $N!$ in counting implies

$$S_{\text{indistinguishable}} - S_{\text{distinguishable}} = -k \ln N!. \tag{3.101}$$

The fact that real identical atoms are not distinguishable results in a loss of entropy by $k \ln N!$. We have fewer microscopic states from among which the required macroscopic state can be chosen. Therefore, the macroscopic state is the less probable than it was when we could distinguish the atoms apart. We can view entropy as a measure of our 'uncertainty' as to the microscopic state of a thermodynamic system*. In the *information theory* formulation of statistical thermodynamics we can say that in this case our a priori uncertainty of the state of the gas has been reduced.

In general, let us consider a set of events. It may consist of values of a property, or states of a system. Initially we may know only that $Y$ of them occur with equal probability, $P° = 1/Y$, and the rest are impossible. A *bit* (binary unit of information) is defined *as the amount of information that allows one to make a choice between equally probable events.* In the case considered, after one bit is received, only $\frac{1}{2}Y$ events are known to be possible, equally so; after 2 bits $\frac{1}{4}Y = (\frac{1}{2})^2 Y$ events are known to be possible, equally so; after $J$ bits, $(\frac{1}{2})^J Y$ events are known to be possible, equally so. Each bit received reduces the field of uncertainty by half. Thus, after posing $J$ binary questions, the field of possible events is reduced from $Y$ to $X$, each with probability $P' = 1/X$. The probability of the remaining $Y - X$ events becomes zero. We can write

$$(\tfrac{1}{2})^J\, Y = X, \quad \text{or} \quad Y/X = 2^J, \tag{3.102}$$

*See, for example, E.T. Jaynes, *Phys. Rev.* **106**. 620, 1957; and **108**, 171, 1957.

$$\ln (Y/X) = J \ln 2, \tag{3.103}$$

$$J = -\frac{\text{bit}}{\ln 2} \ln X + \frac{\text{bit}}{\ln 2} \ln Y,$$

$$= \frac{\text{bit}}{\ln 2} \ln P' - \frac{\text{bit}}{\ln 2} P^{\circ}. \tag{3.104}$$

Because each of the nonzero probabilities has the same value, either $P_i^{\circ} = 1/Y$ or $P_i' = 1/X$, we get

$$\ln P_i' = \ln P_i' \sum_i P_i' = \sum P_i' \ln P_i'. \tag{3.105}$$

From (3.104) and (3.105),

$$J = \frac{\text{bit}}{\ln 2} \sum_i P_i' \ln P_i' - \frac{\text{bit}}{\ln 2} \sum_i P_i^{\circ} \ln P_i^{\circ}$$

$$= \mathcal{N}' - \mathcal{N}^{\circ} = \Delta \mathcal{N}$$

$$= \mathcal{S}^{\circ} - \mathcal{S}' = -\Delta \mathcal{S}, \tag{3.106}$$

where $\mathcal{S}$ is *Shannon's measure of uncertainty*, and $\mathcal{N}$ is the negative of Shannon's measure,

$$\mathcal{S} \equiv -\frac{\text{bit}}{\ln 2} \sum_i P_i \ln P_i, \quad \mathcal{N} \equiv \frac{\text{bit}}{\ln 2} \sum_i P_i \ln P_i. \tag{3.107}$$

Meaning of (3.106) is that the information received in bits is the difference between two functions of the state of knowledge of the system. The result obtained here for ensembles describing equally probable events is true for other ensembles as well.

For equally probable events,

$$P_1^{\circ} = \tfrac{1}{2}, \quad P_2^{\circ} = \tfrac{1}{2}, \tag{3.108}$$

a choice reduces the probability distribution (3.108) to certainty

$$P_1' = 1, \quad P_2' = 0, \tag{3.109}$$

with

$$J = \frac{\text{bit}}{\ln 2}(1 \ln 1 + 0 \ln 0) - \frac{\text{bit}}{\ln 2}(\tfrac{1}{2} \ln \tfrac{1}{2} + \tfrac{1}{2} \ln \tfrac{1}{2})$$

$$= 1 \text{ bit}, \tag{3.110}$$

as expected. The value of $\mathcal{S}$ is the number of binary questions that must be answered to reduce the uncertainty in knowledge to certainty. If we have two ensembles, the one with greater $\mathcal{S}$ has less information.

For the set of states $i$, we can find the probabilities $P_i$ corresponding to maximum ignorance (or minimum information) about the liklihood of those states by maximizing $\mathcal{S}$. It amounts to finding the extremum of

$$\mathcal{J} = \sum_{i=1}^{Y} P_i \ln P_i. \tag{3.111}$$

subject to the normalization

$$\sum_{i=1}^{Y} P_i = 1. \tag{3.112}$$

Due to (3.112), all the $P$'s are not independent. Suppose it fixes one called $P_Y$,

$$P_Y = 1 - \sum_{i=1}^{Y-1} P_i. \tag{3.113}$$

Now we can vary all the other $P$'s independently to maximize

$$\mathscr{I} = \sum_{i=1}^{Y-1} P_i \ln P_i + P_Y \ln P_Y. \tag{3.114}$$

Thus

$$\begin{aligned} 0 = \frac{\partial \mathscr{I}}{\partial P_j} &= \frac{\partial}{\partial P_j}(P_j \ln P_j + P_Y \ln P_Y) \\ &= 1 + \ln P_j + 1\frac{\partial P_Y}{\partial P_j} + \frac{\partial P_Y}{\partial P_j} \ln P_Y. \end{aligned} \tag{3.115}$$

Using $\partial P_Y/\partial P_j = -1$ from (3.113), we finally get

$$P_j = P_Y, \quad \text{for all } j. \tag{3.116}$$

All the $P$'s have the same value when $\mathscr{S}$ is maximum. The $P_i = 1/Y$ for all $i$ maximizes the uncertainty. This is understandable because to weigh $P_i$ differently, from, say $P_j$ in the ensemble, some more information is needed.

Above discussion shows that the entropy $S$ is proportional to Shannon's measure of uncertainty $\mathscr{S}$. It is a measure of uncertainty regarding the quantum state of the system. We can write

$$S = -k \sum_i \ldots, \quad \mathscr{S} = -\frac{\text{bit}}{\ln 2} \sum_i \ldots,$$

$$S = \frac{k \ln 2}{\text{bit}} \mathscr{S}, \tag{3.117}$$

where $\mathscr{S}$ is the number of binary choices to be made to find the quantum state of the system.

As a simple illustration of the information theory concept of entropy consider the dependence of $S$ on $V$ for an ideal gas. Let $N$ molecules, initially in volume $V_1$, finally occupy a volume $V_2$, where $V_2 < V_1$. The field of uncertainty in the location of each molecule is reduced by a factor of $V_2/V_1$. The gain in information is given by

$$J = \frac{\text{bit}}{\ln 2} \ln \frac{Y}{X} = \frac{\text{bit}}{\ln 2} \ln \frac{V_1}{V_2}. \tag{3.118}$$

The total gain in information is $NJ$, and the entropy change is

$$\Delta S = \frac{k \ln 2}{\text{bit}} \Delta \mathscr{S} = -\frac{k \ln 2}{\text{bit}} J = -Nk \ln \frac{V_1}{V_2}. \tag{3.119}$$

It agrees with the thermodynamic calculation.

*PROBLEMS*

3.1 Find the equation of state for an ideal gas enclosed in a box made of insulating walls and a movable piston loaded with weight $w$.

3.2 A system of $N$ independent harmonic oscillators enclosed in a volume $V$ satisfies a microcanonical distribution with an energy $E$. Find the phase volume $\Gamma$ and the entropy $S$.

3.3 Show that the absolute temperature is given by

$$\beta = \frac{1}{kT} = \frac{\partial \ln \Omega(E)}{\partial E}.$$

Does it imply that for any ordinary system $T > 0$? Estimate $\beta$ by assuming $\Omega(E) \propto (E - E_0)^f$, where $E_0$ is the ground state energy and $f$ the number of degrees of freedom of the system.

3.4 Verify that the absolute temperature of any system is an increasing function of its energy.

3.5 The atoms displaced from the lattice sites inside the crystal can migrate to the crystal surface forming Schottky defects. Consider a crystal of $N$ atoms with $n$ Schottky defects. If $w$ is the energy of formation of a Schottky defect, show that $S(n) = k \ln [(N+n)!/(n!\,N!)]$ and that in the equilibrium state at temperature $T$,

$$n/N = (e^{w/kT} - 1)^{-1}.$$

Estimate $n/N$ at $T = 290$ K and $10^3$ K, for $w = 1$ eV.

3.6 Consider a perfect crystal of $N$ atoms. We get $n$ Frenkel defects if $n$ atoms are replaced from the regular lattice sites to interstitial sites. The number $N'$ of interstitial sites into which an atom can enter is of the same order as $N$. Let $w$ be the energy required to move an atom from a lattice site to an interstitial site. Show that

$$S(n) = k \ln \frac{N!}{n!\,(N-n)!} - \frac{N'!}{n!\,(N'-n)!}$$

and that in the equilibrium state at temperature $T$, $w \gg kT$,

$$\frac{n^2}{(N-n)\,(N'-n)} = \exp(-w/kT).$$

3.7 Consider the following one-dimensional model of rubber. A chain consists of $n$ ($\gg 1$) elements, each of length $a$. The ends of the chain are distance $x (< na)$ apart. With origin at the left-end of the chain, if $n_+$ ($n_-$) is the number of links directed to the right (left), then show that

$$n_\pm = \frac{na \pm x}{2a}, \quad \Omega(x) = \frac{n!}{n_+!\,n_-!}, \quad \eta_t \approx \frac{kT}{na^2}\,x \text{ for } x \ll na,$$

where analogous to the pressure we have for the tension

$$\eta_t = -T\,(\partial S/\partial x)_E$$

and all orientations of the elements have the same energy.

3.8 Calculate the value of the thermal de Broglie wavelength $\lambda$ in A for an electron, a nucleon, and a $^4$He atom at room temperature.

[Ans. 745 $T^{-1/2}$; 17.4 $T^{-1/2}$; 8.7 $T^{-1/2}$]

3.9 Show that Shannon's measure of uncertainty based on the joint probability $P_{ij}$ of two events is given by the sum of the uncertainties of each event separately (based on $P_i$ and $P_j$), when the events $i$ and $j$ are uncorrelated.

3.10 Let $\rho_i\,(E_i)$ be the probability density that system $i$ is in a state with energy $E_i$. For two statistically independent systems

$$\rho\,(E_1 + E_2) = \alpha\rho_1\,(E_1)\,\rho_2\,(E_2).$$

If $\rho_1$, $\rho_2$ and $\rho$ have the same functional dependence on energy, then show that $\rho(E) = Ae^{-\beta E}$, when there are many states available to the system, irrespective of the assumptions of classical or quantum physics.

[Hint: Dropping the subscripts on $\rho$,

$$\frac{1}{\rho(E_1 + E_2)}\frac{d}{d(E_1 + E_2)}\rho(E_1 + E_2) = \frac{1}{\rho(E_1)}\frac{d}{dE_1}\rho(E_1) = \frac{1}{\rho(E_2)}\frac{d}{dE_2}\rho(E_2),$$

$$\frac{1}{\rho(E)}\frac{d}{dE}\rho(E) = \text{constant} = -\beta.]$$

# 4
# CANONICAL AND GRAND CANONICAL ENSEMBLES

## 4.1 CANONICAL ENSEMBLE

The microcanonical ensemble provides a general basis for statistical mechanics. It has a fundamental importance because (1) it deals with the simplest system known, that is, an isolated system, (2) the postulate of equal a priori probability is strictly applicable in this case, and (3) other ensembles can be deduced from the microcanonical ensemble. With its help we have shown that the basic postulates of statistical mechanics lead to correct thermodynamic relations. However, it is inconvenient in practice for two reasons: (1) We do not know how to specify the width of the ergodic shell between $E$ and $E + \Delta E$ in any given case. (2) It deals with systems, isolated from the rest of universe, with a given $E$, that do not occur in the laboratory. In thermodynamics we usually deal with systems kept in contact with a heat reservoir at a given temperature. The temperature is introduced in a artificial way, or through (3.31, 51). It would be useful to construct an ensemble based on this fact. In a microcanonical ensemble we expect the temperature to differ from element to element.

We have seen that an ensemble is stationary if $P_l = P_l(E)$. For the microcanonical ensemble we made the simple choice $P_l = P_l(E_l \text{ only}) =$ constant. We can make other choices. For example, we can replace the constant energy constraint with the constant temperature constraint. Thus for a system ($N$, $V$, $T$) in thermal equilibrium with a large heat reservoir (Fig. 4.1a) we can allow the energy to vary from element to element in the representative *canonical ensemble* (Fig. 4.1b). This is possible because the heat reservoir is such a large source of energy that the exchange of energy with a system in contact does not change it appreciably. We now determine a suitable functional dependence of $P_l$ on $E_l$ in such a case.

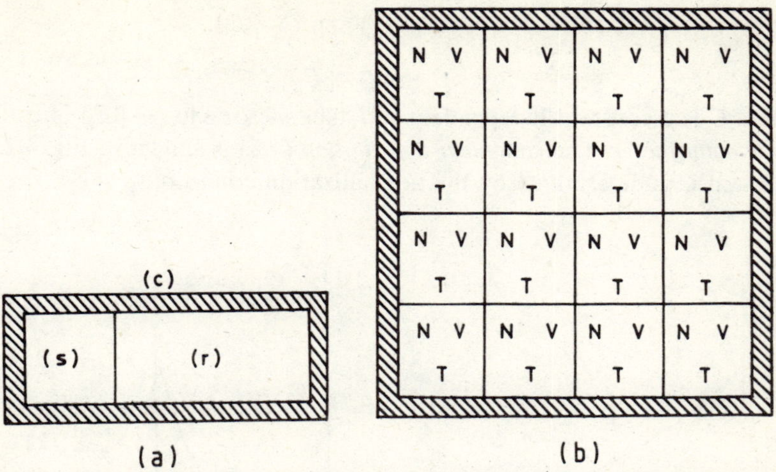


**Fig. 4.1** (a) Isolated composite system ($c$) consisting of the system ($s$) and the reservoir ($r$) in thermal contact. (b) Canonical ensemble of replicas in thermal contact (exchange of energy across the conducting walls) forming a lattice.

The composite system ($c$), consisting of system of interest ($s$) and reservoir ($r$), can be regarded as a single isolated system to which the method of microcanonical ensemble can be applied. If $P_c$ is the probability that the composite system is in $\Delta\Gamma_c\,(E_c, E_i)$, then by (3.5, 6),

$$P_c\,(E_c) \propto \Delta\Gamma_c\,(E_c, E_i) = \Delta\Gamma_s\,(E_i)\,\Delta\Gamma_r\,(E_r) \quad \text{for } E_c < E < E_c + \Delta$$
$$= 0 \qquad \text{otherwise,} \tag{4.1}$$

where $E_i$ is the unperturbed energy of $i$th quantum state of the system, $E_r$ is the energy of the reservoir ($E_r \gg E_i$), and $E_c$ is the constant total energy of the isolated composite system, $E_c = E_i + E_r =$ constant.

Let the system ($s$) be in one definite state $i$. Then the number of states accessible to the composite system ($c$) must simply be the number of states that are accessible to the reservoir ($r$). Thus

$$P_i \propto \Delta\Gamma_r\,(E_c - E_i), \tag{4.2}$$

where $P_i$ is the probability that the system is in the quantum state $i$. As $E_i \ll E_c$, we can expand the slowly varying function $\ln \Delta\Gamma_r\,(E_c - E_i)$ about the value $E_r = E_c$,

$$\ln \Delta\Gamma_r\,(E_o - E_i) = \ln \Delta\Gamma_r\,(E_c) - \left(\frac{\partial \ln \Delta\Gamma_r\,(E_r)}{\partial E_r}\right)_{E_r = E_c} + \ldots$$
$$= \ln \Delta\Gamma_r\,(E_c) - \beta E_i, \tag{4.3}$$

$$\beta = \left(\frac{\partial \ln \Delta\Gamma_r\,(E_r)}{\partial E_r}\right)_{E_r = E_c} = \frac{1}{kT}, \tag{4.4}$$

where $T$ is the constant temperature of the reservoir*. We can express (4.3) as

*According to R.C. Tolman (*The Principles of Statistical Mechanics*, OUP London 1938, p. 501) a canonical ensemble is the representative of a system in equilibrium (thermal contact) with a heat reservoir. The temperature $T$ is not mentioned and so the zeroth law of thermodynamics is not presupposed. Without the zeroth law one can still infer the existence of functions $\theta$ and $s$ of the coordinates of the system such that $dQ = \theta ds$ (see H.A. Buchadel, *J. Phys. A* **16**, 111, 1983).

$$\Delta\Gamma_r\,(E_c - E_i) = \Delta\Gamma_r\,(E_c)\exp(-\beta E_i), \tag{4.5}$$

or

$$P_i = C\exp(-\beta E_i), \tag{4.6}$$

where $C$ is a constant independent of $i$. The factor $\exp(-\beta E_i)$ is called the *Boltzmann factor*. The *canonical distribution* (4.6) is shown in Fig. 4.2a. The constant $C$ is determined by the normalization condition

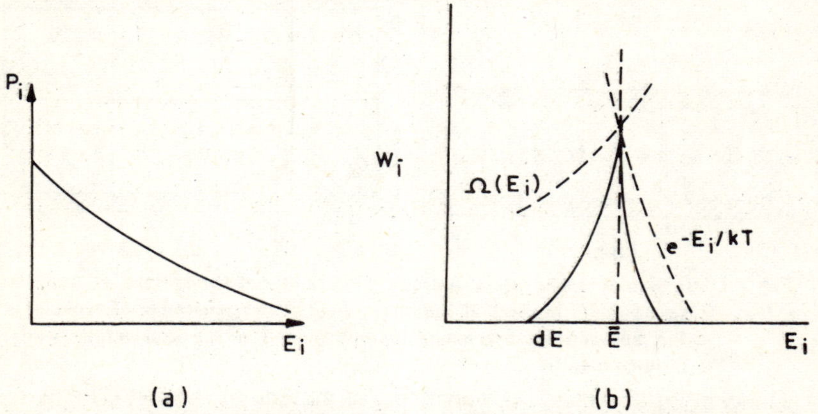


**Fig. 4.2** **(a) Canonical distribution. (b) A typical plot of probability $W_i$ that the system has a particular energy $\bar{E}$ as a function of energy $E_i$. It is the product of the functions $\Omega(E_i)$ and $P_i$. At $\bar{E}$ the $W_i$ has its maximum value. The total area under the curve is unity.**

$$\sum_i P_i = C\sum_i \exp(-\beta E_i) = 1. \tag{4.7}$$

Then the probability distribution is given by

$$P_i\,(N, V, T) = \frac{\exp[-\beta E_i(N, V)]}{Z\,(N, V, T)}, \quad Z = \sum_{\substack{i \\ \text{(states)}}} \exp[-\beta E_i(N, V)], \tag{4.8}$$

where $Z$ is called the *partition function of the system.*

The microscopically sharp energy requirement of the microcanonical distribution is replaced by one where the system may have any energy. The higher the energy of a quantum state, the less likely the system is to be in that state, (4.6). Note that all the $N$-particle quantum states with the same energy have the same probability. In this sense a canonical ensemble is constructed from thin slices of microcanonical ensembles, one for each different possible energy level, weighted as (4.6).

In spite of the form (4.6), Fig. 4.2a, the canonical distribution can predict average quantities with sharply defined means. The reason is that the probability $W_i$ that the system will be found with a particular energy $E_i$ is the sum of probabilities (4.8) of all states of this energy. The number of such states is the degeneracy $\Omega\,(E_i)$ of the energy level $E_i$. Since all the degenerate states have equal probabilities,

$$W_i = P_i\,\Omega\,(E_i) = \frac{\exp(-\beta E_i)\,\Omega\,(E_i)}{Z}$$

Being a product of an exponentially decreasing curve, $\exp(-\beta E_i)$, and a rapidly rising curve, $\Omega\,(E_i)$ of Fig. 2.1, the $W_i$ is very sharply peaked over a small range of energies (Fig. 4.2b). For macroscopic systems, Fig. 4.2b is similar to Fig. 1.9a.

Dropping the suffix $i$ in (4.6), we get $\ln P = \ln C - \beta E$. This quantity is *additive* for two subsystems in thermal contact. Thus, $\ln P_1 = \ln C_1 - \beta E_1$ and $\ln P_2 = \ln C_2 - \beta E_2$ give $\ln P = \ln P_1 P_2 = \ln C - \beta E$, where $C = C_1 C_2$ and $E = E_1 + E_2$.

Usually ($s$) is a large (macroscopic) system and therefore distinguishable. The distribution (4.6) gives the probability $P_i$ of finding the system in any one given state $i$ of energy $E_i$. The use of (4.6) can be extended to the case of molecules that are distinguishable. The gas is then said to be *non-degenerate* (or classical) in the sense that the mean separation of the molecules is larger than the typical thermal de Broglie wavelength of a molecule.

Suppose the gas containing $N$ molecules is in equilibrium at temperature $T$. The molecules being distinguishable, we can select a particular molecule and imagine it to be a small system in thermal contact with a heat reservoir consisting of all the other molecules of the gas. Applying (4.8) to the molecule,

$$p_{\mu i} = \frac{\exp(-\beta\epsilon_i)}{\sum_i \exp(-\beta\epsilon_i)} = \frac{\bar{n}_i}{N}, \quad \text{(molecule).} \tag{4.9}$$

Here $p_{\mu i}$ is the probability of finding the molecule in any of its state $i$, where its energy is $\epsilon_i$.

Consider a canonical ensemble of $M(M \to \infty)$ replicas of the system of interest (Fig. 4.1b). The statistical weight $\Omega_M$ of the ensemble associated with a particular macrostate $\{m_i\}$ of the canonical ensemble is

$$\Omega_M\{m_i\} = \frac{M\,!}{\prod_i m_i\,!}, \tag{4.10}$$

where $m_i$ is the number of *systems* in the state $i$.

If we maximize $\ln \Omega_M \{m_i\}$ subject to the constraints

$$\sum_i m_i = M, \qquad \sum_i m_i E_i = E, \tag{4.11}$$

we arrive at (4.8),

$$\frac{m_i}{M} = P_i = \frac{\exp(-\beta E_i)}{\sum_i \exp(-\beta E_i)} = \frac{\exp(-\beta E_i)}{Z}, \tag{4.12}$$

as it should be. We can, as before, identify $\beta = 1/kT$. Note that

$$\overline{m_i}/\overline{m_j} = \exp(-\beta E_i)/\exp(-\beta E_j). \tag{4.13}$$

The useful features of the Gibbs canonical ensemble are:

(1) The probability density depends on $E$ and $T$.

(2) The additive property of $\ln P_i$ allows us to couple two canonical ensembles with the same $T$ so that the resulting ensemble is again a canonical ensemble.

(3) Canonical ensembles apply equally to macroscopic systems or microscopic (atomic) subsystems.

(4) There is a strong resemblance between the probability density of a canonical ensemble and the distribution function of classical statistics. This is not accidental, because in the latter case we have an example of a subsystem (one molecule) in an isolated system (the gas).

(5) As the system of interest is in thermal equilibrium with the heat reservoir, fluctuations do not occur in temperature but appear in energy.

## 4.2 ENTROPY OF A SYSTEM IN CONTACT WITH A HEAT RESERVOIR

The entropy of the *ensemble* is given by

$$S_M = k \ln \Omega_M = k(\ln M! - \sum_i \ln m_i!)$$
$$\simeq k(M \ln M - \sum_i m_i \ln m_i).$$

The average entropy of one of the elements in the ensemble is

$$S = S_M/M. \tag{4.14}$$

To determine $S$, note that

$$\begin{aligned}(1/M) \ln \Omega_M &= (1/M) (\sum_i m_i \ln M - \sum_i m_i \ln m_i)\\ &= - \sum_i (m_i/M) \ln (m_i/M)\\ &= - \sum_i P_i \ln P_i,\end{aligned} \tag{4.15}$$

where $P_i = m_i/M$ is the probability that a *single system* chosen at random is in the state $i$. Thus

$$S = - k \sum_i P_i \ln P_i, \tag{4.16}$$

in agreement with (3.87).

The energy $E_i$ of the system is a function of $V$ and $N$. Therefore, the partition function of the system,

$$Z = \sum_i \exp(- \beta E_i) = \exp(- \beta E_1) + \exp(- \beta E_2) + \ldots \tag{4.17}$$

is a function of $V$, $N$ and $\beta$. The terms in the summation of $Z$ indicate how the systems are partitioned among the various energy states, (4.13).

As $\bar{R} = \sum_i P_i R(E_i)$ for the average value of a quantity $R(E_i)$, we can write

$$\bar{E} = \sum_i P_i E_i = Z^{-1} \sum_i (\exp(- \beta E_i)) E_i$$
$$= - Z^{-1} \sum_i \frac{\partial}{\partial \beta} [\exp(- \beta E_i)] = - Z^{-1} \frac{\partial}{\partial \beta} [\sum_i \exp(- \beta E_i)]$$

$$= -\frac{1}{Z}\frac{\partial Z}{\partial \beta} = -\frac{\partial \ln Z}{\partial \beta} \tag{4.18}$$

Substituting (4.8) in (4.16),

$$S = -k\sum_i P_i(-\beta E_i - \ln Z)$$

$$= k\beta\bar{E} + k \ln Z \sum_i P_i$$

$$= k\beta\bar{E} + k \ln Z, \tag{4.19}$$

$$dS = k\bar{E}d\beta + k\beta d\bar{E} + kd(\ln Z). \tag{4.20}$$

Noting that $Z = Z(\beta, V, N)$ and $E_i = E_i(V, N)$, we have

$$\frac{\partial(\ln Z)}{\partial V} = \frac{\partial}{\partial V} \ln \{\sum_i \exp[-\beta E_i(V, N)]\}$$

$$= -\beta \frac{\sum_i \frac{\partial E_i}{\partial V}\exp(-\beta E_i)}{Z} = -\beta \sum_i P_i(\partial E_i/\partial V), \tag{4.21}$$

$$\frac{\partial \ln Z}{\partial N} = -\beta \sum_i P_i(\partial E_i/\partial N), \tag{4.22}$$

$$d(\ln Z) = \frac{\partial \ln Z}{\partial \beta}d\beta + \frac{\partial \ln Z}{\partial V}dV + \frac{\partial \ln Z}{\partial N}dN$$

$$= -\bar{E}\,d\beta - \beta[\sum_i P_i(\partial E_i/\partial V)_N]\,dV - \beta[\sum_i P_i(\partial E_i/\partial N)_V]\,dN. \tag{4.23}$$

Therefore, on writing $\bar{E} = U$, (4.20) becomes

$$dS = k\beta dU - k\beta[\sum_i P_i(\partial E_i/\partial V)_N]\,dV - k\beta[\sum_i P_i(\partial E_i/\partial N)_V]\,dN$$

$$= \left(\frac{\partial S}{\partial U}\right)_{V,N} dU + \left(\frac{\partial S}{\partial V}\right)_{U,N} dV + \left(\frac{\partial S}{\partial N}\right)_{U,V} dN, \tag{4.24}$$

whence

$$1/T = (\partial S/\partial U)_{V,N} = k\beta, \tag{4.25}$$

$$P/T = (\partial S/\partial V)_{U,N} = -k\beta\sum_i P_i(\partial E_i/\partial V)_N, \tag{4.26}$$

$$-\mu/T = (\partial S/\partial N)_{U,V} = -k\beta\sum_i P_i(\partial E_i/\partial N)_V.$$

Substituting $\beta = 1/kT$ in (4.19),

$$S = (U/T) + k \ln Z(V, N, T), \tag{4.27}$$

and so the Helmholtz free energy $F(V, N, T)$ is

$$F(V, N, T) \equiv U - TS = -kT \ln Z(V, N, T), \tag{4.28}$$

$$Z(V, N, T) = \sum_i \exp[-E_i(V, N)/kT]. \tag{4.29}$$

This provides a complete thermodynamic description of the system. In particular,

$$\bar{E} = U = -\frac{\partial \ln Z}{\partial \beta} = -\frac{\partial \ln Z}{\partial T}\frac{\partial T}{\partial \beta} = kT^2 \frac{\partial \ln Z}{\partial T}, \tag{4.30}$$

$$C_V = \left(\frac{\partial U}{\partial T}\right)_V = \frac{k}{T^2}\frac{\partial^2 \ln Z}{\partial (1/T)^2}. \tag{4.31}$$

Dropping the suffix $i$, we can combine (4.6) and (4.28) to write the canonical distribution as

$$P(E) = \exp\,[(F - E)/kT], \tag{4.32}$$

because $C = 1/Z = \exp(F/kT)$.

## 4.3 IDEAL GAS IN CANONICAL ENSEMBLE

For an ideal (Boltzmann) gas consisting of $N$ molecules of mass $m$,

$$E = \sum_{i=1}^{3N} p_i^2/2m, \tag{4.33}$$

$$\exp(-E/kT) = \prod_i \exp(-p_i^2/2mkT), \tag{4.34}$$

$$Z = [\sum_{p_1} \exp(-p_1^2/2mkT)]\,[\sum_{p_2} \exp(-p_2^2/2mkT)]$$
$$\ldots [\sum_{p_{3N}} \exp(-p_{3N}^2/2mkT)] = [\sum_{p_i} \exp(-p_i^2/2mkT)]^{3N}. \tag{4.35}$$

We can replace the sum over discrete states by an integral over the phase space,

$$\sum_i \to \frac{1}{N!}\int \cdots \int \frac{dq_1 \ldots dq_{3N}\, dp_1 \ldots dp_{3N}}{h^{3N}}, \tag{4.36}$$

where $1/N!$ comes from the Boltzmann counting. Thus the canonical partition function is

$$Z = \frac{1}{N!\,h^{3N}}\int \exp\,[-E(p)/kT]\, d\Gamma$$
$$= \frac{1}{N!\,h^{3N}}\, V^N \left[\int_{-\infty}^{\infty} \exp(-p_i^2/2mkT)\, dp_i\right]^{3N}. \tag{4.37}$$

The integral can be evaluated by using $\int_{-\infty}^{\infty} e^{-ax^2}\, dx = (\pi/a)^{1/2}$,

$$Z = \frac{V^N}{N!}\left(\frac{2\pi mkT}{h^2}\right)^{3N/2} = \frac{z^N}{N!}, \tag{4.38}$$

where $z$ is the *single-particle partition function*, $N = 1$. We have

$$z = \sum_i \exp(-\beta\epsilon_i) = V\left(\frac{2\pi mkT}{h^2}\right)^{3/2} = \frac{V}{\lambda^3} = n_Q V = n_Q/n, \tag{4.39}$$

where $\beta\epsilon_i = p_i^2/2mkT$, $n_Q = 1/\lambda^3$ is called the *quantum concentration* and $n = 1/V$ is the concentration. The $n_Q$ is the concentration associated with one particle in a cube of side $\lambda$. Note that, in general, $Z \neq z^N$.

Using the Stirling formula $n! = (n/e)^n$, the Helmholtz free energy is given by

$$\begin{aligned} F = -kT \ln Z &= -NkT \ln \left[\frac{Ve}{N}\left(\frac{2\pi mkT}{h^2}\right)^{3/2}\right] \\ &= -NkT \ln (ez/N) \\ &= -NkT \ln (z/N) - NkT. \end{aligned} \tag{4.40}$$

Therefore, the *translational* contribution to entropy is

$$\begin{aligned} S = -(\partial F/\partial T)_V &= Nk \ln (ez/N) + \tfrac{3}{2} Nk \\ &= Nk \ln [(V/N)(1/\lambda^3)] + \tfrac{5}{2} Nk, \end{aligned} \tag{4.41}$$

in agreement with the Sackur-Tetrode equation.

We find that the knowledge of $Z$ allows us to calculate $S$ very easily. The entropy (4.41) tends to infinity as $T \rightarrow 0$. *The third law of thermodynamics* assumes that for *any* system

$$\lim_{T \to 0} S = 0. \tag{4.42}$$

This contradiction would not have occurred if we had used the original definition (4.17) of $Z$ where we have to sum over (discrete) states. The replacement of sum by the integral (4.37) is not justified near the absolute zero. At $T = 0$ the lowest state ($p = 0$) becomes important, while its contribution has been excluded altogether in (4.37). In classical statistics, since $p$ is a continuous variable and the size of cell in the phase space is not fixed, we cannot estimate (4.42). For this we have to go to quantum theory.

## 4.4 MAXWELL VELOCITY DISTRIBUTION

The canonical ensemble is applicable both to macroscopic and atomic subsystems. When applied to a single molecule of mass $m$ in a volume $V$, we can write (4.32) as

$$P(E)\, V d\Gamma_p = \exp (F/kT) \exp (-p^2/2mkT)\, (V/h^3)\, dp_x\, dp_y\, dp_z. \tag{4.43}$$

It gives the probability of finding the molecule in the momentum range $dp_x\, dp_y\, dp_z$ at $(p_x, p_y, p_z)$. The probability of finding the molecule in the velocity range $dc_x\, dc_y\, dc_z$ at $(c_x, c_y, c_z)$ can be expressed as

$$\exp (F/kT) \exp [-m(c_x^2 + c_y^2 + c_z^2)/2kT]\, (V m^3/h^3)\, dc_x\, dc_y\, dc_z.$$

Let us now evaluate $\exp (F/kT) \equiv 1/Z$, (4.28). For $N = 1$, (4.38) gives $Z = z = V/\lambda^3$. Therefore, the *Maxwell distribution of velocities* is given by

$$dn\,(c_x, c_y, c_z)\, dc_x\, dc_y\, dc_z = \left(\frac{m}{2\pi kT}\right)^{3/2} \exp[-m(c_x^2 + c_y^2 + c_z^2)/2kT]\, dc_x\, dc_y\, dc_z, \tag{4.44}$$

or the *Maxwell speed distribution* by

$$dn(c)\,dc = \left(\frac{m}{2\pi kT}\right)^{3/2} \exp\left(-mc^2/2kT\right) 4\pi c^2 dc, \tag{4.45}$$

where $c^2 = c_x^2 + c_y^2 + c_z^2$. The quantity $dn(c)dc$ is the probability that a particle has its speed in $dc$ at $c$.

Using $\epsilon = \frac{1}{2}mc^2$, $c^2dc = (2\epsilon)^{1/2}\,d\epsilon/m^{3/2}$, we get the probability distribution that a molecule has translational kinetic energy between $\epsilon$ and $\epsilon + d\epsilon$,

$$dn(\epsilon) = \frac{2}{\pi^{1/2}} \exp\left(-\epsilon/kT\right) \left(\frac{\epsilon}{kT}\right)^{1/2} d\left(\frac{\epsilon}{kT}\right)$$

$$\equiv f(\epsilon/kT)\, d(\epsilon/kT).$$

In Fig. 4.3 we show the plot of this energy distribution. The Boltzmann

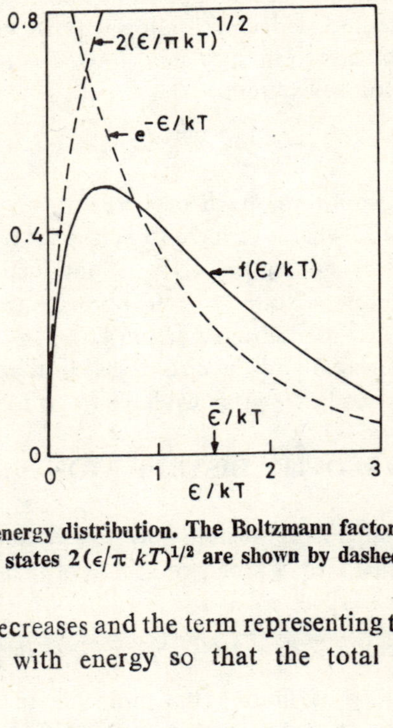


**Fig. 4.3** **Maxwell energy distribution. The Boltzmann factor $e^{-\epsilon/kT}$ and the density of states $2(\epsilon/\pi kT)^{1/2}$ are shown by dashed curves.**

factor $\exp(-\epsilon/kT)$ decreases and the term representing the density of states, $2(\epsilon/\pi kT)^{1/2}$, increases with energy so that the total distribution has a maximum.

## 4.5 EQUIPARTITION OF ENERGY

From (4.18, 38), for the case $N = 1$,

$$U = \boldsymbol{E} = \left(-\frac{1}{Z}\frac{\partial Z}{\partial \beta}\right)_{N=1} = \tfrac{3}{2}\,kT. \tag{4.46}$$

Thus the average energy associated with each variable, like $p_i (i = 1, 2$ or $3)$, which contributes a *quadratic* term to the energy, has the value $1/kT$. This can be verified by direct calculation. Suppose the Hamiltonian of a system of particles is a quadratic function of the $q$'s and the $p$'s,

$$H = \sum_i (a_i p_i^2 + b_i q_i^2).$$

For the particular term $a_i p_i^2$ the average energy is

$$a_i \int_0^\infty p_i^2 \exp(-a_i p_i^2/kT)\, dp_i \Big/ \int_0^\infty \exp(-a_i p_i^2/kT)\, dp_i$$

$$= -a_i kT \frac{\partial}{\partial a_i} \int_0^\infty \exp(-a_i p_i^2/kT)\, dp_i \Big/ \int_0^\infty \exp(-a_i p_i^2/kT)\, dp_i$$

$$= -a_i kT\,(\partial/\partial a_i) \left[\ln \int_0^\infty \exp(-a_i p_i^2/kT)\, dp_i\right]$$

$$= -a_i kT\,(\partial/\partial a_i)\,[\ln (\pi kT/a_i)^{1/2}] = \tfrac{1}{2}\,kT. \tag{4.47}$$

The same result is obtained for a term like $b_i q_i^2$. Thus each term in the $H$ which depends quadratically on a $p_i$ or a $q_i$ contributes a mean energy of $kT/2$ (*theorem of equipartition of energy*).

## 4.6 GRAND CANONICAL ENSEMBLE

In going from the microcanonical ensemble to the canonical ensemble, we relaxed the condition of constant energy $E$. This simplified the calculations in thermodynamics where the exchange of energy is a common phenomenon.

In chemical processes the number of particles varies. In quantum processes also particles are being created and destroyed. Therefore, it would be useful to relax the condition of constant total number of particles $N$ as well.

In the canonical ensemble the subsystem could exchange energy, but not particles, with the reservoir. We now consider the grand canonical ensemble in which the subsystem ($s$) can exchange energy, as well as particles, with the reservoir ($r$) (Fig. 4.4). The variable $N$ is replaced by the variable $\mu$, the chemical potential per particle. The composite system ($c$) is again represented by a microcanonical ensemble, Fig. 4.4b, because the total energy $E_c$ and the total number of particles $N_c$ are fixed,

$$E_c = E_s + E_r, \tag{4.48}$$

$$N_c = N_s + N_r, \tag{4.49}$$

$$\Delta\Gamma_c\,(E_c, N_c, E_s, N_r) = \Delta\Gamma_s\,(E_s, N_s)\,\Delta\Gamma_r\,(E_r, N_r). \tag{4.50}$$

The phase space now depends on the number of particles $N_i$ in the quantum state $i$ of the system, because it affects the number of dimensions. A particular quantum state of the system is denoted by $\psi_{N\,i}$.

We wish to find $P_{Ni}$, the probability in the ensemble of finding the system ($s$) in a given state $i$ when it contains $N_i = N$ particles and has an

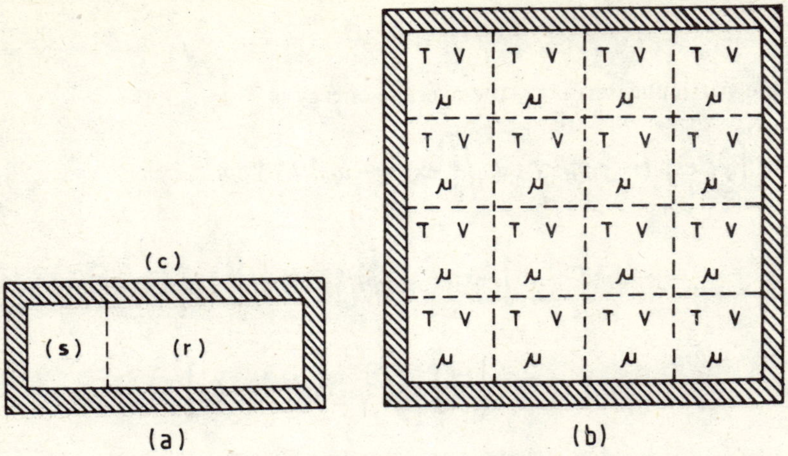


**Fig. 4.4** (a) Isolated composite system ($c$) consisting of the system ($s$) and the reservoir ($r$) in thermal and material contact. (b) Grand cononical ensemble of replicas in thermal and material contact (exchange of energy and particles across the conducting and porous walls) forming a lattice.

energy $E_i = E_{Ni}$. Following the arguments similar to those leading to (4.2),

$$P_{Ni}(E_{Ni}, N) \propto \Delta\Gamma_r (E_c - E_{Ni}, N_c - N). \tag{4.51}$$

Since ($s$) is very small compared to ($r$), $E_{Ni} \ll E_c$ and $N \ll N_c$.

$$\ln \Delta\Gamma_r (E_c - E_{Ni}, N_c - N) = \ln \Delta\Gamma_r (E_c, N_c) - \left(\frac{\partial \ln \Delta\Gamma_r (E_r, N_r)}{\partial E_r}\right)_{E_r = E_c} E_{Ni}$$
$$- \left(\frac{\partial \ln \Delta\Gamma_r (E_r, N_r)}{\partial N_r}\right)_{N_r = N_c} N. \tag{4.52}$$

The derivatives are evaluated for $E_r = E_c$ and $N_r = N_c$ and so are constants characterizing the reservoir ($r$). We can denote them as, (3.29, 39),

$$\beta = \left(\frac{\partial \ln \Delta\Gamma_r}{\partial E_r}\right)_{E_r = E_c}, \quad -\beta\mu = \left(\frac{\partial \ln \Delta\Gamma_r}{\partial N_r}\right)_{N_r = N_c}, \tag{4.53}$$

where the chemical potential $\mu$ represents Gibbs free energy per particle, (3.64). Then (4.52) reads

$$\Delta\Gamma_r (E_c - E_{Ni}, N_c - N) = \Delta\Gamma_r (E_c, N_c) \exp[-\beta(E_{Ni} - N\mu)]. \tag{4.54}$$

Since $\Delta\Gamma_r(E_c, N_c)$ is just a constant independent of $i$ and $N$, (4.51) can be written as

$$P_{Ni}(E_{Ni}, N) = C \exp[-\beta(E_{Ni} - N\mu)], \tag{4.55}$$

$$C = B\, \Delta\Gamma_r (E_c, N_c), \tag{4.56}$$

where $C$ and $B$ are constants independent of $i$ and $N$. This is called the *grand canonical distribution.*

The constant $C$ in (4.55) is determined by the normalization condition

$$\sum_{N,i} P_{Ni}(E_{Ni}, N) = C \sum_{N,i} \exp[-\beta(E_{Ni} - N\mu)] = 1. \tag{4.57}$$

Then

$$P_{Ni}\,(E_{Ni}, N) = \frac{\exp\left[-\beta(E_{Ni} - N\mu)\right]}{\mathscr{Z}}, \; \mathscr{Z} = \sum_{N,i} \exp[-\beta(E_{Ni} - N\mu)], \quad (4.58)$$

where $\mathscr{Z}$ is called the *grand partition function.* It is the sum of the canonical partition functions $Z(N)$ for ensembles with different $N$'s, with weighting factors $\exp(\beta N\mu)$,

$$\mathscr{Z} = \sum_{N=0}^{\infty} Z(N) \exp(\beta N\mu), \quad Z(N) = \sum_{i} \exp(-\beta E_{Ni}). \quad (4.59)$$

Consider a grand canonical ensemble of $M(M \rightarrow \infty)$ elements (Fig. 4.3b). The state of each element is characterized by the energy $E_{Ni}$ and the number $N$ of particles in it. The statistical weight $\Omega_{gM}$ of the ensemble associated with a particular macrostate $\{m_{Ni}\}$ is

$$\Omega_{gM}\,\{m_{Ni}\} = \frac{M!}{\prod_{N}\prod_{i} m_{Ni}!}. \quad (4.60)$$

To find the most probable macrostate $\{m_{Ni}\}$ we maximize $\Omega_{gM}\,\{m_{Ni}\}$ subject to the constraints which are generalizations of (4.48, 49),

$$\sum_{N,i} m_{N_i} = M,$$

$$\sum_{N,i} m_{N_i} E_{N_i} = E_c,$$

$$\sum_{N,i} N\, m_{N_i} = N_c, \quad (4.61)$$

where $N_c$ is the total number of particles in the ensemble. The result is

$$\frac{\overline{m}_{Ni}}{M} = P_{Ni} = \frac{\exp\left[-\beta(E_{Ni} - N\mu)\right]}{\sum_{N,i} \exp\left[-\beta(E_{Ni} - N\mu)\right]}, \quad (4.62)$$

as expected, with the identifications $\beta = 1/kT$, and $\alpha = -\mu/kT$ for the Lagrange multipliers. Here $\alpha$ (or $\mu$) is determined by the last condition in (4.61). The identification of $\beta$ follows from the fact that we get back the canonical distribution if assume $N$ to have a fixed value.

We can again define entropy by

$$S = -k \sum_{N,i} P_{Ni} \ln P_{Ni}. \quad (4.63)$$

Substituting (4.58) in (4.63) and noting that $E_{Ni}$ is a function of $V$ alone,

$$S = k\beta\overline{E} + k \ln \mathscr{Z} - k\beta\overline{N}\mu, \quad (4.64)$$

$$dS = k\beta d\bar{E} - k\beta \left[\sum_{N,i} P_{Ni}(\partial E_{Ni}/\partial V)\right] dV - k\beta\mu d\bar{N},$$

$$\beta = 1/kT,$$

$$P = -\sum_{N,i} P_{Ni}\,(dE_{Ni}/dV). \tag{4.65}$$

We can now rewrite (4.64) as

$$\Omega_g = U - TS - \mu\bar{N} = -P(T, V, \mu)V = -kT \ln \mathscr{Z}(T, V, \mu), \tag{4.66}$$

where $\bar{E} = U$ and $\Omega_g$ is the *grand canonical potential*, which determines the entire thermodynamics. In particular,

$$F \equiv U - TS = \Omega_g + \mu\bar{N},$$

$$G \equiv U - TS + PV = \mu\bar{N},$$

$$S = -(\partial\Omega_g/\partial T)_{V,\mu}, \quad \bar{N} = -(\partial\Omega_g/\partial\mu)_{V,T}. \tag{4.67}$$

Note that from (4.58, 66), dropping the suffixes,

$$P(N) = \exp\left[(\Omega_g - E + N\mu)/kT\right], \tag{4.68}$$

$$\mathscr{Z} = \exp(-\Omega_g/kT) = \sum_N \exp(N\mu/kT) \int \exp[-E(N)/kT]\, d\Gamma(N). \tag{4.69}$$

## 4.7. IDEAL GAS IN GRAND CANONICAL ENSEMBLE

We can write for an ideal gas

$$Z(N) = \frac{1}{N!\,h^{3N}} \int \exp[-E(N)/kT]\, d\Gamma(N)$$

$$= \frac{V^N}{N!}\left(\frac{2\pi mkT}{h^2}\right)^{3N/2} = \frac{z^N}{N!}, \tag{4.70}$$

$$\mathscr{Z} = \sum_N e^{N\mu/kT} Z(N) = \sum_{N=0}^{\infty} \frac{(e^{\mu/kT}\, z)^N}{N!} = \exp(ze^{\mu/kT}) = \exp(z\eta_a) \tag{4.71}$$

where we have used the series expansion $e^x = \sum_{n=0}^{\infty} (x^n/n!)$ and introduced the convenient notation $\eta_a$ for the *absolute activity* (or *fugacity*),

$$\eta_a \equiv e^{\mu/kT}. \tag{4.72}$$

It follows that

$$\Omega_g = -kT \ln \mathscr{Z} = -kTz\eta_a = -kT\, e^{\mu/kT}\left(\frac{2\pi mkT}{h^2}\right)^{3/2} V, \tag{4.73}$$

whence

$$S = -\left(\frac{\partial\Omega_g}{\partial(kT)}\right)_{V,\mu} = V\,\frac{(2\pi mkT)^{3/2}}{h^3}\, e^{\mu/kT}\left(\frac{5}{2} - \frac{\mu}{kT}\right), \tag{4.74}$$

$$\bar{N} = -\left(\frac{\partial \Omega_g}{\partial \mu}\right)_{V,\,T} = V\frac{(2\pi mkT)^{3/2}}{h^3}\, e^{\mu/kT} = -\,\Omega_g/kT. \tag{4.75}$$

The relation (4.74) is the Sackur-Tetrode equation. From (4.75) we get the chemical potential per particle for an ideal gas,

$$\mu = -\,kT\ln\left[(V/\bar{N})\,(2\pi mkT)^{3/2}/h^3\right] = -\,kT\ln(z/\bar{N}) = kT\ln(n/n_Q), \tag{4.76}$$

where $n = \bar{N}/V$ is the concentration of particles and $n_Q = 1/\lambda^3$, (4.39). Thus *$\mu$ increases with increase in n.* We see from (4.72, 76) that *for an ideal gas $\eta_a$ is directly proportional to the concentration.* The pressure is given by (4.73, 75),

$$P = -\,(\partial\Omega_g/\partial V)_{T,\,\mu} = kT\bar{N}/V, \tag{4.77}$$

which is the perfect gas law.

From (4.58, 69, 70) for the ideal gas

$$P_{Ni}\,(E_{Ni}, N) = \frac{1}{N!\,h^{3N}}\,\frac{\exp\left[-\,(E_{Ni} - N\mu)/kT\right]}{\mathscr{Z}}. \tag{4.78}$$

Let us take the sum of $P_{Ni}$ over $i$ for a given $N$, (4.70). This gives the probability $P_N$ that a volume $V$ of the ideal gas at equilibrium will happen to have $N$ molecules in it, irrespective of the energy of the subsystem,

$$\begin{aligned} P_N &= \frac{e^{N\mu/kT}}{\mathscr{Z}}\cdot\frac{\sum_i \exp(-\,E_{Ni}/kT)}{N!\,h^{3N}}. \\ &= \frac{e^{N\mu/kT}}{\mathscr{Z}}\cdot\frac{1}{N!\,h^{3N}}\int \exp\left[-\,E(N)/kT\right] d\Gamma(N) \\ &= (1/\mathscr{Z})\, e^{N\mu/kT}\, Z(N), \end{aligned} \tag{4.79}$$

or

$$\begin{aligned} P_N &= \exp\left[(\Omega_g + N\mu)/kT\right] Z(N) \\ &= \exp(-\,\bar{N})\left(\frac{\bar{N}}{z}\right)^N \frac{z^N}{N!} = \frac{1}{N!}\exp(-\,\bar{N})\,(\bar{N})^N, \end{aligned} \tag{4.80}$$

where in the end we have used (4.75, 76). This is the *Poisson distribution*, which exhibits a maximum near $N = \bar{N}$. Thus the bar can be dropped from $N$ and (4.66) can be taken as a proper definition of $\Omega_g$.

## 4.8 COMPARISON OF VARIOUS ENSEMBLES

We have shown that all the three ensembles, microcanonical (Fig. 1.9), canonical (Fig. 4.1) and grand canonical (Fig. 4.4) are applicable, in principle, to the determination of the thermodynamic properties of a system. The three ensembles are compared in Table 4.1. As far as thermodynamic calculations are concerned, it is simply a matter of convenience which method is followed All of them give equivalent results. Usually, the most convenient from the point of view of factorizability of the partition func-

**Table 4.1. Comparison of the various ensembles**

| Ensemble (macroscopic constraints) | Distribution | Partition function | Thermodynamic function | Contact with surrounding |
|---|---|---|---|---|
| Microcanonical ($V, N, E$) | $P(E) = \delta(E - E_0)$ | $\Delta\Gamma$ or $\Omega(n_i)$ | $-TS = -kT \ln \Omega$ | none |
| Canonical ($V, N, T$) | $P_l = C\, e^{-\beta E_l}$ | $Z = \int e^{-E/kT}\, d\Gamma$ | $F = E - TS$ $= -kT \ln Z$ | thermal |
| Grand canonical ($V, \mu, T$) | $P_{Ni} = C\, e^{-\beta(E_{Ni} - N\mu)}$ | $\mathscr{Z} = \sum_N e^{N\mu/kT} \cdot \int e^{-E/kT}\, d\Gamma(N)$ | $\Omega_g = E - TS - N\mu$ $= -kT \ln \mathscr{Z}$ | thermal + material |
| $T-P$ distribution ($T, P, N$) | $P(dV, i) = C\, e^{-[E_i(V)+PV]}\, dV$ | $Y = \int_0^\infty e^{-PV/RT} . Z(T, V, N) dV$ | $G = -kT \ln Y$ | thermal + pressure (movable wall) |

tion is the grand canonical ensemble. It is possible to construct other ensembles as the need arises. An example is of $T$-$P$ distribution with independent parameters $(T, P, N)$, and volume variable.

The relation between the grand canonical ensemble and the canonical ensemble is in some sense similar to the relation between the canonical and microcanonical ensembles. The description of a subsystem by means of the microcanonical distribution ignores fluctuations in its total energy while the canonical distribution takes it into account. However, the latter ignores the fluctuations in the number of particles (that is, it is microcanonical with respect to the number of particles), whereas the grand canonical distribution takes this into account (that is, it is canonical both as regards energy and the number of particles). If we neglect fluctuations in $N$, we have $\Omega_g + \mu N = F$ and the distribution (4.68) reduces to (4.32).

## 4.9 QUANTUM DISTRIBUTIONS USING OTHER ENSEMBLES

### Canonical Ensemble

The canonical partition function (4.17, 37) for a system is

$$Z = \int \exp\,[-E(q, p)/kT]\, d\Gamma \qquad \text{(classical)}, \tag{4.81}$$

$$Z = \sum_i \exp\,(-E_i/kT), \qquad \text{(quantum)}. \tag{4.82}$$

For a system $(N, V, T)$ the canonical distribution is

$$P_i = \frac{\exp\,(-\beta E_i)}{Z}, \tag{4.83}$$

where $\beta = 1/kT$, and $P_i$ is the probability for a given system in the ensemble to be in the state $i$.

We can use (4.83) to describe a sytsem $(N, V, T)$ consisting of non-interacting bosons or fermions. The wave functions, occupation numbers and macroscopic constraints are

$$\Psi = \begin{cases} \Psi^{(S)}\,(n_1, n_2, \ldots, n_i, \ldots), & \text{(BE)}, \\ \Psi^{(A)}\,(n_1, n_2, \ldots, n_i, \ldots), & \text{(FD)}, \end{cases} \tag{4.84}$$

$$n_i = \begin{cases} 0, 1, 2, 3, \ldots, & (\text{all } i,\ \text{BE}), \\ 0 \text{ or } 1 \quad , & (\text{all } i,\ \text{FD}), \end{cases} \tag{4.85}$$

$$\sum_i n_i = N, \quad \sum_i n_i \epsilon_i = \sum_i E_i = E, \qquad \text{(BE, FD)}. \tag{4.86}$$

The canonical distribution (4.83) is

$$P(n_1, n_2, \ldots, n_i, \ldots) = \frac{\exp\,(-\sum_i n_i \epsilon_i/kT)}{\sum_{\{n_i\}} \exp\,(-\sum_i n_i \epsilon_i/kT)}, \tag{4.87}$$

where the summation in the denominator is over the set of all $n_i$ satisfying (4.85).

The mean value of $n_l$ is given by the average over the ensemble,

$$\bar{n}_l = \sum_{\{nj\}} n_l P\{n_j\} = - kT \frac{\partial}{\partial \epsilon_l} \ln Z \tag{4.88}$$

where

$$Z = \sum_{\{n_l\}} \exp\left(- \sum_l n_l \epsilon_l / kT\right) = \sum_{\{n_l\}} \prod_l \exp\left(-n_l \epsilon_l / kT\right). \tag{4.89}$$

The evaluation of Z, and therefore of $\bar{n}_l$, is difficult because of the restrictions on $n_l$. We do this here for the FD case.

The constraints $n_l = 0$ or 1 and $\Sigma\, n_l = N$ can be incorporated in the definition of $Z$ by using the Kronecker delta (not to be confused with the Dirac delta function)

$$\begin{aligned} \delta(n) &= 0 \qquad \text{for } n = \pm 1, \pm 2, \ldots, \\ &= 1 \qquad \text{for } n = 0, \end{aligned} \tag{4.90}$$

which can be written as

$$\delta(n) = \frac{1}{2\pi} \int_{-\pi}^{\pi} e^{in\theta}\, d\theta = \frac{e^{\alpha n}}{2\pi} \int_{-\pi}^{\pi} e^{in\theta}\, d\theta, \tag{4.91}$$

where $\alpha$ is some arbitrary real number. Then

$$\begin{aligned} Z &= \sum_{n_1=0}^{1} \ldots \sum_{nj=0}^{1} \ldots \prod_l \exp\left(- n_l \epsilon_l / kT\right) \delta\left(N - \sum_l n_l\right) \\ &= \frac{e^{\alpha N}}{2\pi} \int_{-\pi}^{\pi} e^{iN\theta} \prod_l \sum_{n_l=0}^{1} \exp\{-[(\epsilon_l/kT) + \alpha + i\theta]\, n_l\}\, d\theta \\ &= \frac{e^{\alpha N}}{2\pi} \int_{-\pi}^{\pi} f(\theta)\, d\theta, \end{aligned} \tag{4.92}$$

$$f(\theta) = e^{iN\theta} \prod_l \left(1 + b_l\, e^{-i\theta}\right), \; b_l = \exp\{-[(\epsilon_l/kT) + \alpha]\}. \tag{4.93}$$

The absolute value of $f(\theta)$ has a sharp peak at $\theta = 0$. We can choose $\alpha$ such that $df/d\theta = 0$ at $\theta = 0$. Then the phase of $f(\theta)$ will not change rapidly at $\theta = 0$, and most of the contribution to $Z$ will occur around $\theta = 0$. For convenience we work with the slowly varying function $\ln f(\theta)$, such that

$$\left[\frac{d}{d\theta} \ln f(\theta)\right]_{\theta=0} = 0. \tag{4.94}$$

Then $N$ is determined by

$$N = \sum_l \frac{b_l}{1 + b_l}, \tag{4.95}$$

and

$$f(\theta) = \exp[\ln f(\theta)]$$

$$= \exp\left\{\ln f(0) + \left[\frac{d}{d\theta}\ln f(\theta)\right]_{\theta=0} \theta + \frac{1}{2}\left[\frac{d^2}{d\theta^2}\ln f(\theta)\right]_{\theta=0} \theta^2 + \ldots\right\}$$

$$= f(0)\exp(-aN\theta^2/2), \quad a = -\frac{1}{N}\left[\frac{d^2}{d\theta^2}\ln f(\theta)\right]_{\theta=0}. \tag{4.96}$$

It follows that

$$Z = \frac{e^{\alpha N}}{2\pi} f(0) \int_{-\pi}^{\pi} \exp(-aN\theta^2/2)\, d\theta$$

$$= \frac{e^{\alpha N}}{(\pi a N)^{1/2}} f(0), \tag{4.97}$$

and from (4.88)

$$\bar{n}_l = -kT\frac{\partial}{\partial \epsilon_l}\ln Z$$

$$= -kT\left[N\frac{\partial \alpha}{\partial \epsilon_l} - \frac{1}{2}\frac{\partial}{\partial \epsilon_l}\ln a + \frac{(-1/kT)\, b_l}{1 + b_l} - \sum_i \left(\frac{b}{1 + b_i}\right)\frac{\partial \alpha}{\partial \epsilon_i}\right]$$

$$= \frac{1}{2}kT\frac{\partial}{\partial \epsilon_l}\ln a + \frac{b_l}{1 + b_l} \simeq \frac{b_l}{1 + b_l}, \tag{4.98}$$

neglecting the small quantity $\frac{1}{2}kT\,\partial(\ln a)/\partial\epsilon_l$. Thus*

$$\bar{n}_l = \frac{1}{\exp[\alpha + (\epsilon_l/kT)] + 1}, \qquad \text{(FD).} \tag{4.99}$$

The derivation is very simple and direct if we use the grand canonical ensemble due to greater factorizability of the grand partition function.

**Grand Canonical Ensemble**

The grand partition function (4.59) is

$$\mathscr{Z} = \sum_{N=0}^{\infty} e^{\beta N \mu} Z(N) = \sum_{N=0}^{\infty} \sum_{\{n_i\}} \exp[\beta \sum_l (\mu - \epsilon_l)\, n_l]$$

$$= \sum_{N=0}^{\infty}\left\{ \sum_{\substack{n_1, n_2, \ldots \\ (n_1+n_2+\ldots=N)}} \exp[-\beta(n_1\epsilon_1 + n_2\epsilon_2 + \ldots)] \exp[\beta\mu(n_1+n_2+\ldots)]\right\}. \tag{4.100}$$

*The derivation given here is after G. Speisman. (see, E.A, Desloge, *Statistical Physics*, Holt, Rinehart and Winson, N.Y., 1966, appendix 12). Similar arguments are given by J.S.R. Chisholm and A.H. de Borde, *An Introduction to Statistical Mechanics*, Pergamon Press, London, 1958, pp. 19–22.

In the grand canonical ensemble the limitation to a specific value of $N$ is removed as reflected in the summation $\sum_{N=0}^{\infty}$.

If we first sum over all the $n_1, n_2, \ldots,$ for fixed $N$, then sum over all values of $N$ from 0 to $\infty$, the result of this double summation is equivalent to summing over all values of $n_1, n_2, \ldots,$ independently of each other. Every term in the first case of double sum appears once and once only in the second case of summing each $n_i$ independently. This is easily checked. Therefore,

$$\mathscr{Z} = \sum_{n_1} \sum_{n_2} \ldots \exp[-\beta(n_1\epsilon_1 + n_2\epsilon_2 + \ldots)] \exp[\beta\mu(n_1 + n_2 + \ldots)]$$

$$= \prod_i \{\sum_{n_1} \exp[\beta(\mu - \epsilon_i)n_i]\} = \prod_i \mathscr{Z}_i,\ \mathscr{Z}_i = \sum_{n_i=0}^{\infty} \exp[\beta(\mu - \epsilon_i)n_i]. \quad (4.101)$$

We see that the grand partition function is easily factorized. The reason behind it is that the occupation numbers $n_1, n_2, \ldots$ are not constrained to add up to a fixed number. As a result the statistical distribution for each single-particle state is independent of the presence of other single-particle states.

In (4.101) the sum $\sum_{n_i}$ extends over the values $n_i = 0, 1, 2, 3, \ldots$ for the bosons (symmetric wave function) and the values $n_i = 0, 1$ for the fermions (antisymmetric wave function). Using the expansion $(1-x)^{-1} = \sum_{n=0}^{\infty} x^n, |x| < 1$, we get for the *bosons*

$$\mathscr{Z}^{(S)} = \prod_i \{\sum_{n_i=0} \exp[\beta(\mu - \epsilon_i)n_i]\}$$

$$= \prod_i \frac{1}{1 - \exp[\beta(\mu - \epsilon_i)]},\ \mu < \epsilon_i \text{ for all } i, \quad \text{(BE)}. \quad (4.102)$$

For the *fermions* the result is obtained directly,

$$\mathscr{Z}^{(A)} = \prod_i \left[\sum_{n_i=0 \text{ or } 1} \exp(\beta(\mu - \epsilon_i)n_i)\right]$$

$$= \prod_i \{1 + \exp[\beta(\mu - \epsilon_i)]\}, \quad \text{(FD)}. \quad (4.103)$$

From (4.66, 67) and the definition of grand potential $\Omega_g = E - TS - \mu\overline{N}$, we have for bosons

$$\Omega_g^{(S)} = -kT \ln \mathscr{Z}^{(S)}$$

$$= -kT \sum_i \ln\{1 - \exp[\beta(\mu - \epsilon_i)]\}^{-1} = \sum_i \Omega_{gi}^{(S)}, \quad (4.104)$$

$$\Omega_{gi}^{(S)} = kT \ln \{1 - \exp [\beta(\mu - \epsilon_i)]\}, \tag{4.105}$$

and

$$\bar{N} = -(\partial\Omega_g/\partial\mu)_{V,T} = -\sum_i (\partial\Omega_{gi}/\partial\mu) = \sum_i \bar{n}_i, \tag{4.106}$$

$$\bar{n}_i = -\frac{\partial\Omega_{gi}^{(S)}}{\partial\mu} = -kT\frac{\partial}{\partial\mu}\{\ln[1 - \exp(\beta(\mu - \epsilon_i))]\}$$

$$= kT\frac{\beta \exp[\beta(\mu - \epsilon_i)]}{1 - \exp[\beta(\mu - \epsilon_i)]} = \frac{1}{\exp[\beta(\epsilon_i - \mu)] - 1}, \quad \text{(BE)}. \tag{4.107}$$

This agrees with (2.65), provided $g_i = 1$ (single-particle state), and

$$\alpha = -\beta\mu = -\mu/kT. \tag{4.108}$$

Note that $\alpha$ depends upon temperature.

For fermions the grand potential is

$$\Omega_g^{(A)} = -kT \ln \mathscr{Z}^{(A)} = -kT\sum_i \ln\{1 + \exp[\beta(\mu - \epsilon_i)]\}$$

$$= \sum_i \Omega_{gi}^{(A)}, \tag{4.109}$$

and the distribution is

$$\bar{n}_i = -\frac{\partial\Omega_{gi}^{(A)}}{\partial\mu} = kT\frac{\partial}{\partial\mu}\{\ln[1 + \exp(\beta(\mu - \epsilon_i))]\}$$

$$= kT\frac{\beta \exp[\beta(\mu - \epsilon_i)]}{1 + \exp[\beta(\mu - \epsilon_i)]} = \frac{1}{\exp[\beta(\epsilon_i - \mu)] + 1}, \quad \text{(FD)} \tag{4.110}$$

in agreement with (2.66), provided $g_i = 1$ (single-particle state), and $\alpha = -\beta\mu = -\mu/kT$. We can imagine the $i$th single-particle state to be our system and the remaining single-particle states to be the heat and particle reservoir. Exchange of energy and particles occurs on collision.

The sum in (4.82) is over all microstates or quantum states of the system. The energies of the various single-particle states $E_i$ are not necessarily different. Then the sum will involve many repeated equal terms given by the degeneracy. To include the degeneracy term $g_i$, the partition function (4.82) can be expressed *in an equivalent way* as a sum over the different energy levels and the degeneracies of the levels,

$$Z = \sum_{\substack{i \\ \text{(levels)}}} g_i \exp(-\beta E_i), \qquad (\text{all } E_i \text{ different}). \tag{4.111}$$

The form (4.82) is used for simplicity but for actual final calculations one must remember to include $g_i$, if the sum is over the energy levels.

From (2.72), for an ideal gas

$$e^{\alpha} = \exp(-\mu/kT) = \frac{V/N}{\lambda^3} \gg 1, \quad (\textit{classical limit}), \tag{4.112}$$

$$(V/N)/\lambda^3 = 0.026\, M^{3/2}\, T^{5/2}/P, \tag{4.113}$$

where $P$ is the pressure in atmospheres and $M$ is the molecular weight in atomic mass units. The right side of (4.113) is about $10^6$ for air at NTP, $10^{-4}$ for electrons at room temperature, and 1 for helium gas at 2 K, 1 atm. Thus the classical statistics can be used for air, fails for electrons, and nearly fails for helium gas, under above conditions. When the classical distribution fails, we are in a region where the distribution is degenerate.

## 4.10 THIRD LAW OF THERMODYNAMICS

The entropy defined in terms of the classical phase space, $\sigma = \ln \Delta\Gamma$, gives only entropy differences. It means that classically the entropy is defined up to an arbitrary additive constant only. Therefore, in classical thermodynamics the third law of thermodynamics is needed to fix the absolute value of entropy as $T \to 0$. According to the *Nernst postulate*, the entropy of any system vanishes in the states for which $T = 0$ K.

In quantum statistics we can define the entropy unambiguously as

$$\sigma = \ln \Omega(E), \quad \Omega(E) = \Delta\Gamma(E)/h^f, \tag{4.114}$$

where $\Omega(E)$ is the number of *discrete* quantum states available to the system with degrees of freedom $f$. At $T = 0$ K, both fermion and boson gases approach a unique *ground state*. For a discrete spectrum we can write (4.114) as

$$\sigma_{T=0} = \ln g_{(gr)}, \tag{4.115}$$

where $g_{(gr)}$ is the degeneracy of the ground state. At $T = 0$, $g_{(gr)}$ is unity when there is only one state of the system having the lowest possible energy.

If the ground state is a single pure quantum-mechanical state, $g_{(gr)} = 1$, we at once get

$$\sigma_{T=0} = 0. \tag{4.116}$$

It is generally believed that the ground state of any system is non-degenerate, $g_{(gr)} = 1$. This cannot be proved but is found to be true in all cases checked. As (4.116) depends on the discreteness of energy levels, it is a consequence of quantum mechanics.

## 4.11 PHOTONS

The BE statistics is applicable to photons (spin 1). For photons $p = h/\lambda = h\nu/c$, $dp = h d\nu/c$, and number of states for photons with momenta between $p$ and $p + dp$ is

$$g(p)\, dp = \frac{4\pi p^2\, dp}{h^3/V} = 4\pi V \frac{\nu^2}{c^3}\, d\nu, \tag{4.117}$$

where $V$ is the volume of the enclosure for blackbody radiation. As there are two independent directions of polarization,

$$g(\nu)\, d\nu = 8\pi V(\nu^2/c^3)\, d\nu \tag{4.118}$$

for the total number of states lying in the frequency range $d\nu$ at $\nu$. Then the BE distribution for photons is

$$dn = \frac{g(\nu)\, d\nu}{\exp(\alpha + \beta\epsilon) - 1} = 8\pi V \frac{\nu^2}{c^3} \cdot \frac{d\nu}{\exp(\alpha + \beta\epsilon) - 1}, \tag{4.119}$$

where $\epsilon = h\nu$. The energy density $u d\nu = (dn/V)\epsilon$ in the specified energy range is

$$u d\nu = \frac{8\pi h\nu^3}{c^3} \cdot \frac{d\nu}{\exp(\alpha + \beta h\nu) - 1}. \tag{4.120}$$

This is the *Planck radiation formula* if we put for photons

$$\alpha = 0,\ \beta = 1/kT. \tag{4.121}$$

The requirement $\alpha = 0$ (or $\mu = 0$) simply means dropping the condition $\delta N = \Sigma \delta n = 0$, for the fixed number of particles. Photons differ from other bosons in that their total number is not conserved. Thus, for photons

$$u d\nu = \frac{8\pi h\nu^3}{c^3} \cdot \frac{d\nu}{\exp(h\nu/kT) - 1}, \quad (\textit{Planck law}). \tag{4.122}$$

Such a derivation of the Planck law* was first given by Bose+.

For $h\nu \ll kT$, $\exp(h\nu/kT) - 1 \simeq h\nu/kT$,

$$u d\nu = \frac{8\pi kT}{c^3} \nu^2\, d\nu, \quad (\textit{Rayleigh-Jeans law}). \tag{4.123}$$

For $h\nu \gg kT$, $\exp(h\nu/kT) - 1 \simeq \exp(h\nu/kT)$,

$$u d\nu = \frac{8\pi h\nu^3}{c^3} \exp(-h\nu/kT)\, d\nu, \quad (\textit{Wien law}). \tag{4.124}$$

The total energy density is

$$\frac{U}{V} = \int_0^\infty u(\nu, T)\, d\nu = \frac{8\pi h}{c^3} \int_0^\infty \frac{\nu^3 d\nu}{\exp(h\nu/kT) - 1}$$

$$= \frac{8\pi h}{c^3} \left(\frac{kT}{h}\right)^4 \int_0^\infty \frac{x^3 dx}{e^x - 1} = bT^4,\ b = \frac{8\pi^5 k^4}{15 c^3 h^3}, \tag{4.125}$$

where $x = h\nu/kT$ and the integral is given by (Appendix IV)

$$\Gamma(4)\, \Sigma\, n^{-4} = 6\zeta(4) = \pi^4/15. \tag{4.126}$$

The result (4.125) is called the *Stefan-Boltzmann law*.

With no restraint on $n_i$, the partition function is

$$Z_{ph}(T, V) = \sum_{n_1=0}^{\infty} \sum_{n_2=0}^{\infty} \ldots \exp[-\beta(n_1\epsilon_1 + n_2\epsilon_2 + \ldots)]$$

---

*M. Planck, *Ann. Physik* 4, 553 (1901).

+S.N. Bose, *Z. Physik*, 26, 178 (1924).

$$= \prod_{l=0}^{\infty} \left[ \sum_{n_l=0}^{\infty} \exp(-\beta\epsilon_l n_l) \right] = \prod_{l=0}^{\infty} \frac{1}{1-\exp(-\beta\epsilon_l)}, \tag{4.127}$$

so that

$$\begin{aligned}
\ln Z_{ph} &= -\sum_l \ln[1-\exp(-\beta\epsilon_l)] \\
&= -\int \ln[1-\exp(-\beta h\nu)]\, g(\nu)\, d\nu \\
&= -\frac{8\pi V}{c^3} \int_0^{\infty} \nu^2\, d\nu \ln[1-\exp(-\beta h\nu)],
\end{aligned} \tag{4.128}$$

$$\begin{aligned}
F(T, V) &= -kT \ln Z_{ph}(T, V) \\
&= \frac{8\pi VkT}{c^3} \int_0^{\infty} \nu^2\, d\nu \ln[1-\exp(-\beta h\nu)] \\
&= \frac{8\pi VkT}{c^3\beta^3 h^3} \int_0^{\infty} x^2\, dx \ln(1-e^{-x}) \\
&= -\tfrac{1}{3} bVT^4,
\end{aligned} \tag{4.129}$$

where $x = \beta h\nu$, $b$ is given by (4.125), and the integral is equal to

$$\frac{1}{3}\int_0^{\infty} d(x^3) \ln(1-e^{-x}) = \left[\tfrac{1}{3}x^3 \ln(1-e^{-x})\right]_0^{\infty} - \frac{1}{3}\int_0^{\infty} \frac{x^3\, dx}{e^x - 1} = -\frac{\pi^4}{45}. \tag{4.130}$$

Note that $Z$ can be interpreted as the grand partition function for an ideal Bose gas with $\mu = 0$. From (4.129),

$$\begin{aligned}
S &= -(\partial F/\partial T)_V = \tfrac{4}{3} bVT^3, \\
P &= -(\partial F/\partial V)_T = \tfrac{1}{3} bT^4, \\
E &= F + TS = bVT^4.
\end{aligned} \tag{4.131}$$

The radiation pressure at the surface of the sun ($T = 6000°$C) is

$$P = \frac{8\pi^5}{45} \frac{(kT)^4}{(hc)^3} = \frac{8\pi^5 (1.38 \times 10^{-16} \text{ erg deg}^{-1} \times 6273 \text{ deg})^4}{45 (6.62 \times 10^{-27} \text{ erg. s} \times 3 \times 10^{10} \text{ cm s}^{-1})}$$

$$= 39.2 \text{ dyne/cm}^2.$$

## 4.12 EINSTEIN'S DERIVATION OF PLANCK'S LAW: MASER AND LASER

An atom emits radiation if an electron makes a transition from a higher energy state to a lower state $m \rightarrow n$. The transition can be either *spontaneous*, or *induced* by the presence of some external radiation. If $N_m$ is the number of atoms in the state $m$, the number of spontaneous radiative emissions per second is $N_m A_{mn}$, where $A_{mn}$ is a coefficient of proportionality. On the other hand, the number of induced emissions is $N_m B_{mn}\, u$, being proportional to the energy density $u$ of the external radiation present. The $A_{mn}$ and $B_{mn}$ are called *Einstein coefficients*.

We can also have transitions from the state $n$ to the state $m$, $n \to m$, induced by the external radiation. The corresponding number of absorptions will be $N_n B_{nm}\, u$. While the spontaneous radiation is incoherent, the induced radiation has the same phase as the external radiation.

When thermodynamic equilibrium is obtained

$$N_m A_{mn} + N_m B_{mn}\, u = N_n B_{nm}\, u. \tag{4.132}$$

Using Boltzmann distribution for the energy distribution of the atoms

$$N_m/N_n = \exp(-h\nu/kT), \tag{4.133}$$

where $h\nu$ is the energy difference between the levels $m$ and $n$,

$$u = \frac{A_{mn}}{\exp(h\nu/kT)\, B_{nm} - B_{mn}} = \frac{A_{mn}/B_{mn}}{\exp(h\nu/kT) - 1}. \tag{4.134}$$

In the last step we have put $B_{nm} = B_{mn}$. The $B_{nm}$, being a transition probability, is a matrix element of a Hermitian Hamiltonian, and so symmetric.

For very large $T$, $h\nu \ll kT$,

$$u \simeq (A_{mn}/B_{mn})\,(kT/h\nu).$$

Comparison with the Rayleigh-Jeans law (4.123) gives

$$A_{mn}/B_{mn} = 8\pi h\nu^3/c^3. \tag{4.135}$$

Einstein obtained the Planck radiation law from (4.134, 135).

The radiative transitions from a state $m$ to a state $n$ are given by $(A_{mn} + u\, B_{mn})\, N_m$ and the inverse transitions by $B_{nm}\, u\, N_n$. For atoms kept in a radiation field of density $u$, this incident radiation will emerge from the atoms with intensity given by $(N_m - N_n)\, u\, B_{nm}$. It is either amplified or reduced with respect to the incident radiation, depending upon

$$N_m > N_n \qquad \text{(amplified)}, \tag{4.136}$$

or

$$N_m < N_n \qquad \text{(reduced)}. \tag{4.137}$$

The condition $N_m > N_n$ is known as *population inversion*, or sometimes as a negative temperature condition since use of Boltzmann distribution (4.133) implies $T$ negative. More generally, for degenerate levels, the condition for amplification is $N_m > (g_m/g_n)\, N_n$. The population inversion can never occur in thermal equilibrium. Its non-equilibrium nature gives rise to an interesting effect discussed below.

Let a light beam of intensity $I(\nu)$ and frequency range $\nu$ to $\nu + d\nu$ pass through a gas of atoms contained in a thick cavity, where significant attenuation of the beam occurs. The beam travelling in the $x$-direction with velocity $c$ in the gas, traverses a thickness $dx$ in a time $dt = dx/c$. Since the induced radiation has the same phase as the external radiation, the change in intensity of the beam is

$$-dI(\nu)\, d\nu = h\nu\, B_{nm}\, (dN_{n\nu} - dN_{m\nu})\, I(\nu)\, (dx/c), \tag{4.138}$$

where $N_{n\nu}$ is the number of atoms in level $n$ which can absorb radiation in the frequency interval $d\nu$ at $\nu$, and $N_{m\nu}$ in level $m$ which can emit radiation in the same interval. The experimental Lambert's law for the attenuation of radiation is

$$I(\nu) = I_0 \exp\,(-a_\nu x), \tag{4.139}$$

where $I_0$ is the incident radiation and $a_\nu$ is the absorption coefficient.

From (4.138, 139)

$$\int a_\nu\, d\nu = \frac{h\bar{\nu}}{c}\, B_{nm}\, (N_n - N_m), \tag{4.140}$$

where $\bar{\nu}$ is the center of the absorption line and the integration is over the width of the line. We have thus related the transition rate or $a_\nu$ with $B_{nm}$. When $N_m > N_n$, the integrated absorption coefficient is negative. This means that the incident beam is amplified on coming out of the cavity (matter). For an input beam of only several watts it may produce a highly monochromatic flux of many mega-watts/cm$^2$. The *maser* (microwave amplification by stimulated emission of radiation) and *laser* (light...) devices work on this principle. Note that the population inversion cannot be achieved by resonance absorption of radiation at the transition frequency $\nu = (E_m - E_n)/h$ of the two states. However, it can be achieved for levels $m$ and $n$ if a third level is also involved. Let us call the ground state as level 3, and the first and second excited states of the atom are called 1 and 2, respectively (Fig. 4.5).

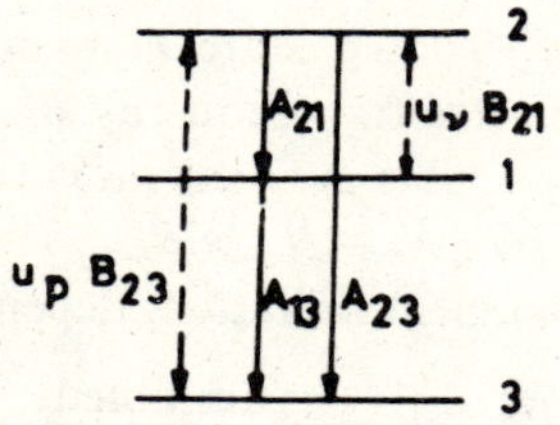


**Fig. 4.5 The atomic energy levels and transition rates for a three level laser.**

A light beam (pump) of energy density $u_p$ excites the transition 3→2, so that $N_2$ atoms are in state 2. The fraction $N_2/N$ of the total number of atoms pumped up like this is very small (about one in a million).

An atom in level 2 can emit light in the transitions 2→1 or 2→3. An atom in level 1 can make the transition 1→3. For appropriate values of the Einstein coefficients, it is possible to achieve the condition $N_2 > N_1$ for the amplification of light of frequency $\nu = (E_2 - E_1)/h$.

The equations for the transition rates for a three-level system are generalizations of those for a two-level system. Besides the pump beam energy $u_p$, we assume that some radiation of energy density $u_\nu$ is present at the frequency $\nu$. For three populated atomic levels

$$N_1 + N_2 + N_3 = N \tag{4.141}$$

the rate equations are

$$\begin{aligned}
dN_2/dt &= -N_2A_{21} - N_2A_{23} + u_pB_{23}(N_3 - N_2) - u_\nu B_{21}(N_2 - N_1),\\
dN_1/dt &= N_2A_{21} - N_1A_{13} + u_\nu B_{21}(N_2 - N_1),\\
dN_3/dt &= N_2A_{23} + N_1A_{13} - u_pB_{23}(N_3 - N_2).
\end{aligned} \tag{4.142}$$

Due to (4.141), the sum of these three rates is zero. In the steady state all the $dN_i/dt = 0$. This gives $N_1$, $N_2$ and $N_3$ in terms of $N$, $u_\nu$, $u_p$. The expressions are lengthy. We bring out the main features of the steady state solutions as follows.

The pumping rate $r$ is defined by

$$r = u_pB_{23}(N_3 - N_2)/N. \tag{4.143}$$

The $rN$ is the net rate at which atoms are excited to state 2 by the pump. The steady state conditions from the last two relations in (4.142) are

$$N_2(A_{21} + B_{21}u_\nu) = N_1(A_{13} + B_{21}u_\nu), \tag{4.144}$$

$$N_2A_{23} + N_1A_{13} = rN. \tag{4.145}$$

We can solve them for $N_1$ and $N_2$. From (4.144), $N_2 > N_1$ if

$$A_{21} < A_{13}, \tag{4.146}$$

that is, the atoms excited into state 2 must decay relatively slowly into state 1, from where they rapidly return to the ground state 3. A gas of atoms, for which (4.146) is satisfied, will amplify the radiation of energy density $u_\nu$.

## 4.13 EQUATION OF STATE FOR IDEAL QUANTUM GASES

The nonrelativistic energy of an elementary particle (no internal degrees of freedom) is just the kinetic energy of its translational motion, $\epsilon = p^2/2m$. Under normal conditions (big box) the translational levels are closely spaced and so the discrete spectrum can be treated approximately as continuous (quasiclassical). Then (4.117) gives

$$g(\epsilon)\,d\epsilon = g_s(2\pi V/h^3)(2m)^{3/2}\epsilon^{1/2}\,d\epsilon,\; g_s = 2s + 1, \tag{4.147}$$

where $g_s$ is the degeneracy associated with the spin $s$ of the particle. The $g(\epsilon)\,d\epsilon$ is the number of quantum states of energy between $\epsilon$ and $\epsilon + d\epsilon$. The $g(\epsilon)$ is the *density of states.* In fact, it is the density of quantum states because it refers to one-particle system and not to the states of $N$ particle system. When multiplied by the *distribution function* $f(\epsilon, T, \mu)$,

$$f(\epsilon, T, \mu) = \exp[-(\epsilon-\mu)/kT], \qquad (\textit{classical}),$$

$$f(\epsilon, T, \mu) = \frac{1}{\exp[(\epsilon-\mu)/kT] \pm 1}, \qquad (\textit{quantum}),$$

it gives the density of occupied quantum states. The total number of particles in a system is then given by

$$N = \int dN = \int f(\epsilon, T, \mu)\, g(\epsilon)\, d\epsilon.$$

In general, if $R_{\mathbf{n}}$ is the value of the quantity $R$ in the quantum state $\mathbf{n}$, then we can write

$$R = \sum_{\mathbf{n}} f(\epsilon_{\mathbf{n}}, T, \mu)\, R_{\mathbf{n}}$$

$$= \sum_{l} g_l f(\epsilon_l, T, \mu)\, R_l, \qquad \text{(all } \epsilon_l \text{ different)},$$

as

$$R = \int f(\epsilon, T, \mu)\, R(\epsilon)\, g(\epsilon)\, d\epsilon. \tag{4.148}$$

Thus the sum over the quantum states can be transformed under appropriate conditions (closely spaced energy levels) to an integral by the substitution

$$\sum_{\substack{\mathbf{n} \\ \text{(states)}}} (\ldots) \text{ or } \sum_{\substack{l \\ \text{(levels)}}} g_l(\ldots) \rightarrow \int (\ldots)\, g(\epsilon)\, d\epsilon. \tag{4.149}$$

Using upper sign for FD statistics and lower sign for BE statistics,

$$N = \int dN = g_s \frac{2\pi V}{h^3} (2m)^{3/2} \int_0^\infty \frac{\epsilon^{1/2}\, d\epsilon}{\exp[(\epsilon-\mu)/kT] \pm 1}, \tag{4.150}$$

$$E = \int \epsilon dN = g_s \frac{2\pi V}{h^3} (2m)^{3/2} \int_0^\infty \frac{\epsilon^{3/2}\, d\epsilon}{\exp[(\epsilon-\mu)/kT] \pm 1}, \tag{4.151}$$

$$\Omega_g = \mp\, kT\, g_s \frac{2\pi V}{h^3} (2m)^{3/2} \int_0^\infty \epsilon^{1/2} \ln\{1 \pm \exp[(\mu-\epsilon)/kT]\}\, d\epsilon. \tag{4.152}$$

Partial integration of the grand potential gives

$$\Omega_g = -g_s \frac{2}{3} \cdot \frac{2\pi V}{h^3} (2m)^{3/2} \int_0^\infty \frac{\epsilon^{3/2}\, d\epsilon}{\exp[(\epsilon-\mu)/kT] \pm 1} = -PV, \tag{4.153}$$

where we have used (4.66), $\Omega_g = -PV$.

From (4.151, 153), for *any* statistics

$$PV = \frac{2}{3} E. \tag{4.154}$$

Putting $x = \epsilon/kT$, write (4.153) as

$$P = g_s \frac{4\pi V}{3h^3} (2m)^{3/2} (kT)^{5/2} \int_0^\infty \frac{x^{3/2}\, dx}{\exp(x-\mu/kT) \pm 1}. \tag{4.155}$$

The equation of state of the gas is determined by (4.150, 155) in terms of the parameter $\mu$.

For $e^{\mu/kT} \ll 1$,

$$\int_0^\infty \frac{x^{3/2}\,dx}{e^{x-\mu/kT} \pm 1} = \int_0^\infty x^{3/2}\,[e^{x-\mu/kT} \pm 1]^{-1}\,dx$$

$$\simeq \int_0^\infty x^{3/2}\,e^{\mu/kT-x}\,[1 \mp e^{\mu/kT-x}]\,dx$$

$$= e^{\mu/kT}\left[\int_0^\infty x^{3/2}\,e^{-x}\,dx \mp e^{\mu/kT}\int_0^\infty x^{3/2}\,e^{-2x}\,dx\right]$$

$$= \frac{3\pi^{1/2}}{4}\,e^{\mu/kT}\left[1 \mp \frac{1}{2^{5/2}}\,e^{\mu/kT}\right], \tag{4.156}$$

$$\Omega_g = -PV = -g_s\,\frac{\pi^{3/2}\,V}{h^3}\,(2m)^{3/2}\,(kT)^{5/2}\,e^{\mu/kT}\,(1 \mp 2^{-5/2}\,e^{\mu/kT}). \tag{4.157}$$

The first term in (4.157) gives just the classical value (4.75) for $g_s = 1$. The next term is the correction to it. We can write

$$\Omega_g = \Omega_g^{MB} \pm g_s\,\frac{\pi^{3/2}\,V}{2^{5/2}h^3}\,(2m)^{3/2}\,(kT)^{5/2}\,e^{2\mu/kT}. \tag{4.158}$$

Using (4.76),

$$e^{2\mu/kT} = \frac{1}{g_s^2}\cdot\left(\frac{h^3\,N}{V(2\pi mkT)^{3/2}}\right)^2, \tag{4.159}$$

we have, to the same approximation, for the *equation of state*

$$PV = NkT\left[1 \pm \frac{Nh^3}{g_s\,2^{5/2}(2\pi mkT)^{3/2}\,V}\right]. \tag{4.160}$$

Thus, the departure from the MB statistics is large when for a given density the temperature is lowered. This is a measure of the degeneracy. In the FD case, the pressure increases due to the correction term, implying the appearance of an effective repulsion between the particles. In the BE case, it leads to a decrease in pressure, implying an effective attraction between the particles.

## PROBLEMS

4.1 If $g(E)$ is the number of microscopic states of the system between $E$ and $E + dE$, then (4.8) implies

$$Z = \int e^{-\beta E}\,g(E)\,dE.$$

Use the Taylor expansion of $\ln \phi(E)$, $\phi(E) \equiv g(E)\, e^{-\beta E}$, to show that

$$\ln Z \simeq \ln g(E_m) - \beta E_m,$$

where $E_m$ is the energy at which $g(E)\, e^{-\beta E}$ has its maximum.

4.2 Find the free energy of a gas in a centrifuge of radius $R$, length $l$ and angular velocity $\omega$. Calculate the mean square distance of a molecule from the axis.

4.3 Consider an ideal gas of $N$ atoms. Assuming canonical distribution at temperature $T$, find the most probable value $E_m$ of the total energy $E$ of the system. Compare it with the mean value $\bar{E}$ in the canonical distribution.

4.4 Show that $Z = 2\pi kT/h\omega$ for an oscillator defined by

$$E = (p^2/2m) + \tfrac{1}{2}m\omega^2 q^2.$$

4.5 Consider a system with given number of particles $N_a$, $N_b$, ... in contact with a bath at temperature $T$ and pressure $p$ through a movable wall. Its volume $V$, like the number of particles in the grand canonical distribution ($T-\mu$ distribution), is also indeterminate. Calculate the results corresponding to (4.58, 59) for this *$T$-$p$ distribution.*

4.6 Approximating the integrand by the largest term use

$$Y = \int_0^\infty Z(N, V) \exp(-\beta p V)\, dV = \int_0^\infty dV \exp(-\beta p V_m)\, Z(N, V_m)$$

$$\exp\left[-\frac{\frac{1}{2}\partial^2 \ln Z(N, V_m)}{\partial^2 V_m}(V - V_m)^2 \ldots\right]$$ to calculate the Gibbs free energy.

[Hint: The integral contributes near $V_m$ to $\ln Y$ a term of the order of $\ln N \ll N$, so put $-kT \ln Y = -kT \ln Z(N, V_m) + pV_m$, where $V_m$ is given by $\partial \ln Z(N, V_m)/\partial V_m = -p/kT$].

4.7 Use (4.16) to show that $P_l$, which makes the entropy under the constraints $\sum_l P_l = 1$, $\sum_l E_l P_l = E =$ constant maximum when the mean energy of the system is $E$, exhibits canonical distribution.
[Hint: Maximize $S + aE + b1$].

4.8 Consider a system of two noninteracting types of particles whose numbers $n_+$ and $n_-$ can vary provided $n_+ - n_- = n =$ constant. Show that

$$\begin{aligned} \mathscr{Z} &= \exp(-\Omega_g/kT) \\ &= \sum_{n_+, n_-} \frac{\exp(\beta n_+ \mu_+)}{n_+!} \cdot \frac{\exp(\beta n_- \mu_-)}{n_-!} \int \exp(-E_N(X)/kT)\, dX \\ &= \sum_{n_-=0}^{\infty} \left(\frac{Z_+}{Z_-}\right)^{n/2} I_n(2\sqrt{Z_+ Z_-}), \end{aligned}$$

where

$$Z_\pm = \int \exp[\beta\mu_\pm - E^\pm(X_l)]\, dX_l,$$

$$E_n(X) = \sum_{i=1}^{n_+} E^+(X_i) + \sum_{j=1}^{n_-} E^-(X_j),$$

and $I_n(x)$ is the modified Bessel function of the first kind of order $n$. Now show that

$$\bar{n}_\pm = \tfrac{1}{2}\tilde{Z}\frac{I_{n\mp1}(\tilde{Z})}{I_n(\tilde{Z})}, \text{ where } \tilde{Z} = (2\sqrt{Z_+ Z_-})_{\mu_+, \mu_- = 0}.$$

4.9 By definition, $Z(\beta)$ is the Laplace transformation of $g(E)$,

$$Z(\beta) = \sum_i \exp(-\beta E_i) = \int_{E_0}^{\infty} e^{-\beta E} g(E)\, dE,$$

where $E_0$ is the lowest energy and the density of state $g(E) \geqslant 0$, $\lim_{E\to\infty} g(E)\, e^{-\alpha E} = 0$ for $\alpha > 0$. The inverse formula is

$$g(E) = \frac{1}{2\pi i}\int_{\beta'-i\infty}^{\beta'+i\infty} Z(\beta)\, e^{\beta E}\, d\beta.$$

Show that for $Z(\beta) = A/\beta^N$, we get $g(E) = AE^{N-1}/(n-1)\,!$.

4.10 Consider a zipper $AB$ with $N$ links. For each link there is a state of energy 0 when it is closed and a state of energy $\epsilon$ when it is open. The zipper can unlink only from the end $A$, and the $n$th link can open only when all the links $1, 2, \ldots, n-1$ beginning from the end $A$ are open. Show that the partition function sums to

$$Z = \frac{1-e^{-(N+1)\epsilon/kT}}{1-e^{-\epsilon/kT}}.$$

Find the average number of open links for $\epsilon \gg kT$.
[Hint: See C. Kittel, Amer. J. Phys. 37, 917(1969)]

# 5
# PARTITION FUNCTION

## 5.1 CANONICAL PARTITION FUNCTION

Even when quantum effects are negligible (classical limit), the appropriate language for the description of a system is provided by the quantum theory. To characterize a system we should therefore specify the energy eigenvalues and the corresponding wave functions for the system as a whole.

Consider a system $A$ composed of two noninteracting distinguishable (localized) atoms $a$, $b$. The one particle wave functions and energy levels are $u_1, u_2, \ldots, u_i, \ldots$ and $\epsilon_1, \epsilon_2, \ldots, \epsilon_i, \ldots$, respectively. The eigenvalue equations are $H_a\, u_i(a) = \epsilon_{ai}\, u_i(a)$, and $H_b\, u_i(b) = \epsilon_{bi}\, u_i(b)$. For the whole system

$$H = H_a + H_b, \tag{5.1}$$

$$\Psi^{(ij)} = u_i(a)\, u_j(b), \tag{5.2}$$

$$E^{(ij)} = \epsilon_{ai} + \epsilon_{bj}. \tag{5.3}$$

The double index $(ij)$ denotes a single state of the composite system.

The one particle canonical partition function is

$$z = \sum_i \exp(\beta\epsilon_i), \tag{5.4}$$

and the canonical partition function for the whole system is

$$Z = \sum_A \exp(-\beta E_A^{(ij)}) = \sum_i \sum_j \exp[-\beta(\epsilon_{ai} + \epsilon_{bj})]. \tag{5.5}$$

We can write $Z$ as

$$Z = [\sum_i \exp(-\beta\epsilon_{ai})] \cdot [\sum_j \exp(-\beta\epsilon_{bj})] = z_a \cdot z_b, \tag{5.6}$$

where the summations extend over all quantum states of the individual atoms $a$ and $b$.

Since $a$ and $b$ are identical, $\epsilon_{ai} = \epsilon_{bi}$ and $z_a = z_b$. Then

$$Z = z^2, \tag{5.7}$$

and generalizing it to a system of $N$ identical but distinguishable particles,

$$Z = z^N, \quad (\textit{identical distinguishable particles}). \tag{5.8}$$

If $a$ and $b$ are indistinguishable particles, the wave function (5.2) must be symmetrized (or antisymmetrized). Then the 2! ways of obtaining $E^{(ij)}$ in (5.3), $\epsilon_{ai} + \epsilon_{bj}$ and $\epsilon_{aj} + \epsilon_{bi}$, correspond to only one wave function $\Psi_A = u_i(a)\, u_j(b) + u_i(b)\, u_j(a)$ for the symmetric case. Consequently in the sum (5.5), for a given $E_A^{(ij)}$, we have 2 ! terms in the summation which differ only in the particle labels. For example, the two terms $\exp[-\beta(\epsilon_{a1} + \epsilon_{b2})]$ and $\exp[-\beta(\epsilon_{a2} + \epsilon_{b1})]$ correspond to the same energy $E_A^{(12)}$. Because our sum should contain only one term for each distinct $\Psi_A$, the summation (5.6) is too large by a factor of 2 !. Therefore, for indistinguishable particles we should replace (5.7) by

$$Z = \frac{1}{2\,!}\, z^2, \tag{5.9}$$

and (5.8) by

$$Z = \frac{1}{N\,!}\, z^N, \quad (\textit{indistinguishable particles}), \tag{5.10}$$

in agreement with (4.38). Thus, the Boltzmann counting appears as a natural consequence of the symmetry of wave functions in quantum theory.

The use of (5.10) describes a 'Boltzmann gas'. The molecules are identical (either bosons or fermions). When many more particle states than particles are available, (2.71), the difference between bosons and fermions can be neglected, and (2.64) used along with (5.10).

## 5.2 MOLECULAR PARTITION FUNCTIONS

For distinguishable particles forming a system we have to use (5.8). The distinguishability for identical particles arises when the classical limit (2.28), $r_{av} > \lambda$, holds and the particles can be treated as localized. Another example is that of particles constrained to occupy fixed lattice sites as in a crystal. Aside from localization any other measurable property of the particles such as their internal state may be used to distinguish the particles or their states. For example, consider a system of two diatomic molecules. The system can be found in two distinct states

$$\Psi_I = u(a, 1)\, u(b, 2), \quad \Psi_{II} = u(a, 2)\, u(b, 1),$$

where $u(a, 1)$ means the molecule occupying the translational state $a$ is in the vibrational state 1. In this case the occupation of distinct translational quantum states provides as much ground for differentiation as was provided by the occupation of distinct spatial positions in a crystal.

In molecular system various internal degrees of freedom are only weakly coupled to the external degrees of freedom and each other. Therefore the total energy $E$ of the system can be expressed as the sum of translational ($t$), vibrational ($v$), rotational ($r$), electronic ($e$) and nuclear ($n$) energies,

$$E = E(t) + E(v) + E(r) + E(e) + E(n). \tag{5.11}$$

To this approximation, we can write the total partition function $Z_T$ for the system as

$$\begin{aligned} Z_T &= \sum_i \exp(-\beta E_i) \\ &= \sum_i \exp\{-\beta [E_i(t) + E_i(v) + E_i(r) + E_i(e) + E_i(n)]\} \\ &= [\sum_i \exp(-\beta E_i(t)]\cdot[\sum_j \exp(-\beta E_j(v))]\cdot[\sum_k \exp(-\beta E_k(r))]. \ldots \\ &= Z(t)\cdot Z(v)\cdot Z(r)\cdot Z(e)\cdot Z(n). \end{aligned} \tag{5.12}$$

Using (5.8,10) appropriately, for a gass of $N$ identical molecules,

$$Z_T = [(1/N!)\ (z(t))^N]\cdot[z(v)]^N\cdot[z(r)]^N\cdot[z(e)]^N\cdot[z(n)]^N, \tag{5.13}$$

where $z$ represents the partition function for a single molecule. For the translational motion we have used (5.10) because each molecule is free to move about in the whole volume and so the states cannot be distinctly labelled as in the case of solids. The factor $(1/N!)$ multiplying $[z(t)]^N$ is appropriate only when there is no degeneracy (all molecules occupy different quantum states). If the occupation number of a given state $\Psi_i$ is much less than 1, effects arising from symmetry imposed limitations on $\bar{n}_i$ are unimportant and classical statistics applies. This requires a large number of available system's states for a given number of particles. Then the use of (5.13) and MB statistics with correct Boltzmann counting for the translational states is a good approximation. Exception will arise, for example, for helium at about 2K, which remains a gas at this low temperature and the number of accessible translational states becomes comparable to the number of atoms $N$. Then we must use the BE statistics.

Under normal conditions

$$z = \sum_{\substack{i \\ \text{(states)}}} \exp(-\beta\epsilon_i), \tag{5.14}$$

where the sum is over *all* the allowed quantum states of the molecule, or equivalently

$$z = \sum_{\substack{i \\ \text{(levels)}}} g_i \exp(-\beta\epsilon_i), \qquad \text{(different } \epsilon_i\text{)}, \tag{5.15}$$

where $g_i$ is the degeneracy of the level $i$ and the sum is over the energy levels $\epsilon_i$ of the molecule. To a good approximation

$$E_{l\ \text{Total}} = \epsilon_i(t) + \epsilon_j(v) + \ldots, \tag{5.16}$$

$$z_T = z(t)\cdot z(v)\cdot z(r)\cdot z(e)\cdot z(n), \tag{5.17}$$

so that (5.13) simply reads

$$Z_T = z_T^N/N!. \tag{5.18}$$

Thus the problem of calculating $Z_T$ reduces to that of calculating $z_T$.

## 5.3 TRANSLATIONAL PARTITION FUNCTION

The classical value of $z(t)$ is given by (4.39),

$$z(t) = V/\lambda^3, \quad \lambda = h/(2\pi mkT)^{1/2}. \tag{5.19}$$

To derive this result in quantum theory consider the translational energy levels of a molecule enclosed in a box,

$$\epsilon_x = \frac{\pi^2\hbar^2 n_x^2}{2mL_x^2}, \quad n_x = 1, 2, 3, \ldots, \tag{5.20}$$

where $L_x$ is the side length of the rectangular box along the $x$-axis. Similar expressions exist for the motion parallel to $y$ and $z$ axes. The translational partition function is

$$z(t) = \sum_{n_x=1}^{\infty} \exp(-\alpha_x n_x^2) \sum_{n_y=1}^{\infty} \exp(-\alpha_y n_y^2) \sum_{n_z=1}^{\infty} \exp(-\alpha_z n_z^2), \tag{5.21}$$

where $\alpha_x^2 = \pi^2\hbar^2/(2mL_x^2kT)$, etc.

For ordinary temperatures and large size of the box the quantity $\alpha_x$ is much less than unity. Hence, $\alpha_x^2 n_x^2$ changes slowly as we vary $n_x$. For $m = 10^{-23}$ g and $L_x = 1$ cm, we get $\Delta\epsilon_x = (\pi^2\hbar^2/2mL_x^2)(2^2-1^2) \simeq 10^{-30}$ erg. The corresponding characteristic temperature for the translational motion is $\Theta_t = \Delta\epsilon_x/k \simeq 10^{-14}$ K. At temperatures $\Theta_t/T \ll 1$ the energy levels are closely spaced, and we can replace the summation by integration

$$\sum_{n_x=1}^{\infty} \exp(-\alpha_x^2 n_{ix}^2) = \int_0^{\infty} \exp(-\alpha_x^2 n_x^2)\, dn_x = \frac{\pi^{1/2}}{2\alpha_x}. \tag{5.22}$$

Using $V = L_xL_yL_z$ and $\alpha_x = \lambda\pi^{1/2}/2L_x$,

$$z(t) = \frac{\pi^{3/2}}{2^3\,\alpha_x\alpha_y\alpha_z} = \frac{V}{\lambda^3}, \quad \lambda = \hbar(2\pi/m)^{1/2}\,\beta^{1/2}, \tag{5.23}$$

in agreement with (5.19). We have $Z(t) = [z(t)]^N/N!$ and $F(t)$ is given by (4.40).

From (4.18), we get for the system

$$E(t) = -\left(\frac{d\ln Z(t)}{d\beta}\right)_V = -\frac{d}{d\beta}\left[\ln\frac{1}{N!}(z(t))^N\right]_V$$

$$= -\frac{d}{d\beta}\left[\ln\left(\frac{1}{\beta}\right)^{3N/2}\right] = \frac{3}{2}N\frac{d\ln\beta}{d\beta} = \frac{3N}{2\beta}. \tag{5.24}$$

The entropy $S(t)$ is given by (4.41).

## 5.4 ROTATIONAL PARTITION FUNCTION

For a diatomic molecule the rotational energy levels are given by*

$$\epsilon(r) = (\hbar^2/2I)\, J(J+1), \quad J = 0, 1, 2, \ldots \tag{5.25}$$

where $I$ is the moment of inertia of the molecule and $J$ is the rotational quantum number. The degeneracy of each level of a two-dimensional rotator is $g(r) = 2J+1$ since there are $2J+1$ values of the $z$ component of angular momentum. Therefore,

$$z(r) = \sum_J (2J+1) \exp[-\beta(\hbar^2/2I)\, J(J+1)] \tag{5.26}$$

For $I = mr^2 \simeq 10^{-23}\,(10^{-8})^2 = 10^{-39}$ g. cm$^2$, $\Delta\epsilon(r) \simeq 10^{-15}$ erg, $\Theta_r = \Delta\epsilon/k \simeq 10$ K. When $\Theta_r/T \ll 1$, we can replace the sum by integration,

$$z(r) = \int_0^\infty \exp[-\beta(\hbar^2/2I)\, J(J+1)]\,(2J+1)\, dJ$$

$$= \frac{2IkT}{\hbar^2} \equiv \frac{T}{\Theta_r}, \quad \Theta_r = \frac{\hbar^2}{2Ik}, \tag{5.27}$$

where we have neglected unity in comparison to $J$ and used

$$\int_0^\infty x e^{-ax^2}\, dx = (1/2a).$$

The rotational motion about the axis (third rotational degree of freedom) is not counted at ordinary $T$ because it involves the excitation to higher states of electronic angular momentum. Such excitations require large amounts of energy as the moment of inertia about this axis is very small.

The total wave function $\psi_T$ of any molecule, including diatomic molecules, can be expressed as

$$\psi_T = \psi_{\text{internal}}\, \psi_t = \psi_e\, \psi_v\, \psi_r\, \psi_n\, \psi_t, \qquad \textit{(molecule)}.$$

The $\psi_e$ specifies an electronic state. The product $\psi_e\, \psi_v$ specifies a vibrational level because such levels are associated with some particular electronic state. Similarly, a rotational level is specified only when the product $\psi_e\, \psi_v\, \psi_r$ is given. We call $\psi_v$ and $\psi_r$, by themselves, the vibrational and rotational eigenfunctions, respectively.

The $\psi_v$ depends only on the magnitude of the internuclear distance and so is not changed by reflection in the origin. For symmetry considerations it is enough to examine

$$\psi = \psi_e\, \psi_r.$$

*See, for example, B.K. Agarwal, *Quantum Mechanics and Field Theory*, 2nd ed., Lokbharti Publications, Allahabad, 1983, p. 59.

The $\psi_e$ is described with respect to a set of coordinates rigidly attached to the nuclei. It is unaltered if the nuclei rotate. We shall ignore nuclear spin here.

For a simple rotator (heteronuclear), the $\psi_r$ are simply spherical harmonics $Y_{JM}$ whose parity is $(-1)^J$. This means, under a reflection in the origin, $\psi_r$ changes sign only for odd $J$. If $\psi_e$ of the ground state is even and does not change in chemical reactions and phase transformations, the $\psi$ is symmetric for $J$ even and antisymmetric for $J$ odd.

For homonuclear molecules the interchange of nuclei can change the sign of $\psi$. Then each rotational level carries two symmetry labels: plus or minus with respect to a reflection in origin of all the particles symmetric or antisymmetric with respect to an interchange of nuclei. A proper quantum treatment of the indistinguishability of like particles gives the rule: *symmetric linear molecules can have either even* (0, 2, 4, ...) *values or odd* (1, 3, 5, ...) *values of J, but not both.*

Therefore, when the atoms in a diatomic molecule are alike, the allowed energy levels will yield a summation term just half as big as if the atoms were different. In such a case we must divide the sum over all states by the *symmetry number* $\sigma_n = 2$. We therefore should write

$$z(r) = T/(\sigma_n \Theta_r), \qquad \textit{(linear molecule)}, \tag{5.28}$$

where $\sigma_n = 1$ for an asymmetrical molecule and $\sigma_n = 2$ for a symmetric linear molecule. The symmetry number is just the number of indistinguishable orientations of the molecule. Any molecule can only exist in $1/\sigma_n$ of the rotational energy levels. Thus, $\sigma_n = 2$ for $H_2O$, 3 for $NH_3$ and 12 for methane.

For an ideal gas of diatomic molecules we have for $T > \Theta_r$,

$$\begin{aligned}
Z(r) &= [z(r)]^N = [T/(\sigma_n\Theta_r)]^N, \\
F(r) &= -NkT \ln [T/(\sigma_n\Theta_r)], \\
U(r) &= kT^2 (\partial \ln Z/\partial T) = NkT, \\
C_V(r) &= d\,U(r)/dT = \frac{d}{dT}\left[ kT^2 \frac{\partial}{\partial T} \ln Z(r) \right] = Nk.
\end{aligned} \tag{5.29}$$

Note that $Z(r)$ does not exist for a monatomic gas. For diatomic gases $F = F(t) + F(r)$, $U = U(t) + U(r) = \frac{3}{2} NkT + NkT = \frac{5}{2} NkT$, and $C_V = \frac{5}{2} Nk$, where $F(t)$ is given by (4.40).

From (4.27),

$$\begin{aligned}
S(r) &= k \ln Z(r) + U(r)/T \\
&= Nk \left( \ln \frac{IT}{\sigma} + \ln \frac{2k}{\hbar^2} + 1 \right),
\end{aligned} \tag{5.30}$$

so that for one mole of gas

$$S(r) = R \ln (IT/\sigma) + 177.68. \tag{5.31}$$

## 5.5 VIBRATIONAL PARTITION FUNCTION

A diatomic molecule has one degree of freedom associated with the vibrational motion of the nuclei along the axis joining them. The vibration of the massive atomic nuclei is driven by the force provided by the electronic molecular distribution which acts as a 'linear spring' yielding a harmonic motion.

The Hamiltonian for a linear harmonic oscillator of frequency $\nu$ is

$$H(x, p_x) = (p_x^2/2m) + \tfrac{1}{2}k_s x^2, \qquad k_s = m(2\pi\nu)^2.$$

The eigenvalues for this system are

$$\epsilon_n = (n + \tfrac{1}{2})h\nu, \qquad n = 0, 1, 2, 3, \ldots \tag{5.32}$$

It is obviously unreasonable to integrate* to find $Z(\nu)$. Therefore the vibration partition function for the molecule is

$$\begin{aligned} z(\nu) &= \sum_{n=0}^{\infty} \exp\,[-\beta(n + \tfrac{1}{2})h\nu] \\ &= \exp\,(-\beta h\nu/2)\,[1 + \exp\,(-\beta h\nu) + \exp\,(-2\beta h\nu) + \ldots] \\ &= \exp\,(-\beta h\nu/2)\,[1 - \exp\,(-\beta h\nu)]^{-1} \\ &= [2\sinh\,(\beta h\nu/2)]^{-1}. \end{aligned} \tag{5.33}$$

If we count our energies above the *zero-point vibrational energy*, $\epsilon_0 = \frac{1}{2}h\nu$ =constant, as it represents vibrational motion of the molecule in its ground stete, we get

$$z(\nu) = \frac{1}{1-\exp\,(-\beta h\nu)}\,. \tag{5.34}$$

At room temperature (300 K), a typical value is $\nu \simeq 6 \times 10^{13}$ s$^{-1}$, $h\nu/k \simeq$ 3000 K, and so $z(\nu) \simeq 1$.

For the whole system, consisting of $N$(distinguishable) independent oscillators,

$$\begin{aligned} Z(\nu) &= [z(\nu)]^N, \\ F(\nu) &= -kT \ln Z(\nu) = N\epsilon_0 + NkT \ln\,[1-\exp\,(-\beta h\nu)], \\ U(\nu) &= kT^2 \frac{d \ln Z(\nu)}{dT} = N\epsilon_0 + \frac{Nh\nu}{\exp\,(\beta h\nu)-1}, \\ C_V(\nu) &= \frac{N(h\nu)^2}{kT^2}\cdot\frac{\exp\,(\beta h\nu)}{[\exp\,(\beta h\nu)-1]^2}. \end{aligned} \tag{5.35}$$

At low temperatures ($kT \ll h\nu$) these quantities tend exponentially to zero,

$$\begin{aligned} F(\nu) &\approx -NkT \exp\,(-\beta h\nu), \\ C_V(\nu) &\approx Nk(h\nu/kT)^2 \exp\,(-\beta h\nu). \end{aligned}$$

---

*For $kT \gg h\nu$ the important quantum numbers $n$ are the large ones (classical limit). Then the summation can be replaced by integration to give $Z(\nu) = (\beta h\nu)^{-1} = kT/h\nu$.

At high temperatures ($kT \gg h\nu$),

$$F(v) \approx -N(\tfrac{1}{2}h\nu) - NkT \ln (kT) + NkT \ln (h\nu),$$

$$C_V(v) \approx Nk.$$

The contribution to entropy is

$$S(v) = k \ln Z(v) + U(v)/T$$

$$= Nk \ln \frac{1}{1-\exp(-\beta h\nu)} + \frac{N\epsilon_0}{T} + Nk\left(\frac{h\nu}{kT}\right)\frac{1}{\exp(\beta h\nu)-1}. \qquad (5.36)$$

In c.g.s. units

$$\Theta_r = \frac{h^2}{8\pi^2 Ik} = \frac{40.3\times 10^{-40}}{I}, \quad \Theta_v = \frac{h\nu}{k} = \frac{\nu}{2.08\times 10^{10}},$$

where $\Theta_v$ is the characteristic temperature of vibration. For oxygen molecule $\Theta_r = 2.07$ K, $\Theta_v = h\nu/k = 2230$ K, so that at 273 K for one g-mole,

$$z(t) : z(r) : z(v) = 3.4\times 10^{30} : 65.3 : 0.16$$

$$S(t) : S(r) : S(v) = 35.9 : 8.5 : 0.004 \quad \text{(cal/g-mole K)}.$$

The values of parameters for a few diatomic molecules are given in Table 5.1.

**Table 5.1. The values of parameters for some diatomic molecules**

| Molecule | $\Theta_r$(K) | $I$(g.cm²) | $\Theta_v$(K) | $2\pi\nu$ (rad/s) |
|---|---|---|---|---|
| $H_2$ | 85.4 | $0\ 047\times10^{-28}$ | 6210 | $8.13\times10^{14}$ |
| HCl | 15.2 | $0.265\times10^{-28}$ | 4140 | $5.42\times10^{14}$ |
| $N_2$ | 2.9 | $1.41\times10^{-28}$ | 3340 | $4.37\times10^{14}$ |
| $O_2$ | 2.1 | $1.94\times10^{-28}$ | 2230 | $2.92\times10^{14}$ |
| $Cl_2$ | 0.3 | $11.6\times10^{-28}$ | 810 | $1.06\times10^{14}$ |

## 5.6 ELECTRONIC AND NUCLEAR PARTITION FUNCTIONS

Molecules can exist with electrons excited to states higher than the ground state. The energy spacings of these states vary in irregular fashion. Therefore, it is not possible to give a general expression for $z(e)$. However, at ordinary temperatures most of the molecules are usually in their ground state whose energy is by definition set equal to zero. Thus,

$$z(e) = g_{(gr)}(e) + g_1 \exp[-\beta\epsilon_1(e)] + \ldots$$

$$\simeq g_{(gr)}(e), \qquad (5.37)$$

where $g_{(gr)}(e)$ is the degeneracy of the electronic ground state.

The nuclear energy can be taken to be zero. Except in atomic explosions, the nuclei are not excited thermally to states above their ground state. Thus,

$$z(n) = g_{(gr)}(n, s), \tag{5.38}$$

where $g_{(gr)}(n, s)$ is the nuclear spin degeneracy. If in a diatomic molecule the nuclei have spins $s_1$ and $s_2$, we have $g_{(gr)}(n, s) = (2s_1 + 1)(2s_2 + 1)$.

## 5.7 APPLICATION OF ROTATIONAL PARTITION FUNCTION

For a heteronuclear molecule (like, HD or HCl), (5.26) gives

$$\begin{aligned} z(r) &= \sum_{J=0}^{\infty} (2J+1) \exp[-J(J+1)\theta] \\ &= 1 + 3e^{-2\theta} + 5e^{-6\theta} + \dots, \end{aligned} \tag{5.39}$$

where $\theta = \Theta_r/T$.

*High Temperature Limit* $\theta \ll 1$: We can use Euler-MacLaurin formula

$$\sum_{n=0}^{\infty} f(n) = \int_0^x f(x)\,dx + \frac{1}{2} f(0) - \frac{1}{12} f'(0) + \frac{1}{720} f'''(0) - \dots \tag{5.40}$$

Putting $f(x) = (2x+1)\exp[-x(x+1)\theta]$, we have $\int_0^\infty f(x)\,dx = 1/\theta$ as shown in (5.27), and

$$f(0) = 1,\ f'(0) = 2 - \theta,\ f'''(0) = -12\theta + 12\theta^2 - \theta^3, \dots$$

so that

$$z(r) = \frac{1}{\theta}\left(1 + \frac{1}{3}\theta + \frac{1}{15}\theta^2 + \dots\right). \tag{5.41}$$

For $\theta = 1/20$, the first few terms in (5.41) are

$$z(r)_{\text{classical}} = 1/\theta = 20,$$

$$(1/\theta)\left(1 + \frac{1}{3}\theta\right) = 20.333\dots,$$

$$(1/\theta)\left(1 + \frac{1}{3}\theta + \frac{1}{15}\theta^2\right) = 20.33666\dots.$$

The thermodynamic functions are

$$\begin{aligned} F(r) &= -NkT \ln z(r) = -NkT \ln\left[\frac{1}{\theta}\left(1 + \frac{1}{3}\theta + \frac{1}{15}\theta^2 + \dots\right)\right] \\ &= NkT \ln\theta - NkT \ln\left(1 + \frac{1}{3}\theta + \frac{1}{15}\theta^2 + \dots\right) \\ &= NkT \ln\theta - NkT\left(\frac{1}{3}\theta + \frac{1}{90}\theta^2 + \dots\right), \end{aligned} \tag{5.42}$$

$$U(r) = -T^2 \frac{\partial}{\partial T}\left(\frac{\partial F(r)}{\partial T}\right) = NkT - NkT\left(\frac{1}{3}\theta + \frac{1}{45}\theta^2 + \dots\right), \quad (5.43)$$

$$C_V(r) = \frac{\partial U(r)}{\partial T} = Nk + Nk\left(\frac{1}{45}\theta^2 + \dots\right), \quad (5.44)$$

where in (5.42) we have used $\ln(1 + x) = x - \frac{1}{2}x^2 + \dots$. For $T \to \infty$, $C_V$ decreases toward the classical value $Nk$. For $T \to 0$, we must have $C_V(r) = 0$. Therefore, the $C_V(r)$ vs. $T$ curve should show a maximum.

*Low Temperature Limit* $\theta \gg 1$: The series (5.39) can be directly summed. In Fig. 5.1 we plot $z(r)$ and $C_V(r)$ for the range $0 \leqslant T/\Theta_r \leqslant 2$. For HD the value of $\Theta_r$ is 64 K.

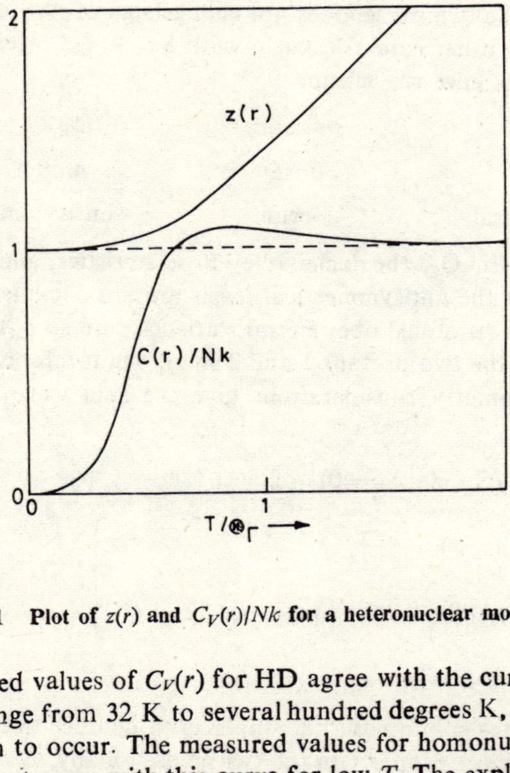


**Fig. 5.1 Plot of $z(r)$ and $C_V(r)/Nk$ for a heteronuclear molecule.**

The measured values of $C_V(r)$ for HD agree with the curve in Fig. 5.1, in the entire range from 32 K to several hundred degrees K, when molecular vibrations begin to occur. The measured values for homonuclear molecules $H_2$ and $D_2$ do not agree with this curve for low $T$. The explanation for this anomaly is to be found in the consideration of indistinguishability of the nuclei in the molecule. The $H_2$ and $D_2$ are the only homonuclear molecules with a low enough boiling point to exhibit this effect of quantum statistics.

## 5.8 HOMONUCLEAR MOLECULES AND NUCLEAR SPIN

So far we have ignored the contribution of nuclear spin to the partition function because it cannot be excited at laboratory temperatures. However, for homonuclear molecules its inclusion can lead to quantum states with different statistical weights even in the ground state of the molecule.

We have already stated that each rotational level $\psi_R = \psi_e \psi_v \psi_r$ is either symmetric or asymmetric with respect to an interchange of nuclei. Here $\psi_R$ is a function of coordinates. If we include the effect of spin of the nuclei,

$$\psi_{RN} = \psi_R \chi_N, \tag{5.45}$$

where the spin wave function $\chi_N$ is not a function of coordinates. The $\psi_{RN}$ is essentially the product of the wave function for the independent rotational and spin modes of energy storage, if we neglect electronic states and consider translation and vibration separately in $\psi_T = \psi_{RN}\psi_t$.

The nuclei can obey either Bose or Fermi statistics depending on whether the spin $S$ of the nucleus is integral or half-integral, respectively. For nuclei with $S = 0, 1, 2, \ldots$ (bosons) the $\chi_S$ is symmetric in the sense that an interchange of the two nuclei does not change sign of the molecular wave function. On the other hand for nuclei with $S = \frac{1}{2}, \frac{3}{2}, \ldots$ (fermions) the sign changes. We thus have the scheme

| $S$ | *statistics* | $\psi_{RN}$ | |
|---|---|---|---|
| integral | Bose | symmetric | |
| half-integral | Fermi | antisymmetric | (5.46) |

If $S = 0$, as in $O_2$, the nuclei obey Bose statistics, and so $\psi_{RN} = \psi_R$ is symmetric. Thus the antisymmetrical levels are not occupied. If $S = \frac{1}{2}$, as in $H_2$, the nuclei (protons) obey Fermi statistics, and so $\psi_{RN}$ must be antisymmetric. For the two protons 1 and 2 in $H_2$ the total spin is $S = S_1 + S_2 = 0, 1$. The symmetry considerations give the spin wave functions $\chi_N$ as follows

$$\chi_N^a\ (S = 0,\ S_z = 0) = 2^{-1/2}\,(\uparrow\downarrow - \downarrow\uparrow)$$

$$\left.\begin{aligned} &\chi_N^s\ (1, 1) = \uparrow\uparrow \\ &\chi_N^s\ (1, 0) = 2^{-1/2}\,(\uparrow\downarrow + \downarrow\uparrow) \\ &\chi_N^s\ (1, -1) = \downarrow\downarrow \end{aligned}\right\} \tag{5.47}$$

where ($\uparrow$) denotes spin up and the superscript denotes the symmetric ($s$) or antisymmetric ($a$) property in the two spins. Finally, for $H_2$ we have the combinations

| $\psi_{RN}$ | $S$ | $\chi_N$ | $\chi_R$ | *spin degeneracy* $(2S + 1)$ | *statistical weight* | |
|---|---|---|---|---|---|---|
| $a$ | 0 | $a$ | $s$(even $J$) | 1 | $1(2J + 1)$ | |
| $a$ | 1 | $s$ | $a$(odd $J$) | 3 | $3(2J + 1)$ | (5.48) |

We find that the antisymmetric rotational levels $\psi_R$ have a statistical weight of $3(2J+1)$ whereas the symmetric ones have a statistical weight of $(2J+1)$.

If $S = 1$, as in $D_2$ or $N_2$, the nuclei again obey Bose statistics and $\psi_{RN}$ is symmetric. The coupling of two spins yields $S = 0, 1, 2$ with $\chi_N$ symmetric for $S = 0$, antisymmetric for $S = 1$, and symmetric for $S = 2$. The allowed combinations are

| $\psi_{RN}$ | $S$ | $\chi_N$ | $\chi_R$ | $2S+1$ | *statistical weight* |
|---|---|---|---|---|---|
| *s* | 0 | *s* | *s* | 1 | $1(2J+1)$ |
| *s* | 1 | *a* | *a* | 3 | $3(2J+1)$ |
| *s* | 2 | *s* | *s* | 5 | $5(2J+1)$ (5.49) |

Thus the symmetric rotational levels $\psi_R$ have a statistical weight of $6(2J+1)$ compared to $3(2J+1)$ for the antisymmetric levels.

**Ortho-and Parahydrogen**

From (5.48) we actually get two types of $H_2$ molecules. The *parahydrogen* molecules with $S = 0$ (antiparallel spins) occupy rotational levels with $J = 0, 2, 4, \ldots$ whereas the *orthohydrogen* molecules with $S = 1$ (parallel spins) occupy levels with $J = 1, 3, 5, \ldots$ . Because each quantum state has equal a priori probability, there exist three times as many odd rotational states (ortho) as even (para), due to the different statistical weights. If thermal equilibrium is established between nuclear spin orientations, we should have

$$z(r)_{\text{equil}} = 3z(r)_{\text{ortho}} + z(r)_{\text{para}}, \tag{5.50}$$

$$z(r)_{\text{ortho}} = \sum_{J=1,3,\ldots} (2J+1)\exp[-J(J+1)\theta]$$
$$= 3e^{-2\theta} + 7e^{-12\theta} + \ldots, \tag{5.51}$$

$$z(r)_{\text{para}} = \sum_{J=0,2,\ldots} (2J+1)\exp[-J(J+1)\theta]$$
$$= 1 + 5e^{-6\theta} + \ldots, \tag{5.52}$$

where

$$\theta = \Theta_r/T = 85/T \quad \text{for} \quad H_2.$$

The MB distribution gives

$$\frac{N_i}{N} = \frac{g_i \exp(-\epsilon_i/kT)}{\sum_i g_i \exp(-\epsilon_i/kT)} = \frac{(2J+1)\exp[-J(J+1)\theta]}{z(r)}. \tag{5.53}$$

The ratio of the number of para to ortho molecules in thermal equilibrium can therefore be expressed as

$$N_{\text{para}} : N_{\text{ortho}} = 3z(r)_{\text{ortho}} : z(r)_{\text{para}}. \tag{5.54}$$

We can easily calculate $N_{\text{para}}/(N_{\text{para}} + N_{\text{ortho}})$ in the equilibrium mixture for various values of $T$ (Fig. 5.2).

At room temperature (300 K) we have one-quarter para and three-quarters orthohydrogen. As the temperature is decreased we expect ortho states (higher energy as odd $J > 0$) to be converted to para state (ground

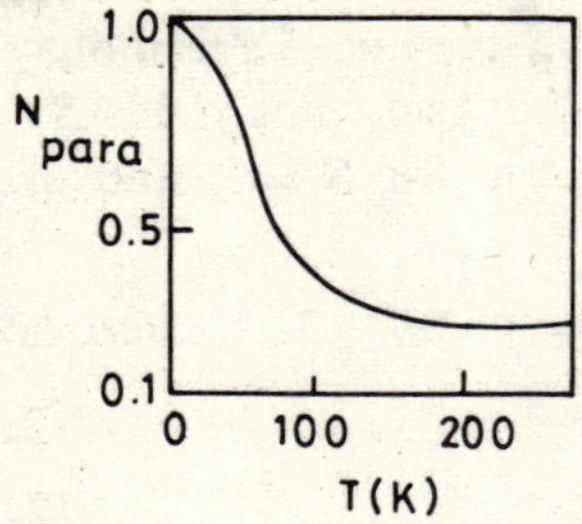


**Fig. 5.2** **Variation of $N_{para}$ with $T$.**

state, $J = 0$), according to Fig. 5.2. Note that (5.50), being the partition function per molecule at equilibrium, will continue to hold even as $T \to 0$, where the gas is entirely para. The specific heat curve ( Fig. 5.3, curve $B$) calculated according to this prescription is *also* found to disagree with experiments. The reason for this anomaly is that the process of establishing thermal equilibrium is very slow. It may take about three years for the process to go half way under ordinary conditions.

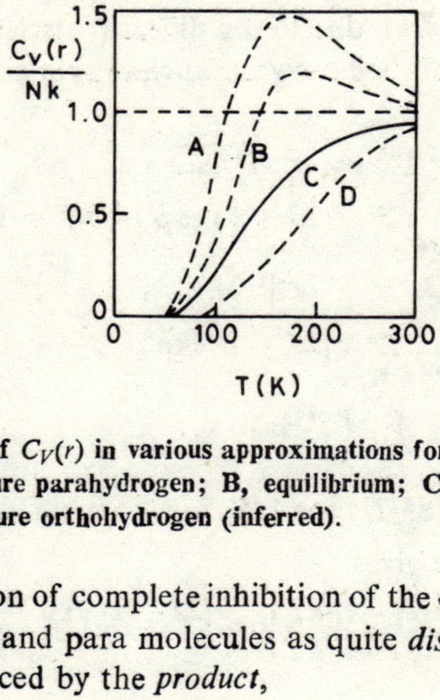


**Fig. 5.3** **Plot of $C_V(r)$ in various approximations for the $H_2$ molecule. A, pure parahydrogen; B, equilibrium; C, $\frac{1}{4}$ para + $\frac{3}{4}$ ortho; D, pure orthohydrogen (inferred).**

Under the condition of complete inhibition of the ortho-para conversion, we must regard ortho and para molecules as quite *distinct*, and the sum in (5.50) should be replaced by the *product*,

$$z(r) = [z(r)_{\text{ortho}}]^{3/4}.\ [z(r)_{\text{para}}]^{1/4}. \tag{5.55}$$

From (5.51, 52, 55)

$$C_V(r)_{\text{ortho}} = Nk\,\frac{\partial}{\partial T}\left[-\frac{\partial \ln z(r)_{\text{ortho}}}{\partial(1/T)}\right]$$

$$= Nk\,\frac{\partial}{\partial T}\left\{-\frac{\partial}{\partial(1/T)}\ln\left[3e^{-2\theta}\left(1+\frac{7}{3}e^{-10\theta}+\ldots\right)\right]\right\}$$

$$= \frac{700}{3} Nk\theta^2 e^{-10\theta} + \ldots, \tag{5.56}$$

$$C_V(r)_{\text{para}} = 180\, Nk\theta^2 e^{-6\theta} + \ldots, \tag{5.57}$$

$$C_V(r) = \frac{3}{4} C_V(r)_{\text{ortho}} + \frac{1}{4} C_V(r)_{\text{para}}, \tag{5.58}$$

with $\theta = 85/T$. The result (5.58) is in agreement with experiment (Fig. 5.3, curve $C$). The curve $A$ (Fig. 5.3) has also been observed by producing pure parahydrogen with the help of a catalyst.

The energy to excite molecules rotationally comes from collisions between molecules in the gas. For $H_2$ the $\Theta_r$ has a relatively large value due to small $I$ (Table 5.1). Therefore, at low temperatures the colliding molecules do not have enough kinetic energy for collisions to provide a whole quantum of rotational energy. In this sense the rotational degrees are 'frozen' and the gas behaves like a classical monoatomic gas.

## 5.9 APPLICATION OF VIBRATIONAL PARTITION FUNCTION TO SOLIDS

Consider a crystal consisting of $N$ atoms. This system has $3N$ degrees of freedom. Einstein assumed that the thermal vibrations of atoms are (1) harmonic oscillations all of the same characteristic frequency $\nu_E$, (2) independent from one atom to another, and (3) independent from one vibrational degree of freedom to another of the same atom. Then the crystal can be simply regarded as an assembly of $3N$ one-dimensional oscillators with the crystalline partition function given by (5.33) as

$$Z = \exp\,[-3N(h\nu_E/2kT)]\,[1-\exp\,(-h\nu_E/kT)]^{-3N}$$
$$= [2\sinh\,(h\nu_E/2kT)]^{-3N}, \tag{5.59}$$

whence

$$U = kT^2 \frac{\ln Z}{\partial T} = 3N \frac{h\nu_E}{2} \coth \frac{h\nu_E}{2kT}$$
$$= 3N \frac{h\nu_E}{2} + 3Nh\nu_E\,[\exp\,(h\nu_E/kT)-1]^{-1},$$
$$U - U_0 = 3Nh\nu_E\,[\exp\,(h\nu_E/kT)-1]^{-1},\ U_0 = \tfrac{3}{2} Nh\nu_E\,, \tag{5.60}$$

$$C_V = \left(\frac{\partial U}{\partial T}\right)_V = 3Nk \frac{x^2 e^x}{(e^x-1)^2}\,, \tag{5.61}$$

$$x \equiv \frac{h\nu_E}{kT} \equiv \frac{\Theta_E}{T}, \quad \Theta_E = \frac{h\nu_E}{k}\,. \tag{5.62}$$

$\Theta_E$ is called the *Einstein temperature* and $\nu_E$ the *Einstein frequency* of the solid. $\Theta_E$ is the only parameter in the theory and can be chosen differently for each solid.

For high temperature, $x = \Theta_E/T \ll 1$, we have $e^x \simeq 1$, $e^x - 1 \simeq x$ and (5.61) reduces to

$$C_V = 3Nk = 3R \simeq 6 \text{ cal/mole K}, \quad (T \gg \Theta_E),$$

in agreement with the experimentally verified Dulong and Petit's law.

For low temperatures, $x = \Theta_E/T \gg 1$, we have $e^x - 1 \simeq e^x$ and (5.61) becomes

$$C_V = 3Nk \left(\frac{\Theta_E}{T}\right)^2 \exp(-\Theta_E/T), \quad (T \ll \Theta_E). \tag{5.63}$$

Thus $C_V \to 0$ as $T \to 0$, in agreement with experiment, but the manner of approach to the zero value given by (5.63) is quite different from the experimental $T^3$ law. The experimental points for $T \leqslant 10$ K lie above the theoretical curve (5.63), Fig. 5.4. The Debye theory explains this in a satis-

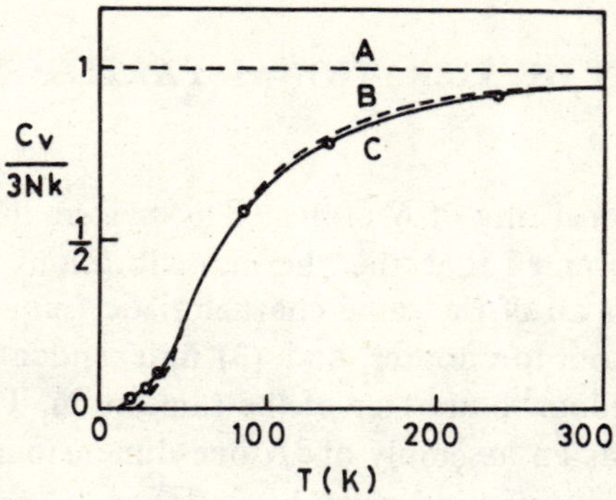


**Fig. 5.4 Specific heat curves for classical (dashed, A), Einstein (dashed, B), and Debye (solid, C) models, using $\Theta_E = 225$ K, $\Theta_D = 310$ K. The experimental points are for copper.**

factory way. Debye noted that the motion of each atom in a solid is not independent of the motions of its neighbours, as assumed by Einstein. He suggested that the single frequency $\nu_E$ of Einstein should be replaced by a spectrum of vibrational frequencies for the crystal.

At low temperatures, when quantum effects begin to be most significant, most substances are in the solid state. The atoms in a solid are arranged in a crystal lattice. The displacement of one atom in a lattice leads to the displacement of neighbouring atoms. This results in a wave propagating along the crystal. The wave nature of the motion of atoms in a crystal lattice enables us to develop a statistical theory of crystals by analogy with the theory of radiation in an enclosure. Complications arise because a lattice is discrete and vibrations of various kinds are possible. We should analyze the vibrations of the whole lattice. The Hamiltonion for a solid can then be approximated by a sum of terms, each representing a harmonic oscillator, corresponding to a normal mode of lattice vibration. Classically, each normal mode is a wave of distortion of the lattice planes, that is, an elastic wave or sound wave. Following de Broglie, a particle (called *phonon*) can be associated with these waves. Thus, the energy of an elastic wave in a solid can be quantized just as the energy of electromagnetic wave in an enclosure is quantized. The quantum energy of a phonon is $h\nu$ for an elastic wave of frequency $\nu$. Like photon, it is also a boson of zero mass.

Enumeration of all the phonons present specifies the solid near its ground state. At very low temperatures, we can regard a solid as a *gas of non-interacting phonons*, enclosed in a volume $V$. Each phonon, being a quantum of a certain harmonic oscillator, has a characteristic frequency $\nu_l$ and energy $h\nu_l$. The state of lattice with one phonon present is described by a sound wave of the form $\mathbf{e} \exp [i(\mathbf{k}.\mathbf{r}-\omega t)]$ where $\mathbf{k}$ is the propogation vector of magnitude $|\mathbf{k}| = \omega/c_s = 2\pi\nu/c_s$, with $c_s$ = velocity of sound, and $\mathbf{e}$ is the polarization vector (not necessarily perpendicular to $\mathbf{k}$). The $\mathbf{e}$ can have three independent directions, corresponding to one longitudinal mode of compression wave and two transverse modes of shear wave.

The harmonic oscillator in an excited state can contain any number of quanta. The phonons obey BE statistics, with no conservation of their total number ($\alpha = 0$). A solid, containing $N$ atoms, has $(3N-6) \approx 3N$ number of modes of vibration. Therefore, we can have $3N$ different types of phonons with frequencies $\nu_1, \nu_2, \ldots, \nu_{3N}$. It is difficult to calculate these characteristic frequencies as their values depend on the lattice of interest. Einstein assumed $\nu_1 = \nu_2 = \ldots = \nu_{3N}$. Improving upon it, Debye assumed that for finding these frequencies the solid can be approximated by an *elastic continuum* of volume $V$. Thus the frequencies required are the lowest $3N$ frequencies of this model system.

An elastic continuum has continuous distribution of normal frequencies. We wish to find the number of normal modes whose frequencies lie between $\nu$ and $\nu + d\nu$. Use of periodic boundary conditions gives $\mathbf{k}=(2\pi/L)\mathbf{n}$, where $L = V^{1/3}$ and $\mathbf{n}$ has the components $n_x, n_y, n_z=0, \pm 1, \pm 2, \ldots$. Since $k_x = (2\pi/L)n_x$, and $n_x$ is integral, the $k_x$ increases by $2\pi/L$ for each additional state. Therefore, the number of $k_x$ values in the interval $dk_x$ is $(L/2\pi)dk_x$. With similar results for other directions, the number we want is

$$g(\nu)\, d\nu = \left(\frac{L}{2\pi}\right)^3 dk_x dk_y dk_z = \frac{V}{(2\pi)^3} 4\pi k^2 dk. \tag{5.64}$$

As $k = \omega/c_s = 2\pi\nu/c_s$,

$$g(\nu)\, d\nu = V \frac{4\pi\nu^2}{c_s^3}\, d\nu. \tag{5.65}$$

Because there are two transverse and one longitudinal types of vibrations,

$$g(\nu)\, d\nu = 4\pi V \left(\frac{1}{c_l^3} + \frac{2}{c_t^3}\right) \nu^2 d\nu. \tag{5.66}$$

The normal modes representing the collective oscillations of the crystal have a continuous spectrum from $\nu = 0$ to $\nu = \nu_D$. Following **Debye**, we require that the *maximum* frequency $\nu_D$, called the *Debye frequency*, is determined from the sum of the degrees of freedom of the $N$ particle system,

$$3N = \int_0^{\nu_D} g(\nu)\, d\nu = \frac{4\pi V}{3}\left(\frac{1}{c_l^3} + \frac{2}{c_t^3}\right)\nu_D^3. \tag{5.67}$$

Such a cut-off in the frequency is associated with the discrete structure of the crystal lattice and was not needed for photons. Writing $c_l = c_t = \bar{c}$, (5.67) gives $\nu_D^3 = (3/4\pi)\,\bar{c}^3\,(N/V)$. This means the minimum phonon wavelength is $\lambda_D = \bar{c}/\nu_D = (4\pi/3)^{1/3}\,(V/N)^{1/3} \approx$ interparticle spacing ($N = 10^{23}$ atoms, per cubic centimeter). For a one-dimensional lattice of atoms, for example, the smallest possible value of the wavelength of a transverse wave is equal to twice the lattice constant when neighbouring atoms vibrate in counterphase (Fig. 5.5). From (5.66, 67)

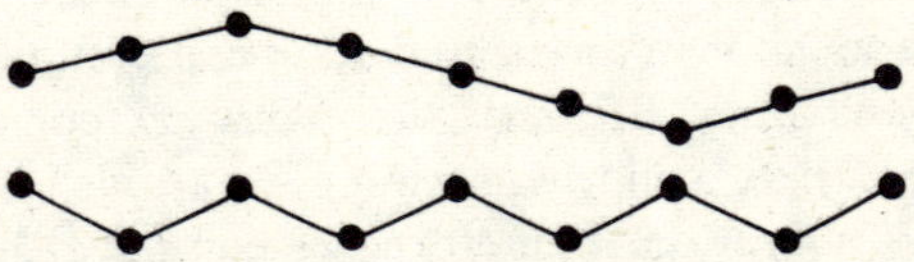

**Fig. 5.5 Transverse waves in a one-dimensional lattice.**

$$g(\nu)\, d\nu = 9N(\nu^2/\nu_D^3)\, d\nu. \tag{5.68}$$

Thus in the Debye approximation $g(\nu) \propto \nu^2$. The frequency spectra of three-dimensional lattice vibrations for various approximations are shown in Fig. 5.6.

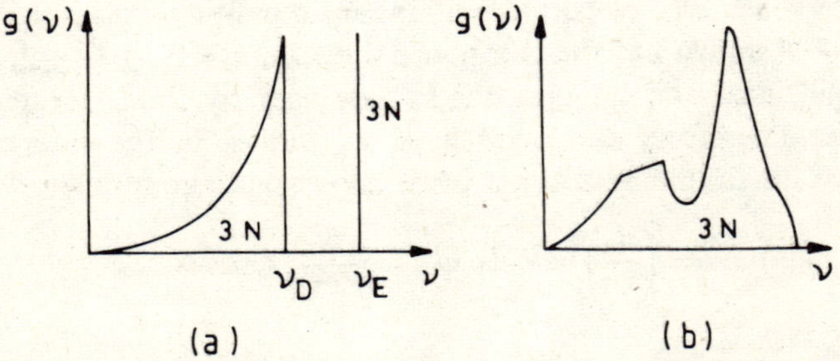


**Fig. 5.6 The density of modes as a function of frequency in (a) Debye and Einstein approximations, and (b) an actual crystal structure(begins as $\nu^2$ for small $\nu$, but develops discontinuities as $\nu$ increases).**

The distribution law for phonons is similar to that for photons,

$$\bar{n}_\nu = \frac{1}{\exp(h\nu/kT) - 1}. \tag{5.69}$$

Therefore, the internal energy of the phonon gas is

$$U-U_0=\int_0^{\nu_D} h\nu\,\bar{n}_\nu g(\nu)\,d\nu$$

$$=\frac{9Nh}{\nu_D^3}\int_0^{\nu_D}\frac{\nu^3 d\nu}{\exp(h\nu/kT)-1}=\frac{9NkT}{x_D^3}\left[\int_0^{x_D}\frac{x^3dx}{e^x-1}\right], \tag{5.70}$$

and

$$C_V=\left(\frac{\partial U}{\partial T}\right)_V=\frac{9Nh^2}{kT^2\nu_D^3}\int_0^{\nu_D}\frac{\nu^4\exp(h\nu/kT)}{[\exp(h\nu/kT)-1]^2}$$

$$=\frac{3Nk}{x_D^3}\left[3\int_0^{x_D}\frac{x^4e^x\,dx}{(e^x-1)^2}\right], \tag{5.71}$$

$$x=\beta h\nu\equiv h\nu/kT,\quad x_D\equiv h\nu_D/kT\equiv\Theta_D/T. \tag{5.72}$$

This defines the *Debye temperature* $\Theta_D=h\nu_D/k$, in analogy to the Einstein temperature $\Theta_E$, and is the only parameter in the theory. If heat capacities of different solids are plotted against $T/\Theta_D$, they should fall on a single curve.

The integral in (5.71) cannot be evaluated analytically. Its tabulated values are available. However, we can obtain the following limits.

*High Temperature Limit* $T\gg\Theta_D$: For $0\leqslant x\leqslant x_D$ and $x_D\ll 1$, we can write

$$\frac{x^4e^x}{(e^x-1)^2}=\frac{x^4}{(1-e^{-x})(e^x-1)}=\frac{x^4}{2(\cosh x-1)}$$

$$=\frac{1}{2}x^4\left(\frac{1}{2!}x^2+\frac{1}{4!}x^4+\ldots\right)^{-1}$$

$$\simeq\frac{1}{2}x^4\left(\frac{1}{2!}x^2\right)^{-1}=x^2, \tag{5.73}$$

so that

$$C_V\simeq\frac{9Nk}{x_D^3}\int_0^{x_D}x^2dx=3Nk,\quad (T\gg\Theta_D), \tag{5.74}$$

in agreement with the Dulong and Petit law.

*Low Temperature Limit* $T\ll\Theta_D$: For $x_D\gg 1$ we can replace the upper limit in (5.70, 71) by $\infty$. Both the integrands, behaving like $x^3e^{-x}$ and $x^4e^{-x}$, go to zero very rapidly for large $x$. Therefore,

$$U-U_0=\frac{9NkT}{x_D^3}\left[\int_0^{\infty}\frac{x^3dx}{e^x-1}\right]=\frac{9NkT^4}{\Theta_D^3}[6\,\zeta(4)]$$

$$=\frac{9NkT^4}{\Theta_D^3}\left[6\sum_{n=1}^{\infty}\frac{1}{n^4}\right]=\frac{9NkT^4}{\Theta_D^3}\left[\frac{\pi^4}{15}\right]$$

$$= \frac{3}{5}\pi^4 NkT^4/\Theta_D^3, \tag{5.75}$$

$$C_V = (12\pi^4 Nk/5)\,(T/\Theta_D)^3, \quad (T \ll \Theta_D), \tag{5.76}$$

where $\zeta(4)$ is the Riemann Zeta function (Appendix VI), $\zeta(4) = \pi^4/90$. Thus for $C_V$ we get the famous $T^3$-law of Debye, valid at low temperatures and in agreement with experiment (Fig. 5.4). Values of $\Theta_D$ for a few elements are Na (160 K), Cu (310 K), Al (380 K) and Diamond (about 1950 K).

## 5.10 VAPOUR PRESSURE

Let us calculate the equilibrium vapour pressure of a monatomic solid at $T \to 0$. The vapour and solid form two subensembles in equilibrium with respect to exchange of particles and energy. This equilibrium is governed by the equality of $\mu$ and $T$. We will calculate $\mu_S$ of the solid and $\mu_G$ of the gas and equate them.

From (4.66),

$$\mu_S = N_S^{-1}\,(U_S - TS_S - \Omega_{gS}), \tag{5.77}$$

where $N_S$ is the number of atoms in the solid phase. The energy $U_S$ of the solid will be the sum of two parts: the energy $U_o$ if there are no phonons present and the energy $U_p$ of the phonons. The $U_o$ is the work required to break up the solid, atom by atom, without exciting any phonon. It gives the latent heat of the sample. If $l_o$ is the latent heat per atom then

$$U_S = U_p - N_S l_o.$$

The entropy of solid is entirely due to phonons, $TS_S = TS_p$. Finally, $\Omega_S = -PV_S \to 0$ as the pressure is very low for $T \to 0$. Thus, (5.77) becomes

$$\mu_S = N_S^{-1}(-N_S l_o + U_p - TS_p). \tag{5.78}$$

As $\mu_p = 0$, from (4.66),

$$\mu_S = -l_o + (\Omega_{gp}/N_S). \tag{5.79}$$

Using $\partial(\beta\Omega_g)/\partial\beta = U$ and (5.75), the low temperature limit of $\Omega_{gp}$ is

$$\Omega_{gp} = -\frac{\pi^4 N_S}{5\beta^4 (k\Theta_D)^3}. \tag{5.80}$$

Therefore,

$$\mu_S = -l_o - \frac{\pi^4 kT^4}{5\Theta_D^3}. \tag{5.81}$$

From (4.76, 77), for the vapour

$$\mu_G = kT \ln\left[\frac{P}{kT}\left(\frac{h^2}{2\pi mkT}\right)^{3/2}\right]. \tag{5.82}$$

Put $\mu_S = \mu_G$ and solve for $P$,

$$P = \left(\frac{2\pi m}{h^2}\right)^{3/2} (kT)^{5/2} \exp\left[-(\pi^4/5)\,(T/\Theta_D)^3\right] \exp\left(-l_0/kT\right). \tag{5.83}$$

Note that in this result $h$ occurs explicitly.

## 5.11 CHEMICAL EQUILIBRIUM

In chemical reactions the number of particles of different species change. A general chemical reaction is described by the formula

$$-\sum_{i} \nu_i M_i \rightarrow \sum_{f} \nu_f M_f. \tag{5.84}$$

It states that a certain number, $-\nu_i$, of molecules of the initial reactants $M_i$ combine to produce a certain number, $\nu_f$, of the final products $M_f$. A simple example is

$$H_2 + Cl_2 \rightleftharpoons 2HCl,$$

for which,

$$M_1 = H_2,\ M_2 = Cl_2,\ M_3 = HCl,$$

$$\nu_1 = -1,\ \nu_2 = -1,\ \nu_3 = +2.$$

The chemical reaction can proceed in either direction, depending upon the conditions. We can express (5.84) as

$$\sum_{j} \nu_j M_j = 0, \tag{5.85}$$

where the negative (positive) $\nu$'s represent initial (final) molecules. When the quantities are measured in moles, the $\nu$'s are called *stoichiometric numbers*.

In a chemical reaction the numbers $N_j$ of moles of the reacting substances will change during the reaction. Therefore, the method of grand canonical ensemble is suitable for the study of chemical equilibrium.

The Gibbs function (3.64) and its change (3.62) in a reaction are

$$G = \sum_{j} N_j \mu_j, \tag{5.86}$$

$$dG = -SdT + Vdp + \sum_{j} \mu_j dN_j, \tag{5.87}$$

where $\mu_j$ is the chemical potential of the $j$th molecular species $M_j$, and $N_j$ is the corresponding number of molecules. If the reaction occurs at constant pressure and temperature, $dp = dT = 0$,

$$dG = \sum_{j} \mu_j dN_j. \tag{5.88}$$

The $N_j$ cannot change arbitrarily because they are related to each other through the chemical reaction (5.85). This requires $dN_j \propto \nu_j$, or

$$dN_j = \nu_j dn, \tag{5.89}$$

$$dG = \sum_{j} \mu_j \nu_j\, dn. \tag{5.90}$$

For equilibrium $G$ is a minimum, that is, a small change in $n$, $\delta n$, corresponds to $\delta G = 0$, or

$$\sum_{j} \mu_j \nu_j = 0, \qquad (\textit{for equilibrium}). \tag{5.91}$$

We require $G$ to be a minimum because in any natural process $G$ can only decrease.

The relation (5.91) is the basic equation for chemical reactions. For the reaction $H_2 + Cl_2 = 2HCl$, it implies

$$\mu_{HCl} = \tfrac{1}{2}(\mu_{H_2} + \mu_{Cl_2}). \tag{5.92}$$

Thus the $\mu$'s of the different components are just added in the same proportions in which they occur in the reaction.

Further progress consists in expressing the $\mu$'s in terms of temperature, pressure, composition and molecular characteristics. We apply (5.91) to a reaction in a gaseous phase. We treat the gases as ideal gases so as to be able to obtain $\mu$'s from the Helmholtz free energy. For ideal gases

$$F(T, V, N_1, N_2, \dots N_r) = \sum_j F_j(T, V, N_j), \tag{5.93}$$

$$Z_j(T, V, N_j) = (N_j\,!)^{-1}\,[z_j(T, V)]^{N_j}, \tag{5.94}$$

$$\begin{aligned} F_j(T, V, N_j) &= -kT \ln Z_j(T, V, N_j) \\ &= -kT[N_j \ln z_j - N_j(\ln N_j - 1)], \end{aligned} \tag{5.95}$$

$$\begin{aligned} \mu_j &= (\partial F/\partial N_j)_{T, V, N_i (i \neq j)} \\ &= -kT\,(\ln z_j - \ln N_j), \end{aligned} \tag{5.96}$$

$$\mathscr{Z}_j = \exp\,[z_j \exp\,(\mu_j/kT)]. \tag{5.97}$$

For a monatomic gas, from (4.39),

$$\begin{aligned} z_j &= V/\lambda_j^3, \quad \lambda_j = (h^2/2\pi m_j kT)^{1/2}, \\ \mu_j &= kT \ln (N_j/V) - \tfrac{3}{2}\, kT \ln (2\pi m_j\, kT/h^2) \\ &= kT \ln P_j - kT \ln kT - \tfrac{3}{2}\, kT \ln (2\pi m_j kT/h^2) \\ &= kT \ln P_j - D_j(T), \end{aligned} \tag{5.98}$$

where $P_j = N_j kT/V$ is partial pressure of the $j$th component and $D_j(T)$ is a function of temperature alone. From (5.91, 98),

$$\sum_j \mu_j \nu_j = kT \sum_j \nu_j \ln P_j + \sum_j \nu_j D_j(T) = 0, \tag{5.99}$$

or

$$\sum_j \nu_j \ln P_j = -(1/kT) \sum_j \nu_j D_j(T). \tag{5.100}$$

Remembering that initial $\nu$'s are negative,

$$\prod_j P_j^{\nu_j} = P_1^{\nu_1} P_2^{\nu_2} \dots = \frac{\prod_f P_f^{\nu_f}}{\prod_i P_i^{\nu_i}} = \exp\,[-(1/kT) \sum_j \nu_j D_j(T)] \equiv K_P(T). \tag{5.101}$$

This relation relates the $P_j$'s to the *equilibrium constant* $K_P(T)$, which is a function of $T$ alone. It is given by

$$\begin{aligned}\ln K_P(T) &= -(1/kT) \sum_j \nu_j D_j(T) \\ &= \sum_j \nu_j [\ln (2\pi m_j/h^2)^{3/2} + \ln (kT)^{5/2}]. \qquad (5.102)\end{aligned}$$

In general, for a molecular species,

$$z_j = z_j(t)\, z_j(\text{int}), \qquad (5.103)$$

where $z_j(t)$ is the translational and $z_j(\text{int})$ is the internal partition function like $z_j(r)$, $z_j(v)$, etc. It is convenient to write

$$z_j\,(\text{int}) = g_{jo} \exp(-\epsilon_{jo}/kT)\, z_{jo}\,(\text{int}), \qquad (5.104)$$

where $\epsilon_{jo}$ and $g_{j0}$ are the energy and degeneracy of the ground state, and

$$z_{jo}\,(\text{int}) = 1 + \sum_{l=1}^{\infty} \frac{g_{jl}}{g_{jo}} \exp(-\epsilon_{jl}/kT). \qquad (5.105)$$

Here $\epsilon_{jl}$ are the internal energy levels relative to the zero ground state energy $\epsilon_{jo}$. As $T \to 0$, $z_{jo}$ (int) $\to 1$, so that at temperatures for which the excitation of the molecule is unimportant,

$$z_j\,(\text{int}) \simeq g_{jo} \exp(-\epsilon_{jo}/kT). \qquad (5.106)$$

The factor $\exp(-\epsilon_{jo}/kT)$ is shown explicitly because often it is the most important term in the $z_j$ (int).

From (5.97) and $-\Omega_{gj} = P_j V = kT \ln \mathscr{Z}_j$,

$$\begin{aligned}P_j &= (kT/V) \ln \mathscr{Z}_j = (kT/V)\,[z_j \exp(\mu_j/kT)] \\ &= (kT/V) \exp(\mu_j/kT)\, z_j(t)\, z_j\,(\text{int}) \\ &= (kT/V) \exp(\mu_j/kT)\, z_j(t) g_{jo} \exp(-\epsilon_{jo}/kT). \qquad (5.107)\end{aligned}$$

We can calculate $K_P(T)$ for any process from (5.101, 107).

For a process of the type

$$M \to B + C, \qquad (5.108)$$

we have

$$\begin{aligned}K_P(T) = \frac{P_B P_C}{P_M} &= \frac{kT}{V} \frac{\exp(\mu_B/kT) \exp(\mu_C/kT)}{\exp(\mu_M/kT)} \cdot V \left(\frac{2\pi m_B m_C kT}{m_M h^2}\right)^{3/2} \cdot \frac{z_B\,(\text{int})\, z_C\,(\text{int})}{z_M\,(\text{int})} \\ &= \left(\frac{2\pi m_B m_C}{m_M h^2}\right)^{3/2} (kT)^{5/2} \frac{z_B\,(\text{int})\, z_C\,(\text{int})}{z_M\,(\text{int})}, \qquad (5.109)\end{aligned}$$

where in the last step we have used the basic equation (5.91), that is,

$$\mu_M = \mu_B + \mu_C, \qquad (5.110)$$

at equilibrium. Since $K_P(T)$ is a function of $T$ only, the ratio $P_B P_C/P_M$ is constant at a given temperature. This is the *law of mass action*.

We can apply the above considerations to the *thermal ionization* of a gas at a very high temperature,

$$M \rightarrow M^+ + e. \tag{5.111}$$

A simple case is the thermal ionization of sodium vapour, $Na \rightarrow Na^+ + e$.

At a given temperature, the ionization reaction attains a state of equilibrium which is quite analogous to the chemical equilibrium for ordinary chemical reactions. Each of these substances $(M, M^+, e)$ behaves like a monatomic gas. Therefore,

$$K_P(T) = \frac{P_{M^+} P_e}{P_M} = \left(\frac{2\pi m_e}{h^2}\right)^{3/2} (kT)^{5/2} \frac{z_{M^+}\,(\text{int})}{z_M\,(\text{int})} g_{eo} \exp\left(-\epsilon_{eo}/kT\right), \tag{5.112}$$

where $g_{eo} = 2$ is the weight function of electron, $\epsilon_{eo}$ is the energy of the released electron, $z_M$ (int) is the internal or electronic partition function of the element $M$, and $z_{M^+}$ (int) is the partition function of the ion. The relation (5.112) was first derived by Saha* and has found several applications in astrophysics.

## 5.12 REAL GAS

In this section we consider imperfect (real) gases, that is, gases for which the interactions between the constituent particles cannot be neglected as was assumed so far.

The Hamiltonian for a real gas is taken to be

$$H(q, p) = E(p) + \sum_{i>j} u_{ij}(r_{ij}), \tag{5.113}$$

where $E(p)$ is the total kinetic energy and $u_{ij}$ the potential energy between the particles $i$ and $j$ We assume that $u_{ij}$ depends only upon the distance $r_{ij} = |\mathbf{r}_i - \mathbf{r}_j|$ between the pairs of particles $i$ and $j$. We can also write the potential energy term as $\frac{1}{2} \sum_{i \neq j} u_{ij}(r_{ij})$, where the factor $\frac{1}{2}$ is included to compensate for counting each interaction twice (once as $ij$ and once as $ji$). For ideal gases $u_{ij} = 0$, and so the potential energy of the ideal gas is

$$\phi(\mathbf{r}_1, \ldots, \mathbf{r}_N) = \sum_{i>j} u_{ij}(r_{ij}) = 0, \qquad (\textit{ideal gas}). \tag{5.114}$$

For a classical imperfect gas we can write the partition function, neglecting internal degrees freedom in molecules, as

$$\begin{aligned} Z &= \frac{1}{N!\,h^{3N}} \int \exp\left[-\beta H(q, p)\right] dq\, dp \\ &= \left[\frac{V^N}{N!} \int \exp\left[-\beta E(p)\right] \frac{dp}{h^{3N}}\right] \cdot \left[\frac{1}{V^N} \int \exp\left[-\beta \phi(q)\right] dq\right] \\ &= Z_p \cdot Z_\phi \end{aligned} \tag{5.115}$$

*M.N. Saha, *Proc. Roy. Soc.* (London), A99, 135 (1921).

where $Z_p$ is the partition function for the perfect gas and

$$Z_\phi = \frac{1}{V^N}\int \exp\left[-\beta\,\phi(q)\right] d\mathbf{q}_1 \ldots d\mathbf{q}_N \tag{5.116}$$

is called the *configurational partition function* of the gas. For an ideal gas $Z_\phi = 1$.

In the approximation (5.113),

$$Z_\phi = \frac{1}{V^N}\int \exp\Big(-\sum_{i>j} \beta u_{ij}\Big) \prod_{i=1}^{N} d^3r_i,$$

$$= \frac{1}{V^N}\int \prod_{i>j} \exp(-\beta u_{ij}) \prod_i d^3r_i. \tag{5.117}$$

In (5.117) each coordinate can vary within the entire volume $V$ of the gas.

Introduce the function

$$\eta_{ij} \equiv \eta_{ij}(r_{ij}) \equiv \exp(-\beta u_{ij}) - 1. \tag{5.118}$$

In the ideal gas limit, $\eta_{ij} \to 0$, and so $\eta_{ij}$ is a measure of the degree of imperfection of the gas. Rewrite (5.117) as

$$Z_\phi = \frac{1}{V^N}\underbrace{\int \ldots \int}_{(3N)} \prod_{i>j} (1+\eta_{ij}) \prod_{i=1}^{N} d^3r_i$$

$$= V^{-N}\int d^3r_1 \ldots d^3r_N (1+\eta_{12})(1+\eta_{13}) \ldots (1+\eta_{N-1,\,N})$$

$$= V^{-N}\int d^3r_1 \ldots d^3r_N \Big(1 + \sum_{i>j} \eta_{ij} + \sum_{\substack{i>j\\ k>l\\ i,j\neq k,l}} \eta_{ij}\,\eta_{kl} + \ldots\Big). \tag{5.119}$$

The complicated integral (5.119) is reduced to a sum of relatively simple but still multiple integrals. Except for the first two of these integrals, the remaining ones are even now very difficult to estimate even approximately.

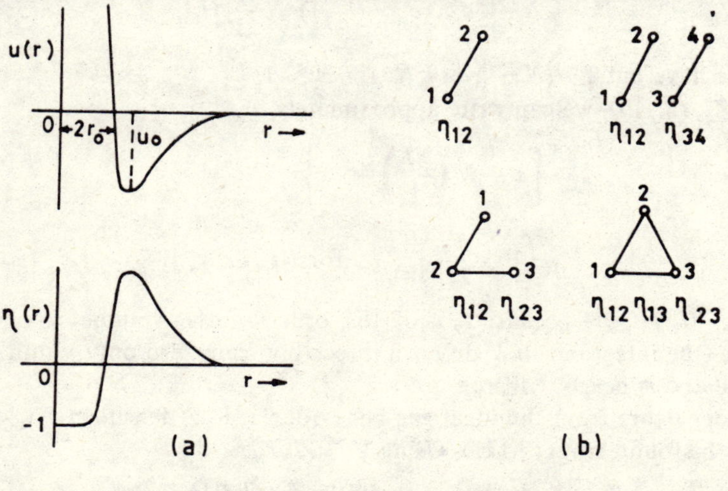


Fig. 5.7 (a) Plot of the intermolecular potential $u(r)$ and the function $\eta(r)$. A usual form is $u(r) = cr^{-m} - dr^{-n}$, $m > n$. (b) Cluster terms.

The usual forms of $u(r)$ and $\eta(r)$ are shown in Fig. 5.7a. It is seen that $\eta(r_{ij})$ is zero except when $r_{ij} \lesssim 2r_0$, where $r_0$ is the radius of one molecule. Thus $\eta_{12}$ is negligible unless molecules 1 and 2 are close together (binary collision). Similarly, $\eta_{12}\eta_{23}$ contributes only when molecules 1, 2, 3 are all simultaneously close together (triple collision), and so on. A graphical representation of such terms is shown in Fig. 5.7b. They represent clusters of molecules. The series (5.119) is called a *cluster expansion*. We shall discuss only the first two terms, 1 and $\eta_{ij}$, in this series.

If we retain the first term, 1, only and neglect all $\eta$'s (that is, neglect all collisions), then (5.119) simply gives $Z_\phi = 1$, $Z = Z_p$ (perfect gas approximation).

Let us consider the second term, $\Sigma\, \eta_{ij}$, in (5.119). This sum contains in all $\frac{1}{2}N(N-1)$ terms determined by the number of ways we can choose pairs of molecules. All these terms are equal because they differ only in the labels of integration variables. Therefore, we can write the contribution of the second term as

$$C_2 = V^{-N}\, \tfrac{1}{2}N(N-1) \int d^3r_1 \ldots d^3r_N\, \eta_{12}$$

$$= V^{-N}\, \tfrac{1}{2}\, N(N-1)\, V^{N-2} \int d^3r_1\, d^3r_2\, [\exp(-\beta\, u(r_{12}))-1],$$

where $u(r_{12})$ is the potential energy between the molecules 1 and 2 as a function of their distance apart $r_{12} = |\mathbf{r}_1 - \mathbf{r}_2|$. If we introduce the new variables of integration,

$$\mathbf{r} = \mathbf{r}_1 - \mathbf{r}_2, \quad \mathbf{R} = \tfrac{1}{2}(\mathbf{r}_1 + \mathbf{r}_2),$$

then the integration over $\mathbf{R}$ gives one more factor $V$, so that

$$C_2 = V^{-1}\, \tfrac{1}{2}\, N^2 \int d^3r\, (e^{-\beta u(r)} - 1) = \frac{N^2}{2V}[I_2], \tag{5.120}$$

where we have put $\frac{1}{2}N(N-1) \simeq \frac{1}{2}N^2$ for $N \gg 1$.

For $Z_\phi$, (5.119), we can write approximately

$$Z_\phi = \left[1 + N\left(\frac{NI_2}{2V}\right) + \ldots\right]$$

$$\simeq \left(1 + \frac{NI_2}{2V}\right)^N, \qquad \text{for } NI_2/2V \ll 1. \tag{5.121}$$

We have $NI_2/2V \ll 1$ because $I_2$ is of the order of the volume $v_0$ of one molecule (the integrand in $I_2$ differs appreciably from zero only within such a volume and is nearly 1 there).

The departure from the ideal gas behaviour at low densities, $Nv_0 \ll V$, can now be found from (5.121). Using $Z = Z_p Z_\phi$,

$$F = -kT \ln Z = -kT \ln Z_p - kT \ln Z_\phi$$

$$= F_p - kTN \ln\left(1+\frac{NI_2}{2V}\right)$$

$$\simeq F_p - kT\frac{N^2I_2}{2V}, \tag{5.122}$$

$$P = -(F/V))_{T,N} = \frac{NkT}{V} - kT\frac{N^2I_2}{2V^2}$$

$$= \frac{NkT}{V}\left(1-\frac{NI_2}{2V}\right), \tag{5.123}$$

remembering that $F_p$ leads to the ideal gas equation.

The equation of state is usually written in the *virial* form

$$P = \frac{NkT}{V}\left[1+\frac{N}{V}B(T)+\left(\frac{N}{V}\right)^2 C(T)+\ldots\right], \tag{5.124}$$

so that (5.123) gives just its beginning. The second virial coefficient $B(T)$ is related to the cluster integral $I_2$ as $B(T) = -I_2/2$.

**Van der Waals Gas**

Let us assume that the molecules are hard spheres of radius $r_0$ and volume $v_0 = (4/3)\,\pi r_0^3$. Then the cluster integral $I_2$ can be divided into two parts. In the first part, $0 < r < 2r_0$, the potential term $u(r)$ is infinite (two hard spheres in contact), so that $\exp(-\beta u(r)) - 1 \simeq -1$. The second part extends from $2r_0$ to $\infty$ (Fig. 5.8). Thus

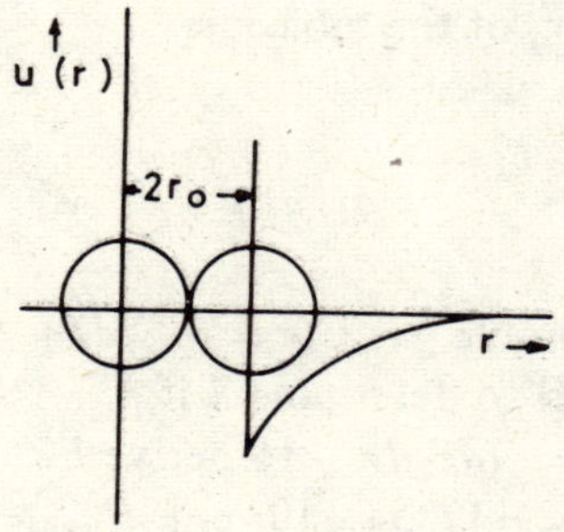


**Fig. 5.8 Hard sphere potential for Van der Waals gas.**

$$B(T) = -\tfrac{1}{2}I_2 = -\tfrac{1}{2}\int_0^\infty d^3r[\exp(-\beta u(r))-1]$$

$$= -\tfrac{1}{2}\int_0^{2r_0}(-1)\,d^3r - \tfrac{1}{2}\int_{2r_0}^\infty d^3r[\exp(-\beta u(r))-1]$$

$$\simeq \tfrac{1}{2}4\pi\int_0^{2r_0} r^2dr + \tfrac{1}{2}\int_{2r_0}^\infty 4\pi r^2\,dr\beta u(r)$$

$$= 2\pi\tfrac{1}{3}(2r_0)^3 + \frac{2\pi}{kT}\int_{2r_0}^{\infty} dr\, r^2 u(r) \equiv b - \frac{a}{kT}, \tag{5.125}$$

where we have put $\exp(-u(r)/kT)-1 \simeq -u(r)/kT$, assuming $u_0$ (Fig. 5.7) to be numerically small in comparison with $kT$, and

$$b = 4v_0,\ a = -2\pi \int_{2r_0}^{\infty} dr\, r^2\, u(r). \tag{5.126}$$

The equation of state up to $B(T)$ is

$$\begin{aligned} P &= \frac{NkT}{V}\left[1 + \frac{N}{V}\left(b - \frac{a}{kT}\right)\right] \\ &= \frac{NkT}{V}\left(1 + \frac{N}{V}b\right) - \frac{N^2}{V^2}a \\ &\simeq \frac{NkT}{V}\left(1 - \frac{N}{V}b\right)^{-1} - \frac{N^2}{V^2}a, \end{aligned} \tag{5.127}$$

which yields the familiar Van der Waals equation

$$\left(P + \frac{N^2 a}{V^2}\right)(V - Nb) = NkT. \tag{5.128}$$

As suggested by Van der Waals, the correction term $a$ comes from the long-range weak attractive force between the molecules (Fig. 5.7), and $b$ comes from the volume $v_0$ of the molecule.

## *PROBLEMS*

5.1 For a monatomic noble gas $C_V = (dE/dT)_V = 12.47$ J/mole-K. Combine this with (5.24) to determine $k$ if $\beta = 1/k\theta$.

5.2 Use (5.23) and $P = -(dF/dV)_T$ to derive $PV = RT$ for an ideal gas.

5.3 Show that $F_{\text{trans}} = -17354\times10^7$ erg, $U_{\text{trans}} = 1.56\times10^6$ erg, $S_{\text{trans}} = 4.6\times10^4$ erg/deg, $\mu_{\text{trans}} = -0.67\times10^{-12}$ erg and $Z_{\text{trans}} = \exp(4.189\times10^{12}) = 10^{10^{20.26}}$ for Argon gas (at wt. 39.94).

5.4 Calculate the rotational energy of para-and ortho-deuterium.

5.5 Determine the Helmholtz free energy, the entropy and the molar heat capacity at constant pressure of $CO_2$ gas at 0°C and 1 atm, using the ideal gas result. Given: molecular weight $=44.010$, moment of inertia (O–C–O) $I = 71.67\times10^{-40}$ gcm$^2$, normal modes of vibration $\nu_1 = \nu_2 = 667.3$, $\nu_3 = 1383.3$, $\nu_4 = 2439.3$ cm$^{-1}$. The $\nu_1$, $\nu_2$ arise from the bending modes, $\nu_3$ from the two C–O bonds oscillating out of, and $\nu_4$ from in phase.

5.6 The kinetic energy of the rotational motion of a diatomic molecule (rigid rotator) is given by

$$\epsilon_{rot} = \frac{1}{2I} p_\theta^2 + \frac{1}{2I \sin^2 \theta} p_\phi^2$$

where $p_\theta$, $p_\phi$ are canonical conjugate momenta. Use it to derive (5.27).

5.7 Show that $\bar{C}_V \simeq 3.44R$ for $NH_3$ at 300 K. Given: Principal moment of inertia $I_a = 4.44 \times 10^{-40}$, $I_b = I_c = 2.816 \times 10^{-40}$ cm²g; normal modes of vibration $\nu_1 = \nu_2 = 3336$, $\nu_3 = \nu_4 = 950$, $\nu_5 = 3414$, $\nu_6 = 1627$ cm$^{-1}$.

5.8 The deformational vibration modes of a linear molecule $ABA$ are displacement of $B$ in one direction normal to the line $ABA$ and the displacement of both $A$ atoms to equal distances in the opposite direction. Find the angle of deviation from the linear form if the masses of $A$ and $B$, the distance $AB$, and the vibrational frequencies are given. [Hint: use $\frac{1}{2}\mu l^2 \omega^2 \overline{\phi^2} = \frac{1}{2} h\nu(\bar{n}_1 + \bar{n}_2 + 1)$, where $\mu$ is the reduced mass and $l$ the distance between $A$ and $B$.]

5.9 Calculate $K_P(1000°\text{C})$ for the dissociation of iodine. Given

$$I = 750 \times 10^{-40} \text{g cm}^2, \quad \nu = 6.41 \times 10^{12} \text{s}^{-1}$$

and dissociation energy $\epsilon_d = 2.466 \times 10^{-12}$ ergs. [Ans. $K_P = 0.175$ atm]

5.10 Use the result of Prob. 5.9 to estimate the enthalpy change, and extend the result to other temperatures.

5.11 Calculate the equilibrium constant $K_P(5000\text{ K}) = P_N P_N / P_{N_2}$ of the dissociative reaction $N_2 \rightleftharpoons 2N$. Given: $\Theta_r = 2.84(\text{K})$, $\Theta_v = 3.35 \times 10^3(\text{K})$, dissociation energy $\epsilon_d = 169.3$kcal mol$^{-1}$, the electronic ground state of the molecule is nondegenerate while that of atom has degeneracy 4 due to electron spin.

5.12 Show that the quantities $1/h^3$, $1/Vh^3$ and $1/V(2\pi)^3$ can be regarded as densities of states in the phase space, momentum space and the wave vector space, respectively.

5.13 Evaluate the entropy of the lattice vibrations of monatomic crystal in (i) Einstein theory, and (ii) Debye theory.

5.14 The surface tension waves of frequency $\nu$ and wavelength $\lambda$ on the surface of a liquid of density $\rho$ and surface tension $\sigma$ are described by

$$\nu = \left(\frac{2\pi\sigma}{\rho\lambda^3}\right)^{1/2}.$$

Use Debye-like approach to calculate the surface energy of the liquid at low temperatures. $\left[\text{Hint: Use } g(\nu)\,d\nu = \frac{4\pi}{3}\left(\frac{\rho}{2\pi\sigma}\right)^{2/3} A\,\nu^{1/3}\,d\nu \text{ where } A \text{ is surface area.}\right]$

5.15 In an imperfect gas the intermolecular potential is

$$\phi(r) = A \exp(-br^2), \quad A \ll 1.$$

Expand the partition function in a series in $A$ and obtain the equation of state to first order in $A$.

# 6
# IDEAL BOSE-EINSTEIN GAS

## 6.1 BOSE-EINSTEIN DISTRIBUTION

For an ideal BE gas of $N$ molecules in a volume $V$, the most probable number of particles with energy $\epsilon_i$ is

$$\bar{n}_i(\epsilon_i) = \frac{g_i}{\exp(\alpha + \beta\epsilon_i) - 1} = \frac{g_i}{\exp[(\epsilon_i - \mu)/kT] - 1}, \tag{6.1}$$

where $\beta = 1/kT$, $\alpha \equiv -\mu/kT$, and $g_i$ = degeneracy of the $i$th level. The parameter $\alpha$ (or $\mu$) is determined as a function of $N$ and $T$ by the condition

$$N = \sum_{\substack{i=0 \\ \text{(levels)}}}^{\infty} \bar{n}_i = \frac{g_0}{\exp[\beta(\epsilon_0 - \mu)] - 1} + \frac{g_1}{\exp[\beta(\epsilon_1 - \mu)] - 1} + \ldots$$

$$= \bar{n}_0 + \bar{n}_1 + \ldots. \tag{6.2}$$

where the sum is over the energy levels as we have included $g_i$ in (6.1). Equivalently, we can replace $g_i$ by 1 in (6.1) and then sum over the quantum states in (6.2).

We must have $\bar{n}_i \geqslant 0$, because the number of particles in a level cannot be negative. Therefore, for a boson gas at all temperatures $(\epsilon_i - \mu)$ must be greater than zero for all $\epsilon_i$, that is,

$$\exp(-\mu/kT) \geqslant 1, \quad \text{or} \quad \mu \leqslant 0. \tag{6.3}$$

We can replace the sum by an integral in (6.2) by using in place of $g_i$ the density of states, (4.149),

$$g(\epsilon)d\epsilon = \frac{2\pi V}{h^3}(2m)^{3/2}\epsilon^{1/2}d\epsilon. \tag{6.4}$$

Then (6.2) becomes

$$N = \int_0^{\infty} \frac{g(\epsilon)d\epsilon}{e^{-\beta\mu}\, e^{\beta\epsilon} - 1}$$

$$= \frac{2\pi V}{h^3} (2m)^{3/2} \int_0^\infty \frac{\epsilon^{1/2} d\epsilon}{\frac{1}{\eta_a} e^{\beta\epsilon} - 1}$$

$$= \frac{V}{\lambda^3} F_{3/2}(\eta_a) = V n_Q \, F_{3/2}(\eta_a), \tag{6.5}$$

where $\eta_a$ = the absolute activity (or fugacity, for a gas), $n_Q = 1/\lambda^3$ = quantum concentration (concentration associated with one atom in a cube of side equal to $\lambda$),

$$e^{\beta\mu} \equiv \eta_a \leqslant 1, \quad \lambda = \frac{h}{(2\pi mkT)^{1/2}}, \tag{6.6}$$

and with $x = \beta\epsilon = \epsilon/kT$,

$$F_{3/2}(\eta_a) = \frac{2}{\pi^{1/2}} \int_0^\infty \frac{x^{1/2} dx}{\frac{1}{\eta_a} e^x - 1}$$

$$= \frac{2}{\pi^{1/2}} \int_0^\infty dx \, x^{1/2} \, \eta_a e^{-x} (1 + \eta_a e^{-x} + \eta_a^2 e^{-2x} + \ldots)$$

$$= \eta_a + \frac{\eta_a^2}{2^{3/2}} + \frac{\eta_a^3}{3^{3/2}} + \ldots = \sum_{n=1}^{\infty} \frac{\eta_a^n}{n^{3/2}}. \tag{6.7}$$

The $F_{3/2}(\eta_a)$ is a special case of the general class of functions (Appendix VI)

$$F_s(\eta_a) = \sum_{n=1}^{\infty} \frac{\eta_a^n}{n^s} \tag{6.8}$$

illustrated in Fig. 6.1. At the limiting value $\eta_a = 1$ (or $\mu = 0$), the derivative of $F_{3/2}(\eta_a)$ diverges but the series (6.7) converges

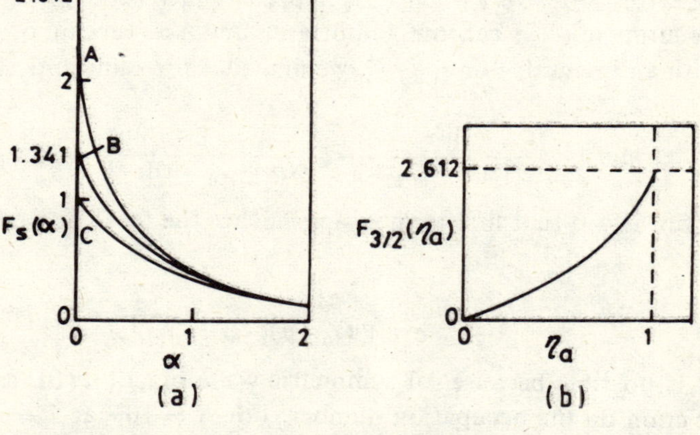


**Fig. 6.1** (a) The functions $F_{3/2}(\alpha)$, curve A; $F_{5/2}(\alpha)$, curve B; and $F_\infty(\alpha) = e^{-\alpha}$, curve *C*. (b) The function $F_{3/2}(\eta_a)$. Note that $\eta_a = e^{-\alpha}$.

$$F_{3/2}(\eta_a = 1) = \sum_{n=1}^{\infty} \frac{1}{n^{3/2}}$$

$$= 1 + \frac{1}{2^{3/2}} + \frac{1}{3^{3/2}} + \ldots = \zeta\left(\frac{3}{2}\right) = 2.612, \tag{6.9}$$

where $\zeta$ is the Riemann zeta function. This is the *maximum* possible value of $F_{3/2}(\eta_a)$ due to (6.3, 6).

We can define a minimum temperature $T_0$, called the *critical temperature*, at which $\eta_a$ has the maximum value 1, (6.5), by

$$N = \frac{V}{\lambda_0^3}\, 2.612 = V\left(\frac{2\pi mkT_0}{h^2}\right)^{3/2} 2.612, \tag{6.10}$$

$$T_0 = \frac{h^2}{2\pi mk}\left(\frac{\rho}{2.612}\right)^{2/3}, \quad \rho = N/V. \tag{6.11}$$

If we have one mole of gas, so that $N$ is Avogadro number,

$$T_0 = \frac{115}{MV_M^{2/3}}\ \text{K}, \tag{6.12}$$

where $M$ is the molecular weight and $V_M$ is the molar volume in $\text{cm}^3\ \text{mol}^{-1}$.

## 6.2 BOSE-EINSTEIN CONDENSATION

We find that (6.5) has no solution for $T < T_0$. This difficulty does not occur in the original sum (6.2). Therefore, it must come from improperly changing the sum (6.2) into the integral (6.5). For low temperatures, $T < T_0$, this causes serious error. Large contribution coming from the first few terms in (6.2) are left out as discussed below.

For small $\eta_a$ (or large $e^{-\beta\mu}$), the terms with the lowest $\epsilon_i$ do not contribute much to the sum, and so replacement of the sum with an integral causes little error. However, when $\eta_a$ is approaching 1 (or $e^{-\beta\mu}$ is small), the first few terms in (6.2) become important, and so we cannot replace the sum with an integral. For $\epsilon_1 > \epsilon_0$, we find that for sufficiently low temperatures,

$$\bar{n}_1 = \frac{1}{\exp[\beta(\epsilon_1 - \mu)] - 1} \ll \frac{1}{\exp[\beta(\epsilon_0 - \mu)] - 1} = \bar{n}_0. \tag{6.13}$$

In fact, for $T \to 0$ that first term $\bar{n}_0$ approaches the total number of particles $N$,

$$\lim_{T\to 0} \bar{n}_0 = N \underset{T\to 0}{\simeq} \frac{1}{\exp[\beta(\epsilon_0 - \mu)] - 1} \simeq \frac{kT}{\epsilon_0 - \mu}, \quad (\bar{n}_0 \text{ large}). \tag{6.14}$$

This is possible because for symmetric wave function (BE case) there is no restriction on the occupation number. For $N = 10^{22}$ at $T = 1\text{K}$, we get $\epsilon_0 - \mu \cong 1.4 \times 10^{-38}$ erg. For such low temperatures $\mu$ is very close to $\epsilon_0$. As $\mu$ is closer to $\epsilon_0$, than to the first excited level, most of the particles tend to occupy $\epsilon_0$ (*Bose-Einstein condensation*) for $T \to 0$. Thus the reason behind the BE condensation is the behaviour of the chemical potential $\mu$ of a boson gas at low temperatures. The BE condensation is a special feature of the BE distribution (6.1), arising from the minus sign of the term unity in the denominator.

For $T \rightarrow 0$, the sum for $N-\bar{n}_0$ can be approximated by an integral without serious error,

$$N-\bar{n}_0 = \sum_{l=1}^{\infty} \bar{n}_l \approx \frac{2\pi V}{h^3} (2m)^{3/2} \int_0^{\infty} \frac{\epsilon^{1/2} \, d\epsilon}{\frac{1}{\eta_a} e^{\beta\epsilon} - 1}$$

$$= (V/\lambda^3) \, F_{3/2}(\eta_a). \tag{6.15}$$

Using (6.10) to eliminate $V$, write (6.15) as

$$N = \bar{n}_0 + N\left(\frac{T}{T_0}\right)^{3/2} \frac{F_{3/2}(\eta_a)}{2.612}. \tag{6.16}$$

We discuss this result in the limits (i) $T < T_0$, and (ii) $T \gg T_0$.

(i) *Below* $T_0$: Without any loss of generality, we can take $\epsilon_0 = 0$ and $g_0 = 1$. Then (6 14) gives

$$\bar{n}_0 = \frac{1}{e^{-\beta\mu} - 1} \approx -\frac{kT}{\mu}, \qquad (\text{large } \bar{n}_0). \tag{6.17}$$

For low temperatures, $\mu$ is very close to zero ($\sim -10^{-38}$ erg),

$$\mu \rightarrow 0, \qquad \text{or} \qquad \eta_a \rightarrow 1, \quad (\textit{quantum region}). \tag{6.18}$$

Therefore, for the energy states above $\epsilon_0$, we can neglect $\mu$ (or put $\eta_a = 1$) and write (6.16) as

$$\bar{n}_0 \equiv N_0 = N - N' = N\left[1 - \left(\frac{T}{T_0}\right)^{3/2}\right], \tag{6.19}$$

where $N'$ is the number of particles in the excited state,

$$N' = N(T/T_0)^{3/2}. \tag{6.20}$$

A plot of $N_0/N$ as a function for $T/T_0$ is shown in Fig. 6.2a. At the *condensation temperature* $T_0$ we have $N'(T_0) = N$. As the temperature is decreased below $T_0$, more and more particles begin to occupy the ground state $\epsilon_0$. The BE gas is then *degenerate* and we are in the quantum region characterized by $\mu \rightarrow 0$. The $T_0$ is also called the *degeneracy temperature* for this reason.

As an alternative to (6.10) defining $T_0$, we can define a *critical volume* $V_0$ such that at a temperature $T$

$$N = \frac{V_0}{\lambda^3} F_{3/2}(1) = V_0 \left(\frac{2\pi mkT}{h^2}\right)^{3/2} 2.612 = V_0 n_Q \, 2.612. \tag{6.21}$$

Using it to eliminate $T$, we can write (6.16) as

$$N = \bar{n}_0 + N \frac{V}{V_0} \frac{F_{3/2}(\eta_a)}{2.612} \simeq N_0 + N \frac{V}{V_0}. \tag{6.22}$$

Thus

$$N_0 = N\left(1 - \frac{V}{V_0}\right), \quad (V \lessdot V_0). \tag{6.23}$$

which is shown in Fig 6.2b. Below $T_0$(or $V_0$) we have BE condensation.

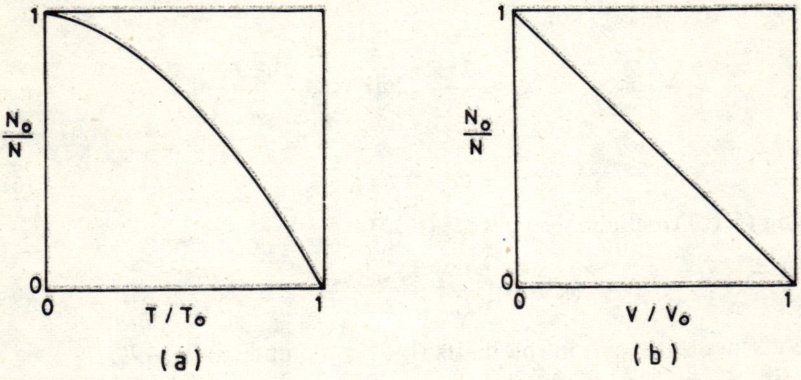


**Fig. 6.2** **Number of BE particles, in the ground state, as a function of (a) temperature, (b) volume.**

We can write (6.10) as

$$\lambda_0^3 = 2.612/\rho. \tag{6.24}$$

At $T_0$ the de Broglie wavelength $\lambda_0$ is of the order of the average particle distance. The wave functions overlap and so the quantum effects are important.

(ii) *Above* $T_0$: For $T > T_0$ we have $\eta_a < 1$ (classical region). In (6.16) the first term $\bar{n}_0$ on the right becomes negligible and the second term increases as $T^{3/2}$ when the BE gas is heated above $T_0$. Thus, (6.15) reduces to $N = (V/\lambda^3)\, F_{3/2}(\eta_a)$ and (6.16) to

$$F_{3/2}(\eta_a) = (T_0/T)^{3/2}\, F_{3/2}(1) = (T_0/T)^{3/2}\, 2.612$$
$$= (\lambda/\lambda_0)^3\, 2.612, \qquad (T > T_0). \tag{6.25}$$

For $T \gg 0$, the ground state is practically empty and most of the particles are in the states with $\epsilon > 0$. We can approximate the BE distribution by the MB distribution. In fact, for $\eta_a \ll 1$, $F_{3/2}(\eta_a) \simeq \eta_a$ from (6.7), and (6.25) becomes

$$\eta_a = (\lambda/\lambda_0)^3\, 2.612 = \rho\lambda^3 = e^{-\alpha}, \quad (\textit{classical limit}).$$

It can now be compared with (4.39), where for the MB gas the single-molecule partition function $z$ is given by

$$e^{-\alpha} = \frac{N}{z} = \frac{N}{V/\lambda^3} = \rho\lambda^3. \tag{6.26}$$

It is instructive to compare the distributions for $T < T_0$ and $T > T_0$ (Fig. 6.3).

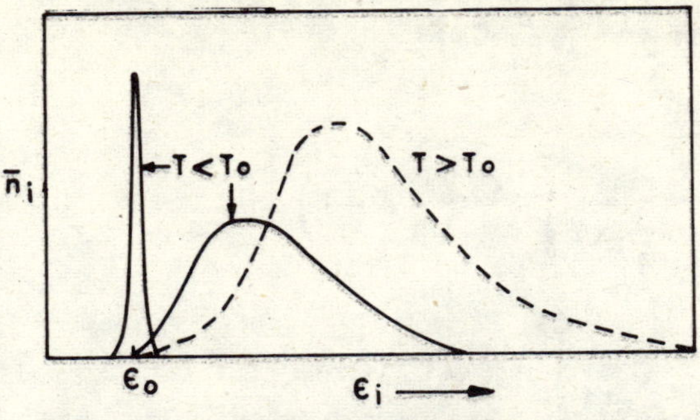


**Fig. 6.3** Schematic diagram of the distribution function for the particles of an ideal BE gas.

## 6.3 THERMODYNAMIC PROPERTIES OF AN IDEAL BOSE-EINSTEIN GAS

### Chemical Potential

The chemical potential $\mu$ is related to the fugacity $\eta_a$ by $\eta_a = e^{\mu/kT}$. We can write (6.15), for $N \to \infty$ but $N/V$ finite, as

$$\rho\lambda^3 = \rho\lambda^3 \; N^{-1} \frac{\eta_a}{1-\eta_a} + F_{3/2}(\eta_a), \tag{6.27}$$

where we have used (6.17), $\bar{n}_0 = \eta_a/(1-\eta_a)$. We know the behaviour of $F_{3/2}(\eta_a)$ (Fig. 6.1). Therefore, we can solve (6.27) graphically to find $\eta_a$ as a function of $\rho$ and $T$ (Fig. 6.4).

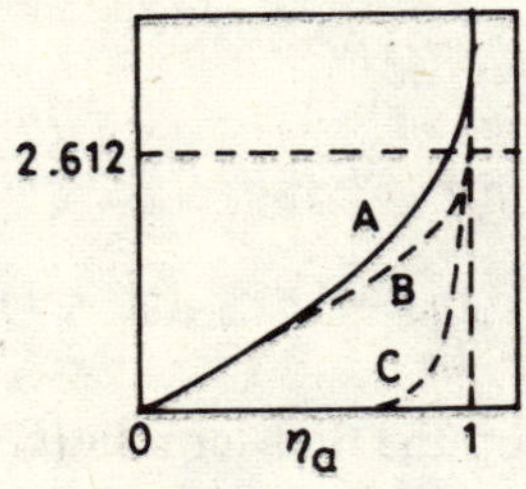


**Fig. 6.4** Graphical solution of (6.27). Curve A, $\rho\lambda^3$; curve B, $F_{3/2}(\eta_a)$; curve C, $\rho\lambda^3\eta_a/(1-\eta_a)$.

### Energy

From (6.1, 4),

$$U = \sum_i \epsilon_i \bar{n}_i = \int_0^\infty \epsilon \, dn = \int_0^\infty \frac{\epsilon g(\epsilon)\, d\epsilon}{e^{\beta(\epsilon-\mu)}-1}$$

$$= \frac{V}{h^3} \, 2\pi \,(2m)^{3/2} \int_0^\infty \frac{\epsilon^{3/2}\, d\epsilon}{e^{\beta(\epsilon-\mu)}-1}$$

$$= \frac{VkT}{\lambda^3} \cdot \frac{2}{\pi^{1/2}} \int_0^\infty x^{3/2} \left(\frac{1}{\eta_a}\, e^x - 1\right)^{-1} dx$$

$$= \frac{3}{2} kT \frac{V}{\lambda^3} F_{5/2}(\eta_a), \tag{6.28}$$

where $x = \beta\epsilon$ and

$$F_{5/2}(\eta_a) = \frac{1}{\frac{3}{4}\pi^{1/2}} \int_0^\infty x^{3/2} \left(\frac{1}{\eta_a} e^x - 1\right)^{-1} dx$$

$$= \frac{1}{\frac{3}{4}\pi^{1/2}} \int_0^\infty x^{3/2} \eta_a e^{-x}(1 + \eta_a e^{-x} + \eta_a^2 e^{-2x} + \ldots)$$

$$= \eta_a + \frac{\eta_a^2}{2^{5/2}} + \frac{\eta_a^3}{3^{5/2}} + \ldots$$

$$= e^{-\alpha} + \frac{e^{-2\alpha}}{2^{5/2}} + \ldots = F_{5/2}(\alpha). \tag{6.29}$$

A plot of $F_{5/2}(\alpha)$ is shown in Fig. 6.1. Note that $F_{3/2}(\eta_a) = \eta_a \frac{\partial}{\partial \eta_a} F_{5/2}(\eta_a)$.

We shall distinguish the two cases: $U_-$ for $T < T_0$ (degenerate gas), and $U_+$ for $T > T_0$ (nondegenerate gas).

For $T < T_0$, $\eta_a = 1$, and so from (6.10, 28),

$$U_- = \frac{3}{2} kT \frac{V}{\lambda^3} F_{5/2}(\eta_a = 1)$$

$$= \frac{3}{2} kT \frac{N\lambda_0^3}{F_{3/2}(\eta_a = 1)\lambda^3} F_{5/2}(\eta_a = 1)$$

$$= \frac{3}{2} NkT \left(\frac{T}{T_0}\right)^{3/2} 0.51 = NkT \left(\frac{T}{T_0}\right)^{3/2} 0.77, \tag{6.30}$$

where we have used $F_{5/2}(\eta_a = 1)/F_{3/2}(\eta_a = 1) = \zeta(\frac{5}{2})/\zeta(\frac{3}{2}) = 1.341/2.612 = 0.5134$, and $\lambda_0^3/\lambda^3 = (T/T_0)^{3/2}$.

For $T > T_0$, $\eta_a \ll 1$, drop the first term in (6.15), and write

$$N = \frac{V}{\lambda^3} F_{3/2}(\eta_a), \quad (T > T_0). \tag{6.31}$$

From (6.28, 31),

$$U_+ = \frac{3}{2} kT \frac{V}{\lambda^3} F_{5/2}(\eta_a) = \frac{3}{2} NkT \frac{F_{5/2}(\eta_a)}{F_{3/2}(\eta_a)}$$

$$= (3/2)\, NkT\, [1 - 2^{-5/2} F_{3/2}(\eta_a) - 2(3^{-5/2} - 2^{-4})\, F_{3/2}(\eta_a)^2 - \ldots]$$

$$= (3/2)\, NkT\, [1 - 0.177\, F_{3/2}(\eta_a) - 0.003\, F_{3/2}(\eta_a)^2 - \ldots]$$

$$= \frac{3}{2} NkT \left[1 - 0.462 \left(\frac{T_0}{T}\right)^{3/2} - 0.023 \left(\frac{T_0}{T}\right)^3 - \ldots\right], \tag{6.32}$$

where $F_{5/2}(\eta_a)/F_{3/2}(\eta_a)$ is given in Appendix VI and in the last step we have used (6.25).

**Specific Heat**

Using $C_V = (\partial U/\partial T)_V$,

$$C_{V-} = \frac{15}{4}\, 0.51\, Nk \left(\frac{T}{T_0}\right)^{3/2} = 1.926\, Nk \left(\frac{T}{T_0}\right)^{3/2}, \tag{6.33}$$

$$C_{V+} = \frac{3}{2} Nk \left[1 + 0.231 \left(\frac{T_0}{T}\right)^{3/2} + 0.045 \left(\frac{T_0}{T}\right)^{3} + \dots\right]. \tag{6.34}$$

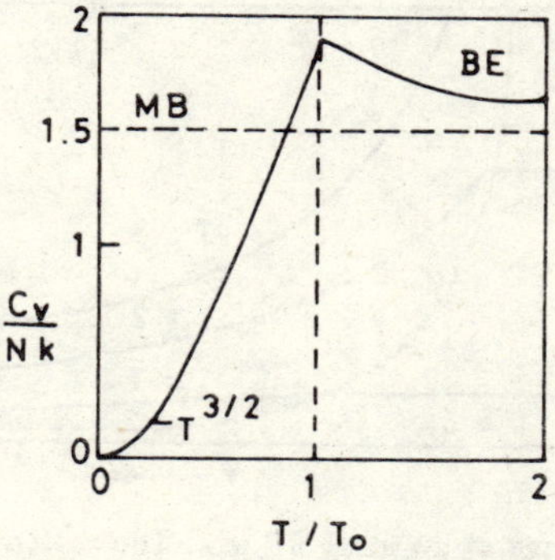


**Fig. 6.5 Specific heat of an ideal BE gas.**

These values are plotted in Fig. 6.5. For $T = T_0$ the values are same, $C_{V-} = C_{V+} = 0.51\,(15/4)\, Nk = 1.926\, Nk$. Therefore, $C_V$ is continuous at $T = T_0$ but shows a kink there. This suggests that the Bose-Einstein condensation is a third order phase transition.

**Entropy**

We have

$$S_-(T) = \int_0^T \frac{C_{V-}}{T}\, dT = \frac{2}{3}\, C_{V-} = \frac{2}{3}\, 1.926\, Nk \left(\frac{T}{T_0}\right)^{3/2}, \tag{6.35}$$

$$S_+(T) = S(T_0) + \int_{T_0}^{T} \frac{CV_+}{T}\, dT$$

$$= S(T_0) + \frac{3}{2} Nk \left[\ln \frac{T}{T_0} + \frac{2}{3}\, 0.231 \left(1 - \frac{T_0}{T}\right)^{3/2} + \dots\right]. \tag{6.36}$$

The entropy shows a sudden drop for $T < T_0$. In (6.35) $S = 0$ at $T = 0$, in accordance with the third law of thermodynamics. This means that for the condensed phase (which exists at $T = 0$) the entropy is zero, that is, all the particles are in one state.

**Pressure**

For all ideal gases $P = 2U/3V$ independently of the statistics, (4.154). Therefore, using $V/V_0 = (T/T_0)^{3/2}$,

$$P_- = \frac{2}{3}\frac{U_-}{V} = Nk\,\frac{T^{5/2}}{VT_0} = Nk\,\frac{T}{V_0}, \tag{6.37}$$

$$P_+ = \frac{2}{3}\frac{U_+}{V} = \frac{NkT}{V}\left[1 - 0.462\frac{V_0}{V} - \ldots\right]. \tag{6.38}$$

Note that $P_-$ is independent of $V$ and a function of $T$ only, as for a condensing gas (Fig. 6.6). In this region further reduction in volume simply condenses more particles into the ground state.

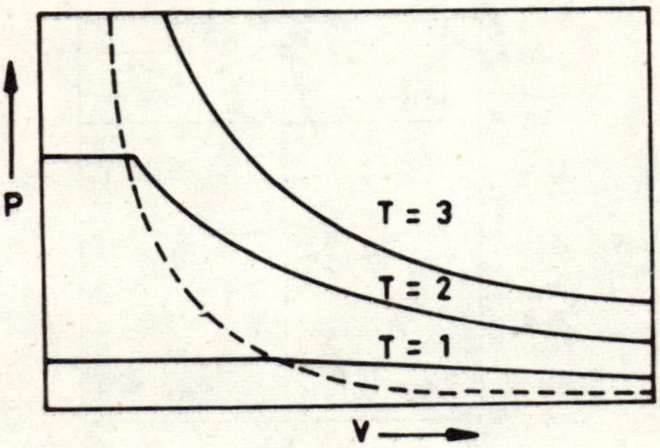


**Fig. 6.6 Isotherms of an ideal BE gas. The region below the dotted curve corresponds to the degenerate state.**

## 6.4 LIQUID HELIUM

A $^4$He atom contains an even number of fermions (2 protons, 2 neutrons, and 2 electrons) and so obeys BE statistics. The phase diagram for helium is shown in Fig. 6.7. The normal boiling point of liquid $^4$He is 4.2 K. The

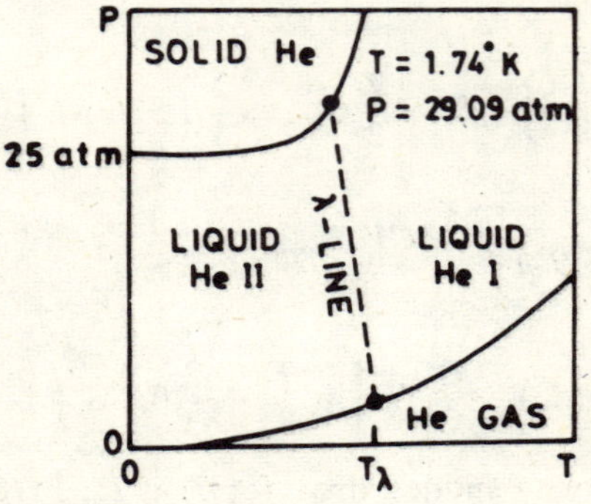


**Fig. 6.7 Schematic phase diagram of $^4$He.**

liquid-gas vapour pressure curve extrapolates to the origin with no sign of the triple point. Thus solid is not formed merely by cooling under its saturated vapour pressure. The solid-liquid curve flattens out to a pressure of 25 atm near 1 K.

As liquid $^4$He in contact with its vapour is cooled, it begins to show dramatic change in properties at $T = T_\lambda = 2.18$ K. For $T > T_\lambda$ its behaviour is that of a normal liquid and is called He I. For $T < T_\lambda$, the liquid helium begins to show remarkable properties, such as zero viscosity (under certain conditions) or superfluidity and zero entropy, and is called He II. There is apparently a further phase transition in the liquid phase, called the *lambda transition*, which divides the liquid state into two phases He I and He II (Fig. 6.7). Evidence for this λ-transition is provided, for example, by the specific heat and entropy measurements (Fig. 6.8). The specific

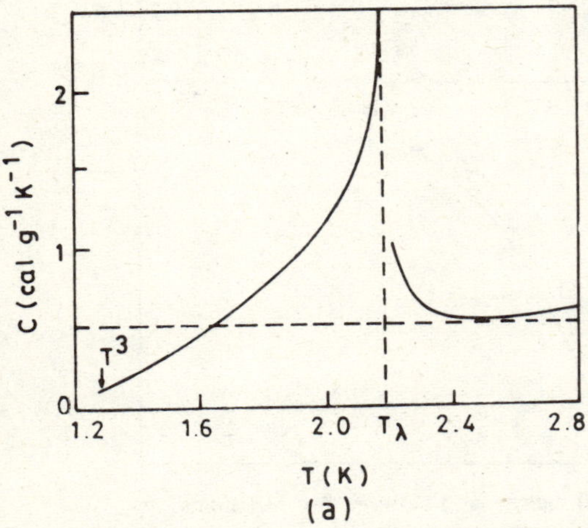


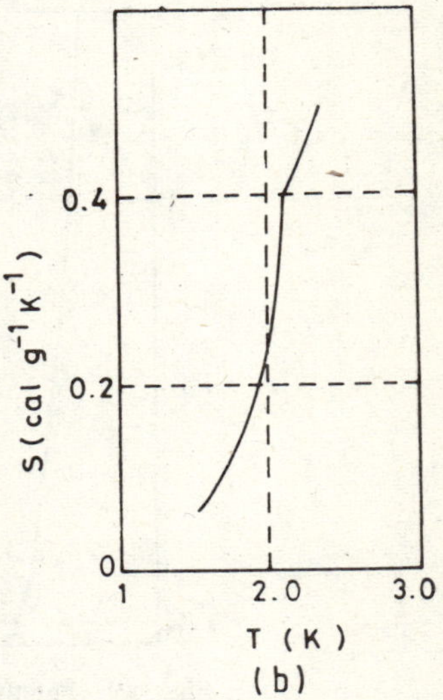


**Fig. 6.8** (a) Specific heat for liquid $^4He$ under saturated vapour pressure.
(b) Entropy curve for liquid $^4He$ under saturated vapour pressure.

heat along the vapour curve becomes logarithmically infinite at the temperature of λ-transition, called the *lambda temperature* $T_\lambda$. The resemblance in the shape of the specific heat curve and the Greek letter λ has given rise to this nomenclature. The λ-transition occurs on the vapour pressure curve at

$$T_\lambda = 2.18 \text{ K}, \; 1/\rho_\lambda = 46.2 \text{ Å}^3/\text{atom}.$$

With $M = 4$, $V_M = 27.6 \text{ cm}^3$, (6.12) gives the condensation temperature $T_0 = 3.14$ K for liquid helium treated approximately as an ideal BE gas. This is close to the observed λ-temperature $T_\lambda = 2.18$ K, considering the approximation involved. This fact, and the similarity of Fig. 6.5 and Fig. 6.8a led London* to suggest that *the lambda transition of $^4He$ is a form of Einstein condensation.*

The fluidity of $^4He$ at low temperatures is mainly due to two reasons:
(1) weak intermolecular force between atoms ($^4He$ is a noble gas), and
(2) small mass of $^4He$ atom.

Small mass makes the thermal wavelength λ and the average distance between the particles $r_0$ to be of the same order of magnitude for $T \leqslant T_0$,

* F. London, *Phys. Rev.* **54**, 947 (1938); *Superfluids* vol. II, John Wiley, New York, 1954.

(6.24). Therefore, it is not possible to specify the position of the helium atom with an uncertainty less than $r_0$. In other words, helium is a liquid and not a solid with atoms localized at lattice sites. To understand this more quantitatively consider the potential energy $\phi(r)$ between two He atoms separated by a distance $r$ (Fig. 6.9).

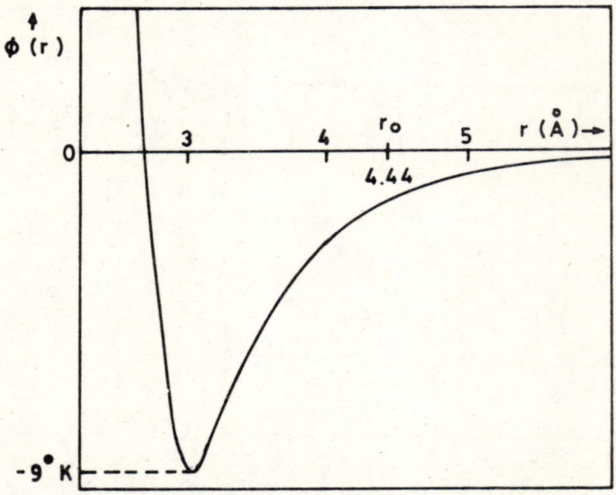


**Fig. 6.9 Potential energy $\phi(r)$ between two He atoms.**

A helium atom will have a well defined location on a crystal site if it can be confined within a distance, say $\Delta x \simeq 0.5$ Å, which is smaller than the range of the potential. The corresponding uncertainty in its energy is of the order of

$$\Delta E \sim \frac{(\Delta p)^2}{2m} = \frac{(h/\Delta x)^2}{2m} = 10 \text{ K}.$$

This energy is enough to allow the atom to escape from the potential well (−9 K), Fig. 6.9. Therefore, localization is not possible. Note that Argon solidifies because of its large mass and hydrogen because of the stronger interaction between the molecules. $^3$He remains liquid, like $^4$He, and forms more complex ordered pairs.

## 6.5 TWO-FLUID MODEL FOR LIQUID HELIUM II

Tisza suggested that He II ($T < T_\lambda$) consists of *two* independent components, a *normal fluid* ($N_n$) and a *superfluid* ($N_s$). We can express the total number of particles $N$ as

$$N = N_s + N_n, \qquad (T < T_\lambda), \tag{6.39}$$

in complete analogy with (6.19) for the BE gas ($T < T_0$).

The superfluid has zero entropy, and vanishing viscosity. The normal fluid ($\epsilon > 0$ for all $N_n$ atoms) contributes to free energy and viscosity.

In the *two fluid model* we can express the mass-density $\rho$ and mass-velocity $u$ as

$$\rho = \rho_s + \rho_n, \tag{6.40}$$

$$\rho u = \rho_s u_s + \rho_n u_n. \tag{6.41}$$

We can use this model to explain the following strange properties of He II.

(1) *Viscosity*: Viscosity measurements made with dynamical methods show that liquid He I ($T > T_\lambda$) behaves like a gas in that it exhibits a lowering of viscosity with decrease in temperature. Usually liquids show the opposite behaviour. In liquid He II ($T < T_\lambda$) the behaviour is even more strange than in He I. Here the viscosity rapidly decreases as the temperature is decreased below $T_\lambda$ (Fig. 6.10).

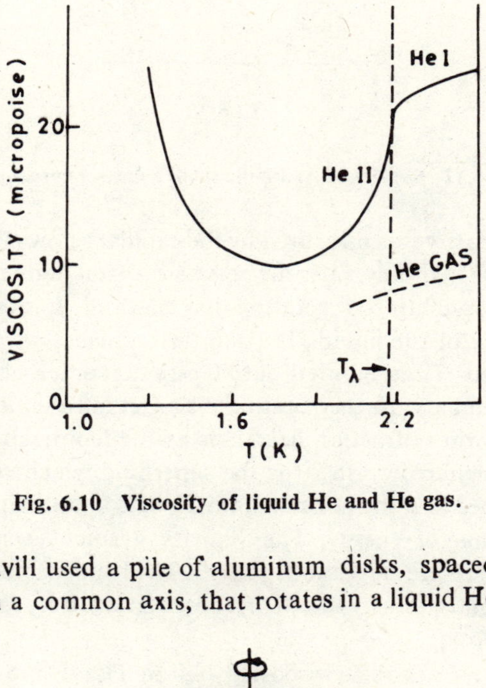


**Fig. 6.10 Viscosity of liquid He and He gas.**

Andronikashvili used a pile of aluminum disks, spaced 0.2 mm apart and mounted on a common axis, that rotates in a liquid He bath (Fig. 6.11).

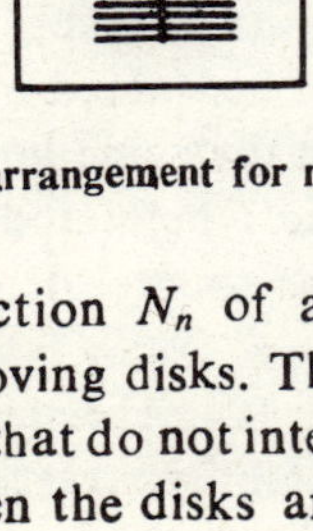

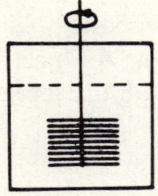

**Fig. 6.11 Andronikashvili's arrangement for measuring $\rho_n/\rho$ of liquid He II.**

In the two-fluid model the fraction $N_n$ of atoms ($\epsilon > 0$) are assumed to interchange energy with the moving disks. The fraction $N_s$ of atoms ($\epsilon = 0$) are regarded as idle spectators that do not interchange any energy. Then only the normal fluid rotates between the disks and thereby contributes to the moment of inertia $I$. For each temperature $T$,

$$\frac{\text{Experimental } I}{\text{Geometrical } I_G} = \frac{\rho_n}{\rho}. \tag{6.42}$$

The observed $\rho_n/\rho$ curve (Fig. 6.12) gives the same qualitative trend for $\rho_s/\rho$ as of the $N_o/N$ curve, (Fig. 6.2a). Hollis-Hallett used a rotating cyclinder viscometer to measure the temperature dependence of $\rho_n/\rho$ in liquid He II.

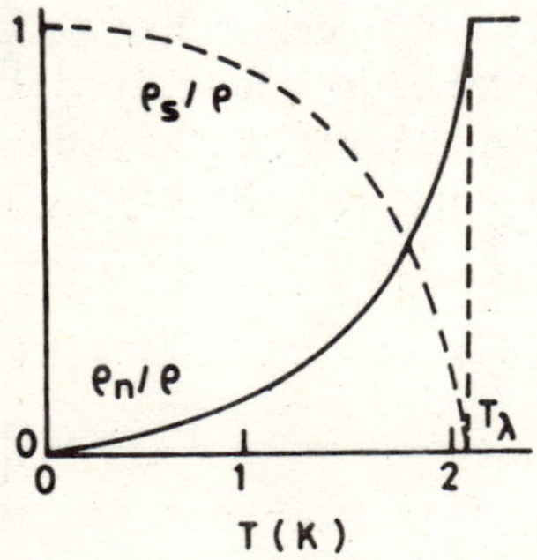


**Fig. 6.12 Results of Andronikashvili's measurements $\rho_n/\rho$.**

When the viscosity was measured by the capillary flow (Poiseuille) method, or by the flow through fine slits, the results did not agree with the values obtained by the oscillating or rotating disk method. It was found that the apparent viscosity of the liquid He II almost vanishes as the width of the channel is reduced. Tisza pointed out an essential difference between the two types of measurements. In the rotating disk method the disk is damped in the normal fluid whose fraction decreases as the temperature is lowered. In fine capillaries, or narrow slits, it is the superfluid which contributes to the measured flow, because the normal fluid is held stationary with the walls due to the exchange of energy. The velocity gradient being absent in the normal fluid, there are no viscous effects. Thus superfluid atoms can move through the normal atoms without friction or viscosity, for small velocities (non-turbulent flow).

For $T \to 0$, $\rho_n \to 0$ and so viscosity $\to 0$ in Tisza's model. However, the observed viscosity increases (Fig. 6.10) as $T \to 0$. This shows that the idea of identifying specific atoms with the superfluid is not correct.

(2) *Thermo-Mechanical Effect*: Daunt and Mendelssohn immersed a small heater coil $H$ in He II contained in the inner bath $A$, which is surrounded by an outer bath $B$ of the same liquid (Fig. 6.13a). They found that there is a transfer of (superfluid) helium due to the formation of a thin liquid film creeping towards the source of heat. This is called the *thermo-mechanical effect*.

The *fountain effect*, analogous to the thermo-mechanical effect, was observed by Allen and Jones. Their apparatus is shown in Fig. 6.13b. One

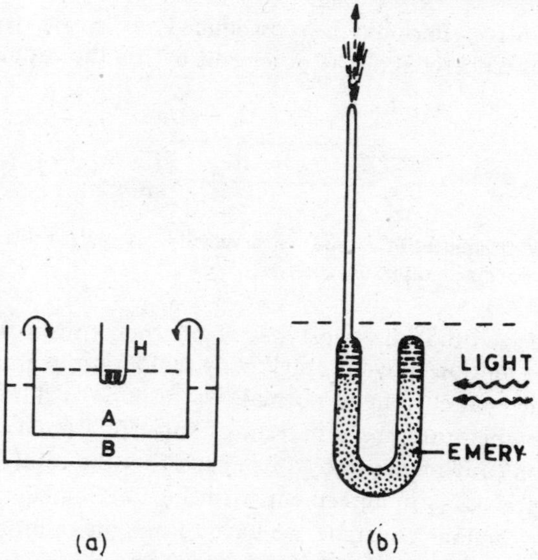


Fig. 6.13 (a) Thermo-mechanical effect, (b) fountain effect.

arm of a U-tube is lengthened in a capillary, the bent portion is packed with emery, and the other arm having an orifice is immersed in He II. When a temperature gradient is created by allowing light to fall on the portion containing emery, a fountain flow is observed at the capillary end.

The *mechano-caloric effect* (reverse of the thermo-mechanical effect) can be observed by connecting two vessels $A$ and $B$ by a very narrow capillary $C$ (Fig. 6.14) through which only the superfluid can pass. When pressure is

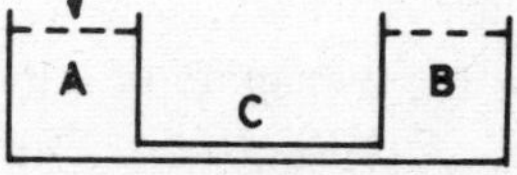


Fig 6.14 Mechano-caloric effect.

exerted on the liquid He II in $A$, some superfluid flows to $B$ and the temperature in $B$ falls.

These experiments can be explained on the two-fluid model by assuming that *the superfluid component has zero entropy in addition to its zero viscosity and being cold* ($\epsilon = 0$) *readily moves towards the source of heat* (Fig. 6.13). Then there is no entropy transport from $A$ to $B$ (Fig. 6.14). The entropy per unit mass $S$ increases in $A$ and decreases in $B$, $S_A > S_B$. According to the equation $C_V dT = TdS$, $T_A$ will rise and $T_B$ will fall.

(3) *Second Sound*: The familiar (first) sound is a pressure wave. Unlike it, the ***second sound*** is a temperature or entropy wave. It was predicted by Tisza* and first observed by Peshkov† using a continuous wave resonance

* L. Tisza, *J. Phys. Radium* **1**, 165 (1940).

† V.P. Peshkov, *J. Phys.* (Moscow) **8**, 131, 381(1944).

technique (Fig. 6.15). Periodic heat produced at $A$ was found to travel through He II and detected at the other end by the thermometer coil at the end $B$.

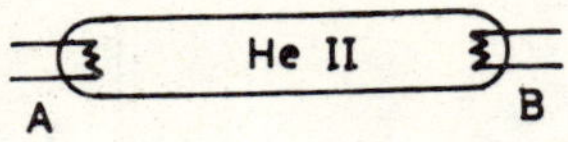


**Fig. 6.15 Measurement of second sound velocity. A, pulsed thermal wave generator; B, receiver.**

Tisza emphasized that the two types of interpenetrating fluids, $\rho_s$ and $\rho_n$, do not resist the motion of each other. Any temperature gradient tends to be smoothed out by a current of normal (warm, $\epsilon > 0$) fluid in the direction of falling temperature and a current of superfluid (cold, $\epsilon = 0$) in the *opposite* direction (out of phase by 180°) such that the total density remains constant. This process is in agreement with the Le Chatelier principle. As the total density remains constant, we have to imagine a motion of fluid for which the average velocity $u$ in (6.41) is zero. Then there is no net transport of mass across any plane in the liquid and the flow equation (6.41) reduces to

$$\rho_s u_s + \rho_n u_n = 0. \tag{6.43}$$

This possible mode of motion in which the normal fluid and superfluid oscillate out of phase by 180° leads to thermal waves or *second sound*.

The velocity $u_2$ of the second sound can be easily calculated*. The superfluid component has zero entropy. Therefore, the total entropy $S$ per gram of the liquid He II can be expressed as

$$\rho S = \rho_n S_n. \tag{6.44}$$

For simplicity consider the transport of entropy in one direction only. Imagine a parallelopiped of unit cross-section and length $\Delta x$ inside the liquid with one pair of opposite faces vertical. The amount of entropy entering the volume per second at one face is

$$\rho S u_n = \rho_n S_n u_n, \tag{6.45}$$

and leaving at the opposite face is

$$\rho S u_n + \frac{\partial}{\partial x}(\rho S u_n)\Delta x.$$

Therefore, the net transport of entropy per unit volume per second is

$$\frac{\partial}{\partial x}(\rho S u_n). \tag{6.46}$$

The original entropy within the volume was $\rho S$ and the loss in unit time is $-(\partial/\partial t)(\rho S)$. This gives the equation of continuity

*D. Gogate and P. Pathak, *Proc. Phys. Soc.* (London) 59, 457(1947).

$$-\frac{\partial}{\partial t}(\rho S) = \frac{\partial}{\partial x}(\rho S u_n). \tag{6.47}$$

If $\rho$ is sensibly constant and $u_n$ is small, we can write

$$-\frac{\partial S}{\partial t} = S\frac{\partial u_n}{\partial x}, \tag{6.48}$$

after neglecting the small term $(\partial S/\partial x)u_n$.

Let the opposite faces have temperatures $T$ and $T + \Delta T$. Then the quantity of heat transferred across $\Delta x$ per second is $Q = \rho S u_n T$. By the second law of thermodynamics $W/Q = \Delta T/T$, where $W$ is the reversible work done in the heat transfer. Thus

$$W = Q(\Delta T/T) = \rho S u_n \Delta T. \tag{6.49}$$

This work brings about a change in kinetic energy, $(\partial U/\partial t)\,\Delta x$, within $\Delta x$. There is no net mass flow, (6.43). So

$$\begin{aligned} U &= \tfrac{1}{2}\left(\rho_n u_n^2 + \rho_s u_s^2\right) \\ &= \frac{1}{2}\rho_n u_n\,[1 + (\rho_s/\rho_n)\,(u_s/u_n)^2] \\ &= \frac{1}{2}\rho_n u_n^2\,[1 + (\rho_n/\rho_s)] \\ &= \frac{1}{2}(\rho\rho_n/\rho_s)\,u_n^2, \end{aligned} \tag{6.50}$$

$$\left(\frac{\partial U}{\partial t}\right)\Delta x = \frac{\rho\rho_n}{\rho_s}u_n\frac{\partial u_n}{\partial t}\Delta x = -W = -\rho S u_n\,\Delta T,$$

or

$$\frac{\rho_n}{\rho_s}\frac{\partial u_n}{\partial t} = -S\frac{\partial T}{\partial x}. \tag{6.51}$$

Using $C_V\,dT = TdS$ and $\partial T/\partial S = T/C_V = (\partial T/\partial x)(\partial x/\partial S)$, we have $(\partial T/\partial x) = (T/C_V)(\partial S/\partial x)$, so that

$$\frac{\rho_n}{\rho_s}\frac{\partial u_n}{\partial t} = -S\frac{T}{C_V}\frac{\partial S}{\partial x}. \tag{6.52}$$

Eliminating $u_n$ in (6.47) and (6.52),

$$\frac{\partial^2 S}{\partial t^2} = \left(\frac{\rho_s}{\rho_n}\frac{S^2 T}{C_V}\right)\frac{\partial^2 S}{\partial x^2}. \tag{6.53}$$

The second sound velocity $u_2$ defined by this equation is

$$u_2 = \left(\frac{\rho_s}{\rho_n}\frac{S^2 T}{C_V}\right)^{1/2}. \tag{6.54}$$

It follows that Tisza's two-fluid model predicts $u_2 \to 0$ both when $T \to 0$ and when $T \to T_\lambda$ ($\rho_s \to 0$). The prediction for $T \to 0$ turned out to be wrong. In fact, $u_2 \to u_1/3^{1/2} = 137$ m/s for $T \to 0$, where $u_1$ is the first sound velocity. This shows that Tisza's idea of associating the condensed particles with superfluid, as a natural extension of London's theory of He II, is not correct.

(4) *Third and Fourth Sounds*: When He II flows through a channel, the normal fluid part gets clamped to the substrate, while the superfluid part moves without hindrance (below the critical velocity). In addition, the surface waves are also possible where the superfluid oscillates parallel to the substrate. This mode is called *third sound*. There is a periodic variation in the film thickness attended with temperature fluctuations. The thicker part contains superfluid in excess and has lower temperature. The phase velocity $u_3$ is obtained by considering the velocity $v_{sur}$ of a surface wave on a shallow liquid of depth $d$,

$$v_{sur}^2 = \left(\frac{f\lambda}{2\pi} + \frac{2\pi\gamma}{\rho\lambda}\right) \tanh \frac{2\pi d}{\lambda},$$

where $f$ is the attractive force of the substrate per unit mass of liquid, $\gamma$ the surface tension, the $\rho$ the density, and $\lambda$ the wavelength. For $\lambda \to$ large, second term is negligible, the tanh is replaced by its argument, and it reduces to

$$u_3^2 = fd\,(\rho_s/\rho).$$

The factor $(\rho_s/\rho)$ is inserted because only the superfluid moves.* The measured values of $u_3$ are in the range 1–40 m/s.

The *fourth sound* has features of the first sound (periodic variation in $\rho$) as well as of the second sound (periodic vibrations in $\rho_s/\rho$). As first sound moves through narrow channel (say, packed powder) its velocity is modified because only the superfluid can move freely. We can write†,

$$u_4^2 = (\rho_s/\rho)\, u_1^2 + (\rho_n/\rho)\, u_2^2.$$

The $u_4$ rises from zero at $T_\lambda$ and approaches $u_1$ as $T \to 0$ K, in agreement with this equation.

## 6.6 LANDAU SPECTRUM OF PHONONS AND ROTONS

Landau found the explanation of London and Tisza based on the Einstein condensation of ideal BE gas for the peculiar properties of liquid He II to be unsatisfactory. In particular, it gives wrong predictions for viscosity and second sound velocity as $T \to 0$. According to him all bodies as $T \to 0$ should show solid-like behaviour. The observed specific heat of He II as $T \to 0$ follows the Debye $T^3$-law (5.76) which is valid for solids at low temperatures. This $T^3$ dependence associated with elastic or sound waves in solids can be understood in terms of *phonons*, the quanta of longitudinal

* For detailed treatment: K.R. Atkins and I. Rudnick, in *Progress in Low Temperature Physics*, ed, C.J. Gorter, vol VI, North-Holland, 1970.

† K.A. Shapiro and I. Rudnick, *Phys. Rev.* A **137**, 1383 (1965); R.J. Donnelly, *Experimental Superfluidity*, UCP Chicago, 1967.

sound waves. Thus, for $T < 1$ K, the potential internal motions of the liquid He II are longitudinal sound waves and *the excited state can be regarded as an aggregate of phonons*. The energy of phonons is linear function of their momentum $p$,

$$\epsilon_{ph} = u_1 p, \tag{6.55}$$

where $u_1$ is the first sound velocity.

This does not exhaust all the $3N$ degrees of freedom, if there are $N$ atoms in the system of liquid He II. A liquid is not expected to sustain high energy transverse (shear) wave, as a solid can. Therefore, for $1\text{K} < T < T_\lambda$, Landau* assumed the existence of excitations called *rotons* (quanta of vortex motion) with the energy spectrum

$$\epsilon_{rot} = \Delta + (p^2/2\mu_r), \tag{6.56}$$

where $\Delta$ is the energy gap (minimum energy required to excite a roton at rest), $p$ the linear momentum of the roton, and $\mu_r$ the effective mass of the roton.

Phonons, like photons, obey BE statistics. Rotons are also assumed to obey BE statistics. However, because the $\epsilon_{rot}$ involves $\Delta > kT$ for $T < T_\lambda$, the aggregate of rotons can be treated, to a good approximation, as a Maxwell-Boltzmann gas. Both phonons and rotons behave like quasi-particles that can move about and form the normal fluid. The superfluid has strictly zero entropy.

To improve the agreement with experiments, Landau** finally proposed the energy spectrum

$$\epsilon = \begin{cases} \epsilon_{ph} = u_1 p, & (\textit{phonon}), \\ \epsilon_{rot} = \Delta + \dfrac{(p-p_0)^2}{2\mu_r}, & (\textit{roton}), \end{cases} \tag{6.57}$$

where the parameters $u_1$, $\Delta$, $p_0$ and $\mu_r$ are adjusted to give the best fit to specific heat measurements. Landau finds

$$u_1 = 22.6 \times 10^3 \text{ cm/s}, \; \Delta/k = 9\text{ K}, \; p_0/h = 2\text{Å}^{-1}, \; \mu_r = 0.3\, m, \tag{6.58}$$

where $m$ is the mass of the He atom. The experimental value for $u_1$ is $23.9 \times 10^3$ cm/s. The Landau spectrum is shown in Fig. 6.16. The active portions

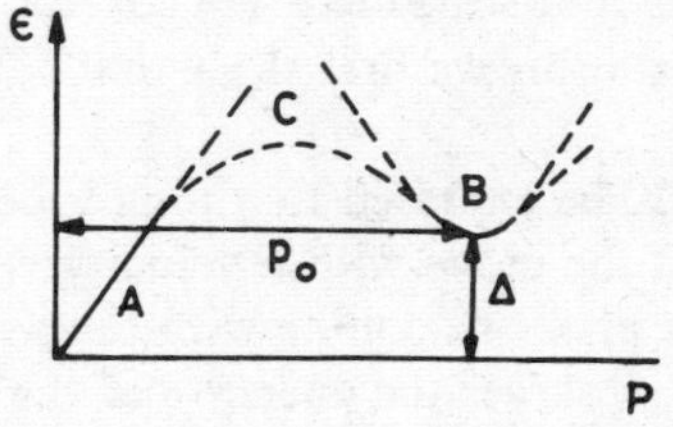


**Fig. 6.16** **Dispersion curve ($\epsilon$ vs. p plot) for liquid $^4$He showing the two branches (A, linear phonon branch, and B, parabolic roton branch) of the Landau spectrum (6.57). The ad hoc curve C smoothly joins the two distinct branches.**

* L.D. Landau, *J. Phys.* (Moscow) **5**, 71 (1941)

** L. D. Landau *J. Phys.* (Moscow) **11**, 91 (1947).

of the spectrum are shown by solid curves. At low temperatures ($0<T<1$K) only the linear part of the spectrum corresponding to phonons becomes active. At about 1 K, both phonons and rotons exist.

If the number of phonons and rotons per unit volume of liquid He II is not large, their aggregate can be regarded as a mixture of two ideal gases—a phonon gas and a roton gas. Then the specific heat will be the sum of the contributions from the two, $C_V = C_{ph} + C_{rot}$.

*Phonon Contribution*: From (5.65), for the longitudinal wave,

$$3N = \int_0^{\nu_m} g(\nu)\, d\nu = \int_0^{\nu_m} 4\pi V \frac{\nu^2}{u_1^3}\, d\nu = \frac{4}{3}\pi V \frac{\nu_m^3}{u_1^3}, \tag{6.59}$$

which relates the cut-off frequency $\nu_m$ with $u_1$. Then (5.76) gives

$$C_V = \frac{12\pi^4 Nk}{5}\left(\frac{kT}{h\nu_m}\right)^3 = V\frac{16\pi^5 k^4}{15h^3u_1^3}\, T^3,$$

that is, the specific heat per unit volume for phonons is

$$C_{ph} = bT^3,\ b = \frac{16\pi^5 k^4}{15h^3 u_1^3}. \tag{6.60}$$

We can express it as per unit mass by using $V = 1/\rho$ for 1 gram. From (6.60),

$$U_{ph} = \int C_{ph}\, dT = bT^4/4, \tag{6.61}$$

$$S_{ph} = \int (C_{ph}/T)\, dT = (b/3)\, T^3, \tag{6.62}$$

where $b$ is a constant. Note that, as for photons, for phonons, $\mu = 0$. It is interesting to compare (4.125) and (6.61), remembering that the former involves two directions of polarization.

In Landau's theory phonons play an important role for $T < 1$ K, where few rotons exist. Unlike the theory of London and Tisza, here they contribute to the normal fluid and so lead to the correct prediction of $u_2$ at low temperatures. To calculate this we first show that $\rho_n^{ph}/\rho = \frac{4}{3}\, U_{ph}/u_1^2$, where $U_{ph} = bT^4/4$.

Let He II at $T < 1$ K be enclosed in a long tube along the $z$ direction and let the walls of the tube move with a velocity $v_z$. The superfluid component remains at rest but phonons collide with the walls and acquire a velocity $v_z$. For a stationary observer the energy $\epsilon$ of the phonon of rest energy $\epsilon_0$ is $\epsilon = \epsilon_0 + v_z p_z = \epsilon_0 + v_z p \cos\theta$, where $\cos\theta = p_z/p$. Then for the phonon gas

$$\bar{p} = \int p\cos\theta\, dN_{ph} = \frac{V}{h^3}\int_0^\infty\int_0^\pi\int_0^{2\pi} \frac{p\cos\theta}{\exp(\epsilon_0/kT)-1}\, p^2 dp\, \sin\theta\, d\theta\, d\phi.$$

By Maclaurin expansion for small $v_z$

$$\frac{1}{\exp(\epsilon_0/kT)-1} = \frac{1}{\exp[(\epsilon - v_z p\cos\theta)/kT]-1} \simeq \frac{1}{\exp(\epsilon/kT)-1} + v_z \frac{(p\cos\theta/kT)\exp(\epsilon/kT)}{[\exp(\epsilon/kT)-1]^2}$$

The zero order term disappears when integrated over the directions, $\int_0^\pi \sin\theta\cos\theta\, d\theta = 0$, and so

$$\bar{p} = \frac{2\pi V}{h^3}\frac{v_z}{kT}\int_0^\infty\int_0^\pi \frac{\exp(\epsilon/kT)}{[\exp(\epsilon/kT)-1]^2} p^4\, dp \sin\theta\cos^2\theta\, d\theta$$

$$= \frac{4\pi V}{3h^3}\frac{v_z}{kT}\int_0^\infty \frac{\exp(pu_1/kT)}{[\exp(pu_1/kT)-1]^2} p^4\, dp,$$

where we have used $\int_0^\pi \sin\theta\cos^2\theta\, d\theta = 2/3$ and $\epsilon = p(u_1 + v_z \cos) \simeq pu_1$ for $v_z \ll u_1$. Integrating by parts and noting that $\frac{kT}{u_1}\left[\frac{p^4}{\exp(pu_1/kT)-1}\right]_0^\infty$ vanishes for both limits,

$$\bar{p} = \rho_{ph} v_z = \frac{4\pi V}{h^3}\frac{v_z}{kT}\frac{4kT}{u_1}\int_0^\infty \frac{p^3 dp}{\exp(pu_1/kT)-1} = \frac{4}{3}\frac{U_{ph}}{u_1^2}v_z,$$

because

$$U_{ph} = \int \epsilon_0 dN_{ph} = \frac{4\pi V}{h^3}\int \frac{pu_1}{\exp(pu_1/kT)-1} p^2 dp.$$

If we write $U_{ph}$ per unit volume then

$$\rho_{ph} = \rho\,\frac{4}{3}\frac{U_{ph}}{u_1^2} = \rho\,\frac{4}{3}\frac{bT^4/4}{u_1^2} = \frac{\rho bT^4}{3u_1^2}. \tag{6.63}$$

In Landau's theory for $T < 1$ K the normal fluid $\rho_n$ is entirely made of $\rho_{ph}$. Therefore, substituting (6.60–63) in (6.54),

$$u_2 = \left(\frac{\rho - \rho_{ph}}{\rho_{ph}}\frac{S^2T}{C_{ph}}\right)^{1/2} = \tfrac{1}{3}(3u_1^2 - bT^4)^{1/2} \simeq \frac{u_1}{3^{1/2}}$$

$$= 13.7\times 10^3 \text{ cm/s}, \quad \text{for } T \to 0. \tag{6.64}$$

The experiment (Fig. 6.17) supports this calculation, whereas Tisza wrongly predicted $u_2 \to 0$ as $T \to 0$.

The reason for the success of Landau's theory is that here not particular particles but *all* excitations, phonons and rotons, are associated with the normal fluid.

*Roton Contribution*: From (4.40), the free energy of a gas of $N$ particles in a volume $V$ is

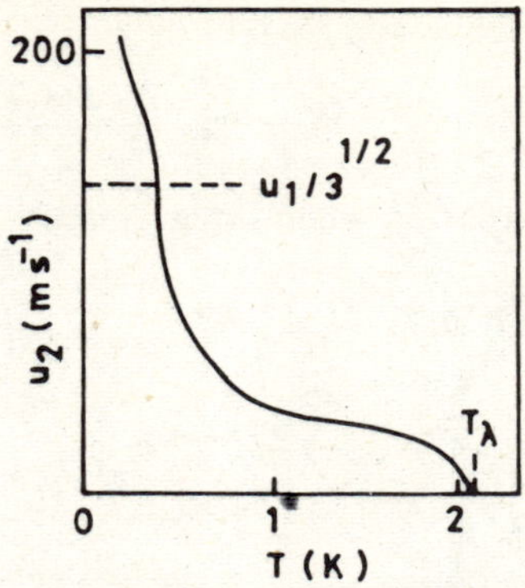


Fig. 6.17 Second sound velocity in liquid He II.

$$F = -kT \ln Z = -NkT \ln z + NkT \ln N - NkT, \tag{6.65}$$

where

$$z = \frac{4\pi V}{h^3} \int_0^\infty \exp(-\epsilon/kT)\, p^2 dp. \tag{6.66}$$

The number of rotons changes with $T$ and is determined by the condition that $F$ be a minimum,

$$0 = \partial F/\partial N = -kT \ln z + kT \ln N,$$

$$N_{rot} = z. \tag{6.67}$$

The corresponding value of the free energy, given by (6.65), is

$$F_{rot} = -kTN_{rot} = -kTz. \tag{6.68}$$

Using (6.57),

$$z = \frac{4\pi V}{h^3} \int_0^\infty \exp\{-[\Delta + (p-p_0)^2/2\mu_r]/kT\}\, p^2 dp.$$

Putting $p - p_0 = q$, $dp = dq$, $p^2 = (q+p_0)^2 \simeq p_0^2$, since $p_0^2 \gg \mu_r kT$, we can write after extending the limits to $-\infty$ and $+\infty$

$$z \simeq \frac{4\pi V}{h^3} \exp(-\Delta/kT)\, p_0^2 \int_{-\infty}^{+\infty} \exp(-q^2/2\mu_r kT)\, dq$$

$$= \frac{4\pi V}{h^3} p_0^2 \exp(-\Delta/kT)(2\pi\mu_r kT)^{1/2} = N_r = -F_{rot}/kT, \tag{6.69}$$

$$S_{rot} = -(\partial F_{rot}/\partial T)_V = kN_{rot}\left(\frac{3}{2} + \frac{\Delta}{kT}\right), \tag{6.70}$$

$$C_{rot} = T(\partial S_{rot}/\partial T)_V = kN_{rot}\left[\frac{3}{4} + \frac{\Delta}{kT} + \left(\frac{\Delta}{kT}\right)^2\right]. \tag{6.71}$$

Because of the factor $\exp(-\Delta/kT)$, very few rotons are present at low temperatures ($N_r \to 0$ as $T \to 0$).

The choice (6.58) of parameters $\Delta$, $p_0$, $\mu_r$ gives a good fit to the measured values of $C_V$, $S$, and $\rho_n/\rho$ for He II. However, Landau's theory has no satisfactory explanation for the existence of $T_\lambda$.

The full dispersion curve (Fig. 6.16) with the phonon (linear) part and roton (parabola) part connected by an ad hoc smooth curve has been observed by the neutron diffraction experiments* (Fig. 6.18). This entire curve is not given by (6.57). Feynman† has derived the full curve from the first principles. However, the quantitative agreement is poor, specially in the roton region where it gives a minimum at about $\epsilon = 19$ K. It is now believed that rotons actually exist in liquid He II, although strictly speaking all evidence is indirect.

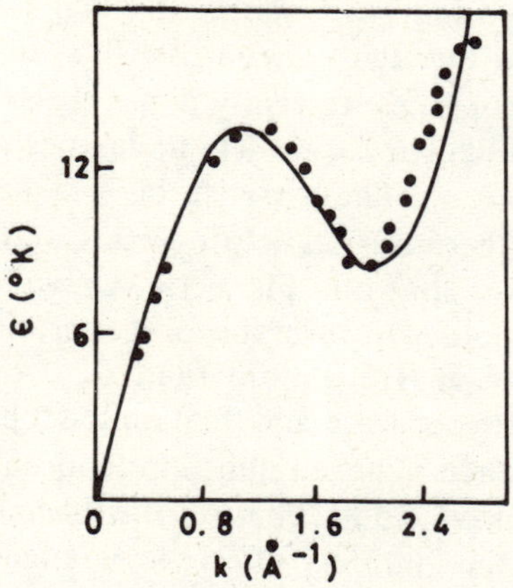


**Fig. 6.18 Full dispersion curve for liquid $^4$He. The points are from the neutron diffraction experiment and the solid curve is a fit according to the linear chain model.****

Several attempts have been made to use approaches other than the Landau model to derive the dispersion curve in agreement the neutron diffraction results. Khanna and Das+ have used a realistic potential for the interaction energy, due to Van der Waals forces between two helium atoms, as their starting point. Agarwal** has regarded He II as an aggregate of long monatomic one-dimensional chains formed under Van der Waals forces. Such a chain supports Debye waves at low temperatures ($T \rightarrow 0$) and develops vacancy waves for temperatures around 1 K and below $T_\lambda$. The resulting dispersion curve is in good agreement with experiment (Fig. 6.18). The melting point of the chain can be associated with $T_\lambda$. It suggests quantized evapora-

*D.G. Henshaw and A.D.B. Woods, *Phys. Rev.* **121**, 1266 (1961); A.D.B. Woods and R A. Cowley. *Rep. Prog. Phys.* **36**, 1135 (1973).

†R.P. Feynman, *Phys. Rev.* **94**, 262 (1954); *Progress in Low Temperature Physics*, ed. by C.J. Gorter, North-Holland, Amsterdam, 1955, Vol. I.

+K.M. Khanna and B.K. Das, *Physica* **69**, 611 (1973).

**B.K. Agarwal, *Lettere al Nuovo Cimento* **17**, 262 (1976).

tion from liquid He II which has been observed.‡ This model‡‡ also shows the possibility of Josephson tunneling of superfluid He atoms through a thin layer of He gas held between two films of liquid He II at $T \lesssim 0.5$ K. This effect has been observed using a level difference in two reservoirs.†

## 6.7 $^3$He-$^4$He MIXTURES

The London-Tisza theory and the Landau theory differ in the relative importance assigned to the role of quantum statistics. It was therefore suggested that a study of the mixture of a BE gas of $^4$He atoms and a FD gas of $^3$He atoms below 2.2 K would throw some light on this issue.

Daunt *et al** showed that $^3$He neither participates in the helium film transfer nor in the flow through very narrow slits. In other words, it does not show superfluid properties. Recently it has been observed** that at a temperature of few millikelvin liquid $^3$He undergoes a superfluid transition connected with spin pairing. The presence of $^3$He decreased the observed value of $T_\lambda$ and other thermodynamic properties were also affected.

The phase diagram is shown in Fig. 6.19. Above 0.86 K we can mix $^3$He and $^4$He in all proportions. However, beyond a certain concentration of $^3$He there is no superfluidity. If $^3$He is more than 6%, then below 0.86 K the $^3$He-rich phase begins to separate and float on the top of the $^4$He-rich phase. There is a visible interface. The equilibrium concentrations are given by points like $P$, $P'$ at, say, 0.5 K. The *tricritical point* $Y$ has received much attention@. As $T \to 0$, the solubility of $^4$He in $^3$He tends to zero, but that of $^3$He in $^4$He to about 6% under the saturated vapour pressure. It means when $^3$He atoms are placed in pure $^4$He at $T = 0$, they have lower energy compared to that in pure $^3$He. As more $^3$He are added, being fermions, they occupy successively higher energy states. After certain concentration it is no more advantageous to be in $^4$He than to be in $^3$He.

Heer and Daunt†† extended London's theory of He II to the mixture by regarding it as an ideal mixture of a degenerate BE gas of $^4$He atoms and a non-degenerate FD gas of $^3$He atoms. They found that $T_\lambda^{mix}$ varies with the $^3$He concentration as

$$T_\lambda^{\text{mix}} = T_0 \left(\frac{N_4 v_4}{N_3 v_3 + N_4 v_4}\right)^{2/3}, \tag{6.72}$$

‡M.J. Baird, F.R. Hope and A.F.G. Wyatt, *Nature*, **304**, 28, 1983.

‡‡B.K. Agarwal, *Lettere al Nuovo Cimento* **21**, 463 (1978).

†P.L. Richards, *Phys. Rev.* A **2**, 1532 (1970).

*J.G. Daunt, R.E. Probst and H.L. Johnston, *Phys. Rev.* **73** 638 (1948).

**For a review see J.C. Wheatley, *Rev. Mod. Phys.* **47**, 415 (1975). Also see O.V. Lounasmaa, *Contemp. Phys.* **15**, 353 (1974).

@G. Ahlers in *The Physics of Liquid and Solid Helium*, eds. K.H. Bennemann and J.B. Ketterson, Willey NY 1976, part I.

††C.V. Heer end J.G. Daunt, *Phys. Rev.* **81**, 447 (1951).

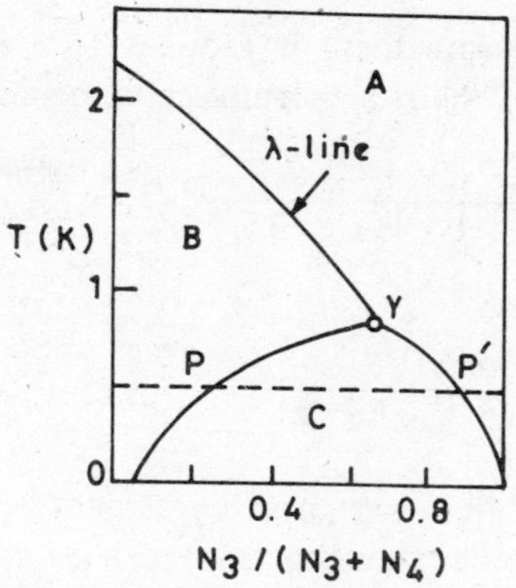


**Fig. 6.19** The phase diagram for liquid $^3He$-$^4He$ mixtures under their saturated vapour pressures. A is the region of normal homogeneous mixture, B of superfluid homogeneous mixture, and C of phase-separated mixture.

where $v_3$ is the volume per atom in $^3$He and $v_4$ in $^4$He, $v_3 > v_4$. This formula is obtained simply on replacing in (6.11),

$$T_0 = \frac{h^2}{2\pi mk}\left(\frac{N_4}{2.612\, V_4}\right)^{2/3},$$

$V_4$ by $V_{\text{mix}} = N_3 v_3 + N_4 v_4$,

$$T_\lambda^{\text{mix}} = \frac{h^2}{2\pi mk}\left(\frac{1}{2.612\, v_4}\right)^{2/3}\left(\frac{N_4 v_4}{N_3 v_3 + N_4 v_4}\right)^{2/3}$$

$$= T_0 X^{2/3}, \quad X = \frac{N_4 v_4}{N_3 v_3 + N_4 v_4}.$$

The agreement with experiment is modest.

Several other phenomenological theories have been given to account for a large body of data on the $^3$He-$^4$He mixture. Nanda* has assumed that $^3$He forms a non-ideal mixture with the $^4$He normal fluid. Mikura** has employed the ad hoc energy spectrum

$$\epsilon = \Delta + p^2/2m^*, \tag{6.73}$$

where $\Delta$ is the energy gap between the zero energy ground state and the lowest excited state, and $m^*$ is the effective mass. His result is

$$T_\lambda^{\text{mix}} = T_0\, \{X \exp\, [(\Delta_0 - \Delta) T_0]\}^{2/3}, \quad \Delta = \Delta_0\, X^{0.4}. \tag{6.74}$$

Agarwal†, following Dingle@, has used the general spectrum

* V.S. Nanda, *Phys. Rev.* **94**, 241 (1954); **97**, 571 (1955).

** Z. Mikura, *Prog. Theoret Phys.* (Japan) **11**, 25 (1954).

† B.K. Agarwal, *Z. Physik* **145**, 515 (1956); **146**, 9 (1956).

@ R.B. Dingle, *Advances in Phys.* **1**, 111 (1952).

$$\epsilon = Mp^{1/r}, \tag{6.75}$$

where $M$ and $r$ are parameters. It reduces to $\epsilon = p^2/2m$ for the choice $M = 1/2m$ and $r = 1/2$. This spectrum leads to the BE distribution*.

$$dn(\epsilon) = \frac{CV\, e^{3r-1}\, d\epsilon}{\exp(\alpha + \epsilon/kT) - 1}, \; C = \frac{4\pi r}{h^3 M^{3r}}. \tag{6.76}$$

The $T_\lambda^{\text{mix}}$ is then given by

$$T_\lambda^{\text{mix}} = T_0 X^{1/3r}, \; r = \frac{1}{2}\frac{v_3}{v_4}. \tag{6.77}$$

Both (6.74) and (6.77) give very good agreement with experiment. Predictions from these and other theories have been compared by Daunt**. For a survey of properties see the review by Wheatley.†

A similar variation of $T_\lambda^{\text{mix}}$ also follows from Landau's theory, without emphasizing the role of statistics. It regards $^3$He atoms as impurity atoms in the mixture, which merely increase the normal fluid concentration. Thus the hope of deciding the exact role of statistics in liquid helium phenomenon by the study of $^3$He—$^4$He mixtures has remained unfulfilled.

## 6.8 SUPERFLUID PHASES OF $^3$He

The phase diagram of $^3$He (Fig. 6.20), is quite different from that of $^4$He. As $^3$He obeys the Fermi-Dirac statistics, it is surprising that it shows superfluidity at all.

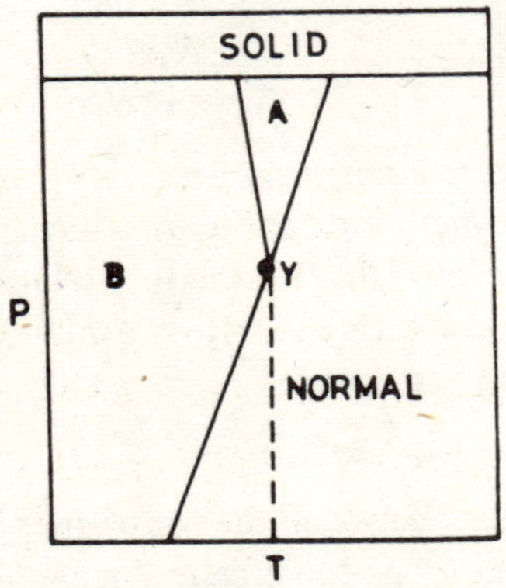


**Fig. 6.20** **Phase diagram of $^3$He when there is no magnetic field. The polycritical point is at about 20 atm pressure and $2.5 \times 10^{-3}$ K temperature.**

*D.S. Kothari and B.N. Singh, *Proc. Roy. Soc.* (London) A **178**, 135 (1941).

**J.G, Daunt, *Low Temperature Physics and Chemistry*, ed. by J.R. Dillinger, Univ. of Wisconsin Press, 1958.

†J.C. Wheatley, *Am. J. Phys.* **36**, 181 (1968).

The superfluidity in $^3$He occurs at 1.0 mK under the saturated vapour pressure and at 2.8 mK near the solidification point. This is very much smaller than $T_\lambda$ for $^4$He.

$^3$He shows two principal superfluid phases, *A* and *B*, rather than one. At the polycritical point *Y* the phases *A* and *B* and the normal liquid co-exist. If a magnetic field **B** is applied, *Y* disappears, *A* stretches down to zero pressure, and a new superfluid phase *A*1 appears between *A* and the normal liquid $^3$He (Fig. 6.21). As field is increased, *B* shrinks towards lower

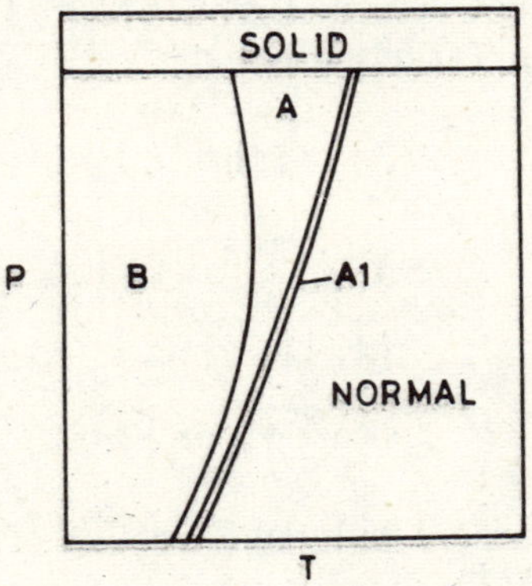


**Fig. 6.21 Phase diagram of $^3$He when the magnetic field is present.**

*T*, while *A* and to a lesser extent *A*1 grow. For very strong fields, *B* is completely replaced by *A*. The superfluid $^3$He is both magnetic and anisotropic. It exhibits three distinct phases when $\mathbf{B} \neq 0$. Various phase transitions are given in Table 6.1.

**Table 6.1 Phase Transitions in $^3$He**

| Magnetic field | Phase transition | Type |
|---|---|---|
| 0 | Normal → *A* | Second order |
| 0, **B** | *A* → *B* | First order |
| **B** | Normal → *A*1 | Second order |
| **B** | *A*1 → *A* | Second order |

The second-order phase transition (normal → *A*) shows a finite discontinuity in the specific heat curve. This curve resembles in shape the normal → superconductivity transition for the electron gas in a metal, rather the λ-transition in liquid $^4$He.

The superfluidity in liquid $^3$He results from the formation of Cooper pairs, analogous to the case of electrons in a superconductor. In latter, the Cooper pairs are formed between electrons with opposite spins ($S = 0$, singlet state) and zero angular momentum ($L = 0$, $s$-state). However, in $^3$He the pairing is between neutral atoms under weak attractive forces with $S = 1$ (triplet state) and $L = 1$ ($p$-wave). Again the overall wave function is antisymmetric, as $L$ is odd. Even in zero magnetic field, the ($S = 1, L = 1$) Cooper pairs can occur with $S_z = 1(\uparrow\uparrow), 0(\uparrow\downarrow + \downarrow\uparrow), -1(\downarrow\downarrow)$, where $\uparrow(\downarrow)$ denotes spin up (down). Various spin states are:

| Phase | Spin | State |
|---|---|---|
| $B$ | $\uparrow\uparrow$, $\uparrow\downarrow + \downarrow\uparrow$, $\downarrow\downarrow$ | $BW$* |
| $A$ | $\uparrow\uparrow$, $\downarrow\downarrow$ | $ABM$† |
| $A1$ | either $\uparrow\uparrow$ or $\downarrow\downarrow$ | ... |

Superfluidity is associated with the formation of Cooper pairs†† and the associated energy gap $\Delta(T)$.

In spite of many differences between liquid $^4$He and liquid $^3$He, the two-fluid model is applicable to $^3$He. The superfluid part in $^3$He is associated with the Cooper pair condensate and the normal fluid part with the unpaired fermions in excited states. Complications arise from the magnetic properties of the liquid and from *texture* (bending of the macroscopic angular momentum vector near the container walls). Due to the liquid $^3$He being both anisotropic and magnetic, the $\rho_s$ depends** on the direction in the liquid and also on the orientation of the external field **B**.

*Effect of* **B** *in A Phase.* In $A$, the vector **L** for every pair tends to be in the same direction. The liquid acquires a macroscopic angular momentum **l**, as if it is an orbital ferromagnet.

It is useful to think of a vector **a** (**k**) such that | **a** (**k**) | signifies the pair condensate amplitude at the intersection point of the momentum direction $\hbar$**k** and the Fermi surface. In $A$, **a**(**k**) is in the same direction for all **k**. For minimum energy **a**(**k**) and **l** are parallel, and **B** (therefore S) is perpendicular to them (Fig. 6.22). The net magnetization $M$ of the liquid is along **B**.

*R. Balian and N.R. Werthamer, *Phys. Rev.* **131**, 1553 (1963).

†P.W. Anderson and P. Morel, *Phys. Rev.* **123**, 1911 (1961); P.W. Anderson and W.F. Brinkman, *Phys. Rev. Lett.* **30**, 1108, (1973).

††E.M. Lifshitz and L.P. Pitaevskii, *Statistical Physics*, Pergamon, 1980, part 2, p. 219.

**J.E. Berthold *et al.*, *Phys. Rev. Lett.* **37**, 1138 (1976).

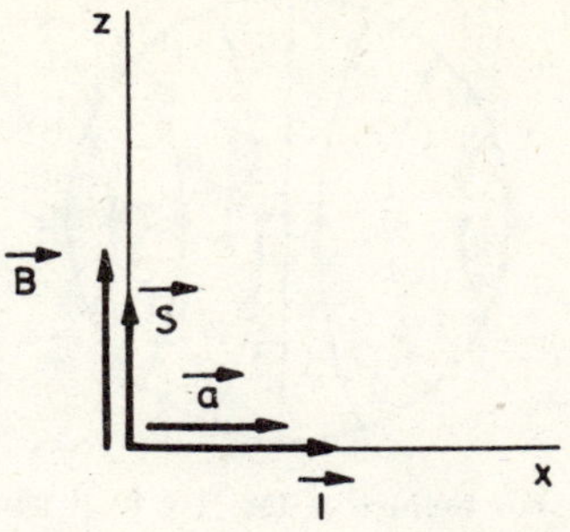


Fig. 6.22 Effect of B in the phase A of $^3$He superfluid.

*Effect of* **B** *in B Phase*: In *B*, the **L** are oriented differently for different Fermi surface points. The resulting net angular momentum in the ground state is zero. Also, | **a** (**k**) | do not depend on **k.** To find how the direction of **a** varies over the Fermi surface, first imagine the **a** vectors to stick out radially over the Fermi surface. To minimize the magnetic dipole energy, rotate all the **a** vectors about an arbitrary axis **n** by the angle $\cos^{-1}(\frac{1}{4}) \simeq 104°$. This adjusts **L** relative to **S** for each pair. The vectors **a** are affected most (least) which lie in a plane *N* normal to **n** (lie on the **n** axis) (Fig. 6.23). The

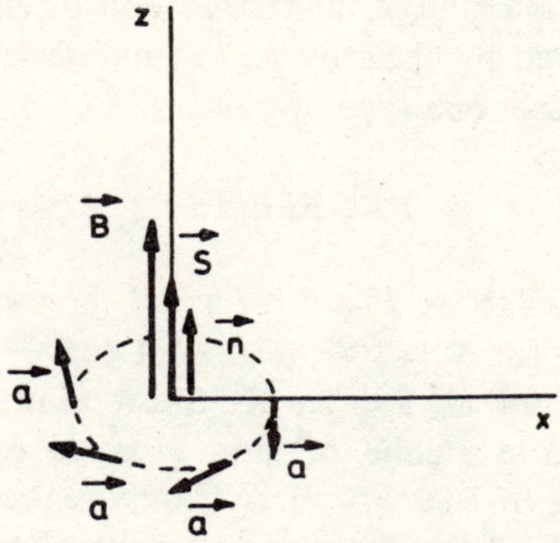


Fig. 6.23 Effect of B in the phase B of $^3$He superfluid.

*N* plane **a** vectors become almost tangential. The dipole energy is minimum when **n** is parallel to the applied field **B**.

*Texture*: The vectors **l, a, n**, are not always arranged throughout the liquid as shown in the ideal pictures given in Figs. 6.22, 6.23. The Cooper pairs tend to orbit in a plane parallel to the surface, when near it. This prevents breaking of pairs by collisions with the surface itself. Thus, **l** in *A* phase is perpendicular to **B** in the interior and bends to be perpendicular to the container walls in the outer region (Fig. 6.24). This gives a *texture* to the liquid.

In the *B* phase, it is the vector **n** which is parallel to **B** and perpendicular to the walls.

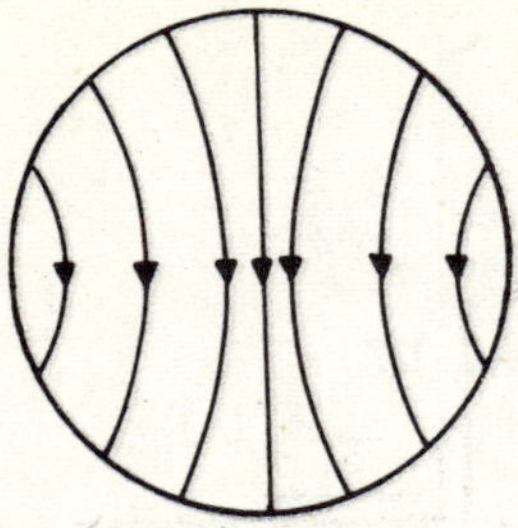

**Fig. 6.24 A possible texture of the $^3$He in $A$ phase inside a sphere.**

*Leggett Effect*: We can associate characteristic frequencies $\omega_A(T)$ and $\omega_B(T)$ with the oscillations of the **a** vector about its equilibrium orientations given in (Figs. 6.22, 6.23). They can be measured by the nuclear magnetic resonance (NMR) method. They can also be measured by the method* of 'parallel ringing'.

When strength of **B** is adjusted in small steps, the magnetization of liquid $^3$He begins to oscillates at $\omega_A$ or $\omega_B$. The weakly damped oscillations last till equilibrium positions are restored for the involved vectors. It can also be understood in terms of tunneling of Cooper pairs between two separate superfluids consisting of $\uparrow\uparrow$ and $\downarrow\downarrow$ pairs. As these fluids are weakly coupled and interpenetrating, it is a kind of internal Josephson effect. It was predicted by Leggett* and observed by Osheroff and Brinkman†. The ringing does not occur in $A$ 1 as it has only one type of pair.

## *PROBLEMS*

6.1 Show that $\mu = -kT/N$ as $T \to 0$ for a BE gas with ground state at $\epsilon = 0$. Show that, for $N = 10^{22}$, $\mu \simeq -1.4 \times 10^{-38}$ erg at $T = 1$ K and $\mu \simeq -1.4 \times 10^{-41}$ erg at $T = 1$m K. Show that the lowest excitation energy of an atom in a cube of side 1 cm is $\Delta\epsilon = \epsilon(211) - \epsilon(111) = 2.48 \times 10^{-30}$ erg or $1.80 \times 10^{-14}$ K. Compare the closeness of $\mu$ to the ground state and of the lowest excited state to the ground state. Justify Fig. P 6.1, with $\epsilon_0 = \epsilon(111) =$ zero of energy

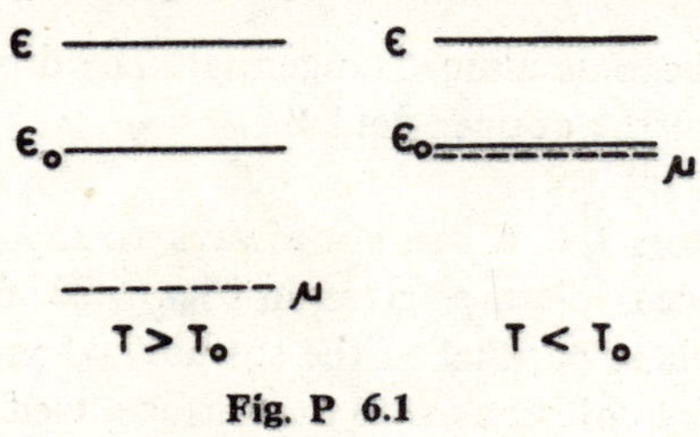


**Fig. P 6.1**

*A.J. Leggett, *J. Phys. C* **6**, 3187 (1973); *Phys. Rev. Lett.* **31**, 352 (1973).
†D.D. Osheroff and W.F. Brinkman, *Phys. Rev. Lett.* **32**, 584 (1974).

6.2 Show that for a two-dimensional ideal Bose-Einstein gas

$$N = \frac{2\pi\, l_x l_y m}{h^2} \int_0^\infty \frac{d\epsilon}{\exp\{(\epsilon-\mu)/kT\} - 1}$$

$$= l_x l_y \frac{2\pi mkT}{h^2} \sum_{s=1}^{\infty} \frac{1}{s} \exp(s\mu/kT)$$

where $l_x l_y$ is the area. Can it undergo Bose-Einstein condensation? (Ans: No, because in this result a value of $\mu$ always exists which is not of order $1/N$. So no levels are occupied by a number of molecules of order $N$.)[Hint: $\epsilon = (\hbar^2/2m)(k_x^2 + k_y^2)$, $k_i = 2\pi n_i/l_i$, $n_i = 0, \pm 1 \ldots$; for large $l_x$, $l_y$ and $k^2 = k_x^2 + k_y^2$,

$$N = \sum_j [\exp\{(\epsilon_j - \mu)/kT\} - 1]^{-1}$$

$$= \frac{l_x l_y}{(2\pi)^2} \int \frac{dk_x\, dk_y}{\exp[\{(\hbar^2 k^2/2m) - \mu\}/kT] - 1}.]$$

6.3 An ideal Bose-Einstein gas consists of particles with an internal degree of freedom. Besides the ground state $\epsilon_0 = 0$, suppose just one excited level $\epsilon_1$ is important. Find the condensation temperature of this gas.

6.4 Calculate $\partial\mu/\partial\theta$, $\theta = kT$, and discuss its sign.

6.5 London suggested an ad hoc energy spectrum for liquid $^4$He,

$$\epsilon_p = \Delta + (p^2/2m^*) = \Delta + \epsilon$$

where $\Delta$ is the energy gap between the ground state $\epsilon_0 = 0$ and the excited state $\epsilon_p$, $\mathbf{p} \neq 0$. Show the $N = \frac{V}{h^3}(2\pi m^* kT)^{3/2} \exp(-\Delta/kT_0)$,

$$N' = N\left(\frac{T}{T_0}\right)^{3/2} \exp(\Delta/kT - \Delta/kT_0).$$

# 7
# IDEAL FERMI-DIRAC GAS

## 7.1 FERMI-DIRAC DISTRIBUTION

For an ideal FD gas of $N$ molecules in a volume $V$, the most probable number of particles with energy $\epsilon_l$ is

$$\bar{n}_l(\epsilon_l) = \frac{g_l}{\exp(\alpha + \beta\epsilon_l) + 1} = \frac{g_l}{\exp[(\epsilon_l - \mu)/kT] + 1}, \tag{7.1}$$

where $\beta = 1/kT$, $\alpha \equiv -\mu/kT$, and $g_i$ = degeneracy of the $i$th level. The parameter $\alpha$ (or $\mu$) is determined as a function of $N$ and $T$ by the condition

$$N = \sum_{\substack{l=0 \\ (levels)}}^{\infty} \bar{n}_l = \sum_l \frac{g_l}{\exp[\beta(\epsilon_l - \mu)] + 1}, \tag{7.2}$$

where the sum is over the energy levels as we have included $g_l$ in (7.1). Because of the factor $+ 1$ in the denominator of (7.1), $\alpha$ need not be restricted to $\alpha \geqslant 0$ as in the BE case. For the FD gas $\alpha$ can be positive or negative. It is more convenient to work with the chemical potential $\mu(T)$, since $\mu$ approaches a finite value $\mu_0$ at $T = 0$ whereas $\alpha$ becomes negatively infinite. The finite value $\mu_0$ at $T = 0$ follows from the Pauli exclusion principle. The fermions in the gas occupy all possible quantum states with energies between zero and the *Fermi energy* $\epsilon_F(T = 0) \equiv \mu_0$ whose value depends on the number of fermions in the gas. The states with energy greater than $\mu_0$ are empty at absolute zero. The *Fermi level* is defined by $\mu(T)$ for $T > 0$.

Once again we can approximate the sum by an integral. In the FD case, due to the Pauli principle, there is no danger that a sizeable fraction of the particles will be in the ground state. If we introduce the Fermi-Dirac distribution function (Fig. 7.1)

$$f(\epsilon) = \frac{1}{\exp[(\epsilon - \mu)/kT] + 1}, \tag{7.3}$$

(7.2) becomes

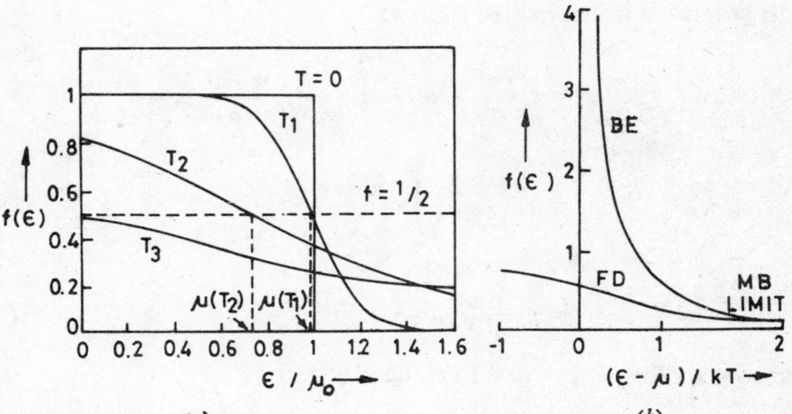


(*a*) (*b*)

**Fig. 7.1 (a) FD distribution function for a gas in three dimensions. At the Fermi level the value of $f$ is $\frac{1}{2}$. Note that $\mu(T_3)$ is negative. As $k = 8.62 \times 10^{-5}$ eV K$^{-1}$, for $\mu/kT \simeq 1$, we have $\mu(T) \simeq$ 1 eV for $T \simeq 10^4$ K. For a metal $\mu_o$ might correspond to $5 \times 10^4$ K as here. (b) Comparison of Fermi-Dirac and Bose-Einstein distribution functions. For $(\epsilon - \mu) \gg kT$ we get the classical limit.**

$$N = \int_0^\infty f(\epsilon)\, g(\epsilon)\, d\varepsilon, \tag{7.4}$$

where

$$g(\epsilon)\, d\epsilon = g_s(2\pi V/h^3)\,(2m)^{3/2}\, \epsilon^{1/2}\, d\epsilon. \tag{7.5}$$

The $g_s = (2s + 1)$ is the spin degeneracy and comes from $(2s + 1)$ different spin orientations possible for the same energy $\epsilon$. For electrons $s = \frac{1}{2}$, $g_s = 2$. The plot of

$$d\bar{n}(\epsilon)/d\varepsilon = g(\epsilon)\, f(\epsilon) \tag{7.6}$$

is shown in Fig. 7.2.

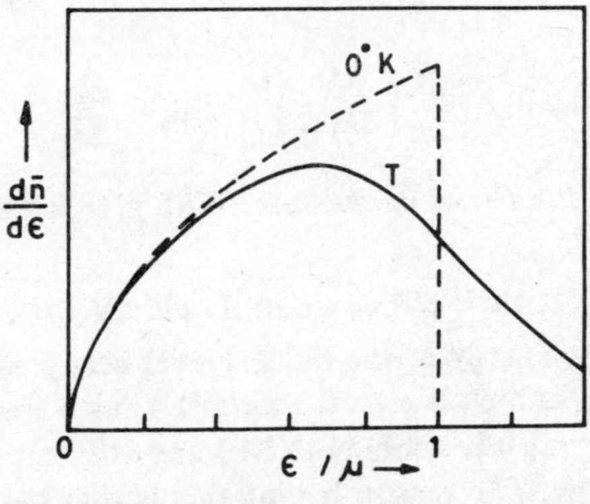


**Fig 7.2 Plot of $d\bar{n}/d\epsilon$ as a function of $(\epsilon/\mu)$, at $T = 0$ K (dashed curve) and $T = \mu/5k$ (solid curve).**

In general, $\mu$ is determined from $N$,

$$N = g_s \frac{2\pi V}{h^3} (2m)^{3/2} \int_0^\infty \frac{\epsilon^{1/2} d\epsilon}{\exp[(\epsilon-\mu)/kT]+1}$$

$$= g_s \frac{V}{\lambda^3} \frac{2}{\pi^{1/2}} \int_0^\infty \frac{x^{1/2} dx}{\frac{1}{\eta_a} e^x + 1}$$

$$= g_s \frac{V}{\lambda^3} G_{3/2}(\eta_a), \tag{7.7}$$

where $x = \epsilon/kT$, $\eta_a = \exp(\mu/kT)$ is fugacity and

$$G_{3/2}(\eta_a) = \frac{2}{\pi^{1/2}} \int_0^\infty dx\, x^{1/2} \left(\frac{1}{\eta_a} e^x + 1\right)^{-1}. \tag{7.8}$$

For small $\eta_a$,

$$G_{3/2}(\eta_a) = \eta_a - \frac{\eta_a^2}{2^{3/2}} + \frac{\eta_a^3}{3^{3/2}} - \ldots = \sum_{n=1}^{\infty} \frac{(-1)^{n+1} \eta_a^n}{n^{3/2}}. \tag{7.9}$$

Thus $G_{3/2}(\eta_a)$ is a monotonically increasing function of $\eta_a$, as is easily verified.

The value of $\epsilon_F(0)$ is determined by noting that at $T = 0$ $f(\epsilon)$ is 1 for $\epsilon < \epsilon_F(0)$ and 0 for $\epsilon > \epsilon_F(0)$,

$$N = \int_0^{\epsilon_F(0)} g(\epsilon)\, d\epsilon = g_s \frac{2\pi V}{h^3} (2m)^{3/2} \int_0^{\epsilon_F(0)} \epsilon^{1/2}\, d\epsilon$$

$$= g_s \frac{2\pi V}{h^3} (2m)^{3/2} \frac{2}{3} [\epsilon_F(0)]^{3/2}, \tag{7.10}$$

or equivalently, with $\epsilon = p^2/2m$,

$$N = g_s \int_0^{p_F} \frac{4\pi p^2 dp}{h^3 h V} = g_s \frac{4\pi V}{3h^3} p_F^3.$$

Thus

$$\epsilon_F(0) = \frac{h^2}{2m} \left(\frac{3N}{4\pi V g_s}\right)^{2/3} = \mu_0 = \frac{p_F^2}{2m}, \tag{7.11}$$

where $p_F$ is called the *Fermi momentum*. With $g_s = 2$ and the density of gas $\rho = (Nm/V)$ kg/m$^3$,

$$\epsilon_F(0) = 0.625 \times 10^{-17} \rho^{2/3} \text{ Joule or } 39\, \rho^{2/3} \text{ eV for electrons.} \tag{7.12}$$

$\epsilon_F(0)$ represents the energy of the highest level occupied at 0 K. For conduction electrons in metals $\rho \simeq 0.1$ kg/m$^3$. Thus a Fermi gas possesses an appreciable energy at 0 K while both MB and BE statistics predict a zero value in the energy. The reason behind this is that the occupation number for each quantum state is 0 or 1 only for the FD statistics.

As the temperature increases above 0 K, the distribution near the $\epsilon_F(0)$ rounds off (Fig. 7.1). If we define a temperature, called the *Fermi temperature*, by

$$T_F = \epsilon_F(0)/k \; (= 4.52\times 10^5 \, \rho^{2/3} \text{ K for electrons}) \tag{7.13}$$

then for $T \ll T_F$, or $kT_F \ll \epsilon_F(0)$, the distribution is called *degenerate*. For $T \gg T_F$, the distribution is nondegenerate and we get the classical limit. The parameter $\alpha = -\mu(T)/kT$ is *negative* in the former case and *positive* in the latter case. It means that $\mu > 0$ at low temperatures, and $\mu < 0$ at high temperatures (Fig. 7.1).

## 7.2 DEGENERACY

We have three cases: (i) $T = 0$, the gas is *completely degenerate*, (ii) $T \ll T_F$, low temperatures, the gas is *degenerate*, and (iii) intermediate temperatures, the gas is *slightly degenerate*.

(i) *Completely degenerate gas*, $T = 0$ K. We have

$$\begin{aligned} dn &= g_s\,(2\pi V/h^3)\,(2m)^{3/2}\,\epsilon^{1/2}\,d\epsilon,\; 0 \leqslant \epsilon \leqslant \epsilon_F(0) \\ &= 0,\; \epsilon > \epsilon_F(0), \end{aligned} \tag{7.14}$$

where $\epsilon_F(0)$ or $\mu_0$ is given by (7.11). The internal energy is

$$\begin{aligned} U_0 &= \int_0^{\mu_0} \epsilon\, dn = g_s\,(2\pi V/h^3)\,(2m)^{3/2} \int_0^{\mu_0} \epsilon^{3/2}\, d\epsilon \\ &= \frac{3}{5} N\mu_0 = \frac{3}{5} N\epsilon_F(0). \end{aligned} \tag{7.15}$$

Other thermodynamic functions are

$$S_0 = 0,$$

$$\Omega_0 = U_0 - S_0 T - \mu_0 N = -\frac{2}{5} N\mu_0, \tag{7.16}$$

$$P = -(\Omega_0/V) = \frac{2}{5}(N/V)\,\mu_0 \; (= 2.71\times 10^7 \,\rho \text{ atm for electrons}).$$

Thus even at 0 K a fermion gas exerts a pressure. If the electrons in a metal were neutral they would exert a pressure of about $10^6$ atm. The Coulomb attraction to the ions counterbalances the pressure. For $T = 0$ K the value of $\epsilon_F(0)$ is positive and large.

(ii) *Degenerate gas*, $T \ll T_F$. The value of $\mu(T)$ is still positive. We have

$$N = \int_0^\infty f(\epsilon)\, g(\epsilon)\, d\epsilon = g_s \frac{2\pi V}{h^3}(2m)^{3/2} \int_0^\infty \frac{\epsilon^{1/2}\, d\epsilon}{\exp[(\epsilon-\mu)/kT] + 1}. \tag{7.17}$$

The series (7.9) does not converge because $1/\eta_a < 1$ with $\mu$ positive. So we have to make use of an approximation. It uses the fact that for the degenerate case $f(\epsilon)$ $[\exp((\epsilon-\mu)/kT)+1]^{-1}$ changes slowly with $\epsilon$ except in the region $\epsilon \simeq \mu$ where it changes very rapidly (Fig. 7.1). In other words, $-(\partial f/\partial\epsilon)$ resembles a delta function (Fig. 7.3) being negligible except when $|\epsilon-\mu|$ Is small and large enough at $\epsilon=\mu$ to give unit area under the peak.

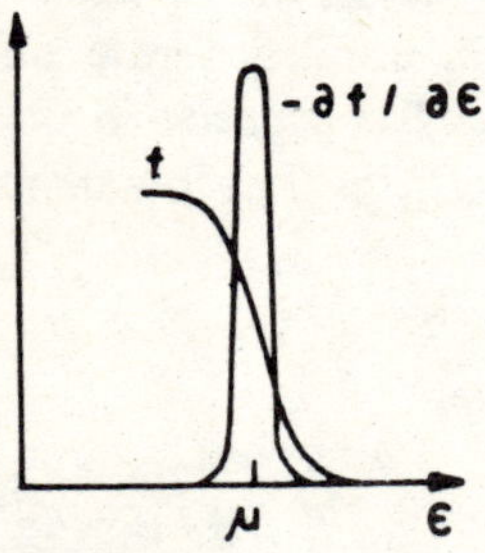


**Fig. 7.3** Plot of $f$ and $-\partial f/\partial\epsilon$ as a function of $\epsilon$.

Introduce the function

$$F(\epsilon)=\int_0^{\epsilon} g(\epsilon)\,d\epsilon, \tag{7.18}$$

so that $F(0)=0$. Now consider the integral (7.17),

$$N=\int_0^{\infty} f(\epsilon)\frac{dF}{d\epsilon}\,d\epsilon. \tag{7.19}$$

Integrating by parts and noting that $F(0)=0$ and $f(\infty)=0$,

$$N=-\int_0^{\infty} F(\epsilon)\frac{df}{d\epsilon}d\epsilon. \tag{7.20}$$

The integrand is appreciable only for $\epsilon=\mu(T)$ (Fig. 7.3). Therefore, expand $F(\epsilon)$ by Taylor's theorem about $\mu$,

$$\begin{aligned}F(\epsilon)&=\sum_{n=0}^{\infty}\frac{(\epsilon-\mu)^n}{n!}\left(\frac{d^nF}{d\epsilon^n}\right)\Bigg|_{\epsilon=\mu}\\&=F(\mu)+(\epsilon-\mu)F'(\mu)+\tfrac{1}{2}(\epsilon-\mu)^2F''(\mu)+\ldots,\end{aligned} \tag{7.21}$$

so that

$$N=I_0F(\mu)+I_1F'(\mu)+I_2F''(\mu)+\cdots, \tag{7.22}$$

where

$$I_0=-\int_0^{\infty} f'(\epsilon)\,d\epsilon,$$

$$I_1=-\int_0^{\infty}(\epsilon-\mu)f'(\epsilon)\,d\epsilon,$$

$$I_2 = -\frac{1}{2}\int_0^{\infty} (\epsilon-\mu)^2 f'(\epsilon)\, d\epsilon. \tag{7.23}$$

At low temperatures the lower limits on the integrals can be replaced by $-\infty$. Then $I_0 = 1$, and, as it is easily seen that $f'(\epsilon)$ is an even power of $\epsilon-\mu$, we have $I_1 = 0$. Writing $y = (\epsilon-\mu)/kT$,

$$I_2 = \tfrac{1}{2}(kT)^2 \int_{-\infty}^{+\infty} \frac{y^2\, e^y dy}{(1+e^y)^2} = \frac{\pi^2}{6}(kT)^2, \tag{7.24}$$

where the definite integral is given in standard tables. Thus

$$N = \int_0^{\infty} f(\epsilon)\, F'(\epsilon)\, d\epsilon = F(\mu) + \frac{\pi^2}{6}(kT)^2 F''(\mu) + \ldots$$

$$= \int_0^{\mu} g(\epsilon)\, d\epsilon + \frac{\pi^2}{6}(kT)^2 g'(\mu) + \ldots \tag{7.25}$$

Since $N$ is also given by (7.10),

$$0 = -\int_0^{\mu_0} g(\epsilon)\, d\epsilon + \int_0^{\mu} g(\epsilon)\, d\epsilon + \frac{\pi^2}{6}(kT)^2 g'(\mu)$$

$$\simeq g(\mu)\,[\mu-\mu_0] + \frac{\pi^2}{6}(kT)^2 g'(\mu). \tag{7.26}$$

Using $g(\epsilon) = (2\pi V/h^3)(2m)^{3/2}\epsilon^{1/2}$,

$$\mu(T) \simeq \mu_0\left[1 - \frac{\pi^2}{12}\left(\frac{kT}{\mu_0}\right)^2\right], \quad T \ll T_F, \tag{7.27}$$

where in the second small term we have replaced $\mu$ by $\mu_0$. Thus $\mu(T)$ is lowered slightly as the temperature is increased (Fig. 7.4).

The internal energy is given by

$$U = \int_0^{\infty} \epsilon f(\epsilon)\, g(\epsilon)\, d\epsilon$$

$$= \int_0^{\mu} \epsilon g(\epsilon)\, d\epsilon + \frac{\pi^2}{6}(kT)^2 \frac{d}{d\epsilon}(\epsilon g)$$

$$\simeq U_0 + [\mu-\mu_0]\,\mu_0\, g(\mu_0) + \frac{3}{2}\frac{\pi^2}{6}(kT)^2 g(\mu_0), \tag{7.28}$$

where we have used (7.15). In view of (7.27),

$$U = U_0 + \frac{\pi^2}{6}(kT)^2 g(\mu_0). \tag{7.29}$$

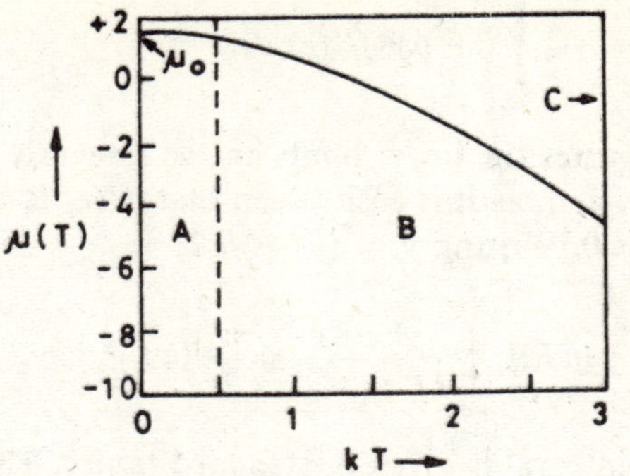


**Fig 7.4** Variation of the chemical potential with $T$ for an ideal FD gas in three dimensions. $\mu_0$ is given by (7.11) and $N/V$ is chosen here to give $\mu_o = (1.5)^{2/3}$, the region $A$ corresponds to the degenerate quantum FD gas, $B$ to slightly degenerate FD gas, and $C$ indicates the region of classical gas.

Using

$$g(\mu_0) = g_s \frac{2\pi V}{h^3} (2m)^{3/2} \mu_0{}^{1/2} = \frac{3N}{2\mu_0} = \frac{5U_0}{2(\mu_0)^2}, \tag{7.30}$$

we get

$$U = U_0 \left[ 1 + \frac{5\pi^2}{12} \left( \frac{kT}{\mu_0} \right)^2 - \ldots \right], \quad U_0 = \frac{3}{5} N\mu_0. \tag{7.31}$$

The heat capacity is

$$C_V = \frac{\pi^2}{3} k^2 T\, g(\mu_0) = \frac{\pi^2 N k^2 T}{2\mu_0} = \frac{\pi^2 N k}{2} \cdot \frac{T}{T_F}. \tag{7.32}$$

This result is due to Sommerfeld (1928).

In Fig. 7.5 we compare the heat capacity of a gas according to the three statistics. The pressure is given by $P = (2U/3V)$.

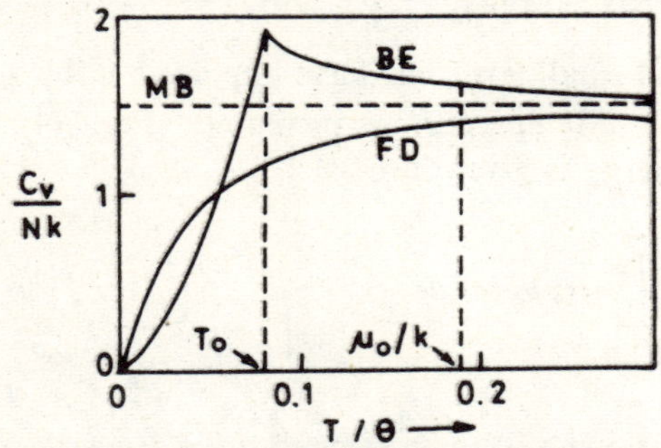


**Fig. 7.5** Comparison of heat capacity of a gas according to the three statistics, $\theta = (h^2/mk)\,(\bar{N}/gV)^{2/3}$.

(iii) *Slight degeneracy*, $kT > \epsilon_F(0)$. In this case $\mu(T)$ is negative or $\alpha$ is positive and $\eta_a < 1$. We can write, as in (7.7),

$$N = g_s \frac{2\pi V}{h^3} (2m)^{3/2} \int_0^\infty \frac{\epsilon^{1/2}\, d\epsilon}{\frac{1}{\eta_a} \exp(\epsilon/kT) + 1}$$

$$= g_s \frac{V}{\lambda^3} \eta_a \left( 1 - \frac{\eta_a}{2^{3/2}} + \frac{\eta_a^2}{3^{3/2}} - \dots \right) \tag{7.33}$$

The series converges for $\exp(\mu/kT) = \eta_a < 1$.

## 7.3 ELECTRONS IN METALS

We have applied FD statistics to an electron gas in which the mutual interaction of electrons is absent. To a first approximation this interaction may be neglected due to the neutralizing effect of positive ions inside a metal.

Drude (1900) was first to suggest that the electrical and thermal behaviour of metals can be correlated by assuming that free electrons exist in thermal equilibrium with the atoms of the metal. Lorentz (1905) gave a logical form to this idea by assuming that the electrons in metals obey MB statistics. If this is so then the free electrons should contribute an amount $\frac{3}{2}k$ per free (valence) electron to the heat capacity, in addition to the contribution from the atomic vibrations. At room temperature, the latter leads to Dulong and Petit's law which gives $3R$ per gram-atom contribution to the heat capacity. For monovalent metals, therefore, the heat capacity should be the sum of two contributions, that is, $9R/2$ per gram-atom. But metals obey Dulong and Petit's law quite accurately. This difficulty is removed if the free electrons in metals are described by the FD statistics, (7.32).

We see from Fig. 7.1 that the FD distribution depends only slightly on temperature. As the temperature is raised from 0 K to $T$, each free electron does not gain energy by an amount $kT$ because most of them are occupying states of energy less than $\mu_0$. By the Pauli principle they cannot be excited to these states as they are already fully occupied. It is only a small fraction of electrons with energy close to $\mu_0$ that can be excited to empty states lying in the range $kT$ about $\mu_0$. This number of excited electrons $N_{exc}$ is given by

$$\begin{aligned} N_{exc} &\simeq g(\mu_0)kT \\ &= \frac{3N}{2\mu_0} kT = \frac{3}{2} N \frac{T}{T_F} . \end{aligned} \tag{7.34}$$

Thus only a small fraction ($\sim 3T/2T_F$) of the conduction electrons are excited. For $T_F \sim 10^4$ K, $T = 300$ K, only a few per cent are excited.

The electronic energy is given by

$$U(T) \simeq N_{exc}\, kT = \frac{3}{2}\, Nk\, \frac{T^2}{T_F}, \tag{7.35}$$

and the electronic heat capacity is

$$C_V(T) = \partial U/\partial T = 3Nk \frac{T}{T_F}, \tag{7.36}$$

which is close to the correct result (7.32).

Thus at room temperature the electronic heat capacity per electron, $3k\,(T/T_F)$, calculated according to FD statistics is very small compared to the atomic specific heat of about $3k$ per atom, in agreement with experiments.

At very low temperatures the lattice heat capacity (due to phonons) is proportional to $T^3$, (6.60), while the electronic heat capacity varies only linearly with $T$. For very low temperatures, the former decreases very rapidly and the latter begins to dominate. This is again in agreement with experiments.

## 7.4 THERMIONIC EMISSION

Electrons with sufficiently large kinetic energy within a conductor can escape from the surface. This is called *thermionic emission.* For interatomic distances of the order of Å, and one conduction electron per atom, (7.13) gives $T_F/T \sim 10$ at room temperature. Thus the electron gas within a metal is highly degenerate.

The *work function* $\phi$ of the metal is defined by

$$\phi = \mu_B - \mu(T), \tag{7.37}$$

where $\mu_B$ is the binding energy (energy required to remove to infinity a zero-energy electron from the metal). The free-electron model for a metal is shown in Fig. 7.6, where $\phi$ is the energy required to remove an electron of energy $\mu(T)$.

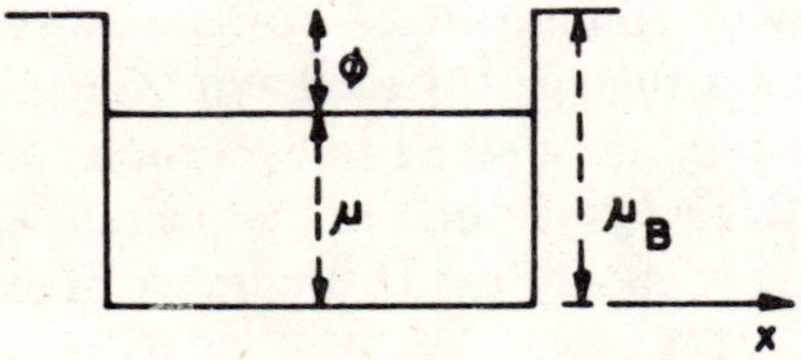


**Fig. 7.6 Free electron model of a metal.**

Let the $x$-axis be normal to the surface of the metal. Let $p_x$ be the momentum of an electron in the $x$ direction. It can leave the metal if

$$p_x^2/2m > \mu_B.$$

If $j$ is the electron current per square centimeter in the $x$ direction, and if $v_x$ is its velocity in that direction, then

$$j = e\frac{g_s}{h^3}\int_{-\infty}^{\infty} dp_y \int_{-\infty}^{\infty} dp_z \int_{\sqrt{2m\mu_B}}^{\infty} v_x f(\epsilon)\,dp_x$$

$$= \frac{2e}{h^3}\int_{-\infty}^{\infty}\int_{-\infty}^{\infty}\int_{B}^{\infty} \frac{dp_y\,dp_z\,d\epsilon}{\exp\,[(\epsilon-\mu)/kT]+1}$$

$$= \frac{2ekT}{h^3} \int_{-\infty}^{\infty}\int_{-\infty}^{\infty} \ln[1 + e^{-\theta}]\, dp_y dp_z, \tag{7.38}$$

where $E = \phi + \mu(T) + (p_y^2 + p_z^2)/2m$ and

$$\theta = [\phi + (p_y^2 + p_z^2)/2m]/kT.$$

At ordinary temperatures $\theta \gg 1$. Put $\ln(x + 1) \simeq x$, if $x$ is small. Thus

$$j = \frac{2ekT}{h^3} \exp(-\phi/kT) \iint \exp[-(p_y^2 + p_z^2)/2mkT]\, dp_y dp_z$$

$$= AT^2 \exp(-\phi/kT), \tag{7.39}$$

where $A = 4\pi mek^2/h^3 = 120$ amp cm$^{-2}$ deg$^{-2}$. This is the *Richardson-Dushman equation*. In general, $\phi$ is of the order of a few electron-volts.

## 7.5 MAGNETIC SUSCEPTIBILITY OF FREE ELECTRONS

The conduction electrons are found to possess a small temperature independent paramagnetic volume susceptibility. It is of the order of $10^{-6}$, in disagreement with the Langevin value $10^{-4}$ at room temperature which varies as $1/T$. Pauli* showed that the use of FD statistics can correct the theory as required.

Each conduction electron has a magnetic moment whose component in the direction of magnetic field is $\pm\, \mu_H$. If there are $n$ conduction electrons per unit volume, the net magnetization is

$$M = \mu_H(n_+ - n_-), \quad n = n_+ + n_-, \tag{7.40}$$

where $n_+$ is the number of electrons parallel to $H$ and $n_-$ the number antiparallel. Regarding the electrons as a fermion gas,

$$n_\pm = \int_0^\infty \tfrac{1}{2}\, g(\epsilon \pm \mu_H H)\, f(\epsilon)\, d\epsilon, \tag{7.41}$$

where $\epsilon$ is the total energy, kinetic plus magnetic, of an electron. The factor $\frac{1}{2}$ is included because the magnetic field $H$ has removed the spin degeneracy. We can write

$$g(\epsilon \pm \mu H) = \frac{2\pi V}{h^3} (2m)^{3/2} \epsilon^{1/2} \left(1 \pm \frac{\mu_H H}{\epsilon}\right)^{1/2}$$

$$\simeq \frac{2\pi V}{h^3} (2m)^{3/2} \epsilon^{1/2} \left(1 \pm \frac{\mu_H H}{2\epsilon}\right). \tag{7.42}$$

$$M = \frac{\mu}{V} \int_0^\infty [\tfrac{1}{2} g(\epsilon + \mu_H H) - \tfrac{1}{2} g(\epsilon - \mu_H H)]\, f(\epsilon)\, d\epsilon$$

$$= \frac{4\pi}{h^3} (2m)^{3/2} \frac{\mu_H^2 H}{2} \int_0^\infty f(\epsilon)\, \epsilon^{-1/2}\, d\epsilon. \tag{7.43}$$

---

*W. Pauli, *Z. Physik* **41**, 81 (1927).

Using (7.25), $\int f(\epsilon)\, F'(\epsilon)\, d\epsilon = F(\mu) + \frac{\pi^2}{6}(kT)^2 F''(\mu)$,

$$M = \frac{4\pi}{h^3}(2m)^{3/2}\mu_H^2 H\left[\mu^{1/2} - \frac{\pi^2}{24}\mu^{-3/2} + \ldots\right]. \tag{7.44}$$

In view of (7.11, 27), at low temperatures,

$$M = \frac{3n\mu_H^2 H}{2\mu_0^{3/2}}\left(\mu^{1/2} - \frac{\pi^2}{24}\mu^{-3/2} + \ldots\right)$$

$$\simeq \frac{3n\mu_H^2 H}{2\mu_0}. \tag{7.45}$$

Therefore, the Pauli result for the susceptibility $\chi$ is

$$\chi = \frac{M}{H} = \frac{3n}{2kT_F}\mu_H^2, \tag{7.46}$$

which is independent of temperature.

The physical reason for this result is again that only a small fraction (7.34) of electrons at the top of the Fermi distribution (Fig. 7.2) has a chance to turn over in the field $H$ and so to contribute to $\chi$. The Langevin result is that the probability that an atom will be lined up parallel to the field exceeds the probability of the antiparallel alignment by a factor $\sim \mu H/kT$. For $n$ atoms this gives a net magnetic moment $\sim n\mu^2 H/kT$. This classical result should be multiplied by $3T/2T_F$ to obtain the correct contribution. Hence $\chi \simeq (n\mu_H^2 H/kT)(3T/2T_F) = 3n\mu^2 H/kT_F$, in agreement with (7.46).

## 7.6 WHITE DWARFS

The white dwarf stars are much fainter, possess smaller diameter, and are very dense, compared to other stars of the same mass. An idealized white dwarf is a mass of helium ($\sim 10^{33}$g $\sim$ mass of sun) with density $\sim 10^7$ g cm$^{-3}$ $\sim 10^7 \times$density of sun, and temperature $\sim 10^7$ K $\sim$temperature of sun. At this temperature the helium atoms get completely ionized. The gas of electrons behaves as an ideal FD gas of density $\sim 10^{30}$ electrons/cc. This gives

$$\mu_0 = \frac{h^2}{2m}\left(\frac{3N}{8\pi V}\right)^{2/3} \sim 20 \text{ MeV}, \; T_F \sim 10^{11} \text{ K}. \tag{7.47}$$

Since $T_F/T \gg 1$, the electron gas is a highly degenerate FD gas. We can regard it as an ideal FD gas at $T = 0$ (that is, in the ground state).

The effect of high electron density is to yield $T_F \gg T$ and also to make the electrons attain relativistic energies due to increase in the mean energy.

We first calculate the pressure exerted by a FD gas of relativistic electrons in the ground state. The single particle enegy levels are given by

$$\epsilon_{ps} = (p^2c^2 + m^2c^4)^{1/2}, \tag{7.48}$$

where the states for a single electron are specified by the momentum **p** and the spin $s$. The ground state energy of the FD gas is

$$E_0 = g_s \sum_{|\mathbf{p}| < p_F} (p^2c^2 + m^2c^4)^{1/2} = \frac{2V}{h^3} \int_0^{p_F} (p^2c^2 + m^2c^4)^{1/2}\, 4\pi p^2 dp, \tag{7.49}$$

where $p_F = h(3N/8\pi V)^{1/3}$, (7.11). Putting $p/mc = x$,

$$E_0 = \frac{8\pi m^4c^5}{h^3} Vf(x_F), \tag{7.50}$$

where $x_F \equiv p_F/mc$ and

$$f(x_F) = \int_0^{x_F} (1 + x^2)^{1/2}\, x^2 dx = \begin{cases} \frac{1}{3} x_F^3 \left(1 + \frac{3}{10} x_F^2 + \ldots\right), & x_F \ll 1, \\ \frac{1}{4} x_F^4 \left(1 + x_F^{-2} + \ldots\right), & x_F \gg 1. \end{cases} \tag{7.51}$$

The $x_F \ll 1$ corresponds to the non-relativistic case, and $x_F \gg 1$ to the relativistic case. The pressure exerted by the FD gas is

$$P_0 = -\frac{\partial E_0}{\partial V} = -\frac{8\pi m^4c^5}{h^3}\left[f(x_F) + V\frac{\partial f(x_F)}{\partial x_F}\frac{\partial x_F}{\partial V}\right]$$

$$= \frac{8\pi m^4c^5}{h^3}\left[\frac{1}{3} x_F^3 ( + x_F^2)^{1/2} - f(x_F)\right]$$

$$\approx \begin{cases} [8\pi m^4c^5/(15h^3)]\, x_F^5, & x_F \ll 1, \\ [2\pi m^4c^5/(3h^3)]\, (x_F^4 - x_F^2), & x_F \gg 1. \end{cases} \tag{7.52}$$

We can now use these results to discuss white dwarfs.

The mass of the star, $M$, and its radius $R$ can be written as

$$M = (m + 2m_P)\, N \approx 2m_P N, \quad R = (3V/4\pi)^{1/3}, \tag{7.53}$$

where $m_P$ ($\gg$ mass of electron $m$) is the proton mass. Then

$$V/N = (8\pi/3)\, (m_P M/R^3), \tag{7.54}$$

$$x_F = \frac{h}{2\pi mc}\cdot\frac{1}{R}\cdot\left(\frac{9\pi}{8}\frac{M}{m_P}\right)^{1/3} \equiv \frac{\bar{M}^{1/3}}{\bar{R}}, \tag{7.55}$$

where $\bar{M} = (9\pi/8)\,(M/m_P)$ and $\bar{R} = R/(h/2\pi mc)$.

The enormous zero-point pressure $P_0$ exerted by the electron gas is balanced off by the gravitational attraction that binds the star. The work done to go from the state of infinite diluteness to a state of finite density to form the star of radius $R$ is

$$-\int_\infty^R P_0 4\pi r^2 dr,$$

where $P_0$ is the pressure of a uniform FD gas. We now introduce the gravitational interaction. The gravitational self-energy, on dimensional considerations, has the form $-aB_GM^2/R$, where $B_G$ is the gravitational constant and $a$ ($\sim 1$) is a number determined by the functional form of density. Differentiation of

$$\int_{\infty}^{R} P_0 4\pi r^2 dr = -aB_G \frac{M^2}{R}, \tag{7.56}$$

with respect to $R$, gives the equilibrium condition

$$P_0 = \frac{aB_G}{4\pi} \cdot \frac{M^2}{R^4} = \frac{a}{4\pi} B_G \left(\frac{8m_P}{9\pi}\right)\left(\frac{2\pi mc}{h}\right)^4 \frac{\bar{M}^2}{\bar{R}^2}. \tag{7.57}$$

We now relate this result with (7.52) expressed in terms of $\bar{M}$ and $\bar{R}$. For $x_F \ll 1$,

$$\frac{4}{5} K \bar{M}^{5/3} / \bar{R}^{5} = K' \bar{M}^2/\bar{R}^2, \quad \textit{(non-relativistic)}, \tag{7.58}$$

where

$$K = \frac{mc^2}{12\pi^2}\left(\frac{2\pi mc}{h}\right)^3, \; K' = \frac{a}{4\pi} B_G \frac{8m_P}{9\pi}\left(\frac{2\pi mc}{h}\right)^4. \tag{7.59}$$

Thus the radius of the star decreases when the mass increases,

$$\bar{M}^{1/3} \bar{R} = 4K/(5K'). \tag{7.60}$$

For $x_F \gg 1$,

$$K\left(\frac{\bar{M}^{4/3}}{\bar{R}^4} - \frac{\bar{M}^{2/3}}{\bar{R}^2}\right) = K' \frac{\bar{M}^2}{\bar{R}^2}, \quad \textit{(extreme relativistic)}, \tag{7.61}$$

or

$$\bar{R} = \bar{M}^{2/3} \, [1 - (\bar{M}/\bar{M}_0)^{2/3}]^{1/2}, \tag{7.62}$$

where

$$\bar{M}_0 = \left(\frac{K}{K'}\right)^{3/2} = \left(\frac{27\pi}{64a}\right)^{3/2} \left(\frac{ch}{2\pi B_G m_P^2}\right)^{3/2}. \tag{7.63}$$

Thus no white dwarf star can have a mass larger than $M_0 = (9/8\pi) m_P \bar{M}_0 \approx 10^{33}$ g $\approx$ mass of the sun, with $a \approx 1$. Refined estimates give $M_0 = 1.4 \times$ mass of the sun. This mass is known as the *Chandrasekhar limit.**

## 7.7 NUCLEAR MATTER

Qualitatively, the nucleons (neutrons and protons) inside a nucleus form a degenerate FD gas. The radius of a nucleus containing $A$ number of nucleons is given by the empirical formula $R \cong aA^{1/3}$, $a = 1.3 \times 10^{-13}$ cm. Therefore, the concentration of nucleons in nuclear matter is

---

*S. Chandrasekhar, *Stellar Structure*, Dover, N.Y., 1957.

$$n \cong \frac{A}{\frac{4}{3}\pi a^3 A} \cong 1.1\times 10^{38}\ \text{cm}^{-3}. \tag{7.64}$$

It is about $10^8$ times higher than the concentration of nucleons in a white dwarf star. From (7.11), the Fermi energy is given by a relation of the form

$$\epsilon_F(0) = \frac{h^2}{2m}\left(\frac{3n}{8\pi}\right)^{2/3}, \tag{7.65}$$

for each type of particle (neutron or proton are not identical particles). For the simple case,

$$n_{\text{neutrons}} \approx n_{\text{protons}} \approx 0.5\times 10^{38}\ \text{cm}^{-3}, \tag{7.66}$$

we get $\epsilon_F(0) \approx 27$ MeV, or $T_F = \epsilon_F/k \approx 3\times 10^{11}$ K, We compare this value with other typical values in Table 7.1.

**Table 7.1. The characteristic values of Fermi temperature for degenerate FD gases**

| Phase of matter | Particles | $T_F$ (K) |
|---|---|---|
| Liquid $^3$He | atoms | 0.3 |
| Metal | electrons | $5\times 10^4$ |
| White dwarf stars | electrons | $3\times 10^9$ |
| Nuclear matter | nucleons | $3\times 10^{11}$ |
| Neutron stars | neutrons | $3\times 10^{12}$ |

## *PROBLEMS*

7.1 Show that the density of states of a free electron gas in two-dimensions is independent of energy, $g(\epsilon) = 4m\pi/h^2$, per unit area of specimen.

7.2 Show that for a FD gas in two dimensions $\mu(T) = kT\ \ln\ [\exp(nh^2/4\pi mkT) - 1]$ for $n$ electrons per unit area.

7.3 Show that the initial curvature of $\mu$ versus $T$ is upward for a FD-gas in one dimension and downward in three dimensions.

7.4 The $^3$He atom is a fermion as its spin is half. Calculate $v_F$, $\epsilon_F$ and $T_F$ for liquid $^3$He treated as a FD-gas at $T = 0$. The density is 0.081 g cm$^{-3}$ for liquid $^3$He. Calculate $C_V$ at $T \ll T_F$ and compare with the value $C_V = 2.89\ NkT$ observed by A. C. Anderson, W. Reese, and J. C. Wheatley (Phys. Rev. **130**, 495, 1963) for $T < 0.1$ K.

7.5 The $^4$He atoms are bosons and $^3$He atoms are fermions. Let liquid $^3$He (lighter) be in equilibrium with a $^3$He—$^4$He mixture (heavier) with more than 6 pct $^3$He. If $^4$He is added to the mixture, the floating $^3$He evaporates from pure $^3$He liquid into the $^4$He rich phase to restore concentration and thereby absorbs heat. Construct a *helium dilution refrigerator* on this fact (see, for example, D.S. Betts, *Contemporary Physics* **9**, 97, 1968; O.V. Lounasmaa, *Scientific American* **221**, 26, 1969).

7.6 Show that the electron gas in copper is degenerate at room temperature (given: concentration $= 8.5 \times 10^{28}$ m$^{-3}$).

7.7 Calculate the pressure of the electrons in a gas discharge at $T = 2000$ K and concentration $10^{18}$ m$^{-3}$.

7.8 Use (7.11, 32, 46) to calculate the electronic specific heat and spin paramagnetic susceptibility of Lithium ($\rho = 0.534$ g cm$^{-3}$)
[Ans: $C_V = 0.26 \times 10^{-4}\, T$ cal/g. deg, $\chi = 1.48 \times 10^{-6}$ (cgs) (emu)/g.]

7.9 Show that for a relativistic completely degenerate electron gas, with $\epsilon = c\,(p^2 + m^2c^2)^{1/2} = mc^2 \cosh \theta$, $\mu = mc^2 \cosh \theta_0$ and $p_F = mc \sinh \theta_0$, we have

$$N/V = \frac{8\pi}{3}\left(\frac{mc}{h}\right)^3 \sinh^3 \theta_0 = \frac{8\pi}{3}\left(\frac{p_F}{h}\right)^3,$$

$$E/V = \frac{\pi}{4}\frac{m^4\, c^5}{h^3}\,[\sinh\,(4\theta_0) - 4\theta_0],$$

$$P = \frac{\pi m^4\, c^5}{h^3}\,[\tfrac{1}{4}\sinh\,(4\theta_0) - 2\sinh\,(2\theta_0) + 3\theta_0].$$

7.10 Consider the nonrelativistic limit in Prob. 7.9 for completely degenerate case.

7.11 Calculate $\mu$ and $C_V$ for an extreme relativistic ($\epsilon = cp$), highly degenerate, ideal electron gas.

7.12 Pulsars are stars consisting of a cold degenerate gas of neutrons. Show that for a neutron star $M^{1/3}\, R \approx 10^{18}$ g$^{1/3}$ cm.

# 8
# SEMICONDUCTOR STATISTICS

## 8.1 STATISTICAL EQUILIBRIUM OF FREE ELECTRONS IN SEMICONDUCTORS

A single electron in a semiconductor is described by a plane wave with wave vector **k**. It is modulated by the periodic field of the lattice. According to quantum mechanical calculations, the corresponding energy levels are grouped into two bands with an energy gap $\epsilon_g$ between them (Fig. 8.1). One can consider regions in space **x** and in **k** space so that Heisenberg's uncertainty relation $\Delta x\ \Delta k \gtrsim 1$ holds. Then one can simultaneously plot the energy

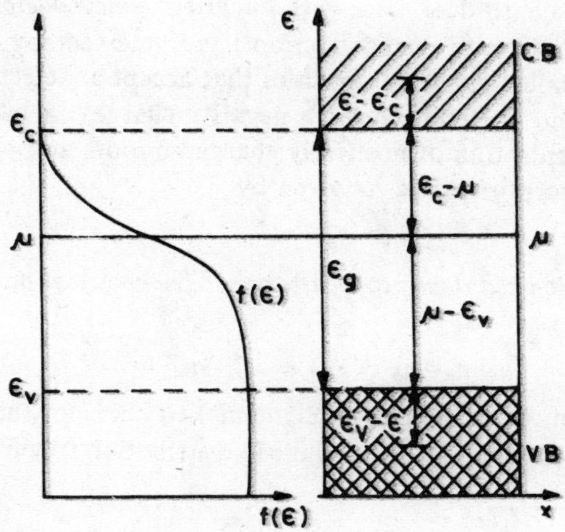


**Fig. 8.1** Energy scale for statistical calculations. The Fermi distribution function is shown for a finite temperature. The conduction band (CB) and the valence band (VB) have their edges at $\epsilon_c$ and $\epsilon_v$, respectively. The Fermi level is marked $\mu$ and is taken to lie well within the energy gap $\epsilon_g$.

levels as a function of spacial coordinate $x$. The ordinate represents the energy of a particle, the abscissa its space coordinate (Fig. 8.1). The upper band is called the *conduction band* and the lower one the *valence band*. In a pure or intrinsic semiconductor (semiconductor without impurities) at $T=0$ all the quantum states are occupied in the valence band and unoccupied in the conduction band. We can write

$$\epsilon_g = \epsilon_c - \epsilon_v, \tag{8.1}$$

where $\epsilon_c$ is the energy of the bottom of the conduction band (conduction band edge) and $\epsilon_v$ of the top of the valence band (valence band edge).

A fully occupied band cannot carry any current. Therefore, at $T = 0$ a pure semiconductor is an *insulator*. For the usual semiconductors $\epsilon_g$ is between 0.1 and 2.5 eV. Because $kT \simeq 1/40$ eV at room temperature ($\sim 300$ K), usually $\epsilon_g \gg kT$. As the temperature is raised, the electrons are thermally excited from the valence band to the conduction band. This creates *conduction electrons* in the conduction band and unoccupied quantum states in the valence band, called *holes*. The conduction electrons and holes can also be created by the presence of *impurities* that change the balance between the number of single-particle quantum states in the valence band and the number of electrons available to occupy them. The conductivity results from the motion of electrons in the conduction band and of holes in the valence band.

If $n_e$ is the concentration of conduction electrons and $n_h$ of holes, then in a pure semiconductor, which is electrically neutral,

$$n_e = n_h. \tag{8.2}$$

In practice, we have to deal more with *impurity semiconductors*. Impurities that supply an electron to the conduction band, and thereby acquire a positive charge, are called *donors*. Impurities that accept an electron from the valence band, and thereby acquire a negative charge, are called *acceptors*. If $n_d^+$ is the concentration of positively charged donors and $n_a^-$ of the negatively charged acceptors, then $\Delta n$ given by

$$\Delta n \equiv n_d^+ - n_a^- \tag{8.3}$$

is called the net *ionized donor concentration*. The electrical neutrality condition can be expressed as

$$n_e - n_h = \Delta n = n_d^+ - n_a^-. \tag{8.4}$$

In the state of statistical (thermodynamic) equilibrium the electron concentration can be calculated from the FD distribution function

$$f_e(\epsilon) = \frac{1}{\exp[(\epsilon-\mu)/kT] + 1}, \tag{8.5}$$

where $\mu = \mu(T)$ is the chemical potential of the electrons (subscript $e$ in $f_e$ refers to electrons). At $T = 0$ K, $\mu(0) \equiv \mu_0 =$ *Fermi energy* $\epsilon_F(T = 0)$. The electron chemical potential $\mu(T)$ is also called *Fermi level*.

For known $\mu$ and $T$, the number of conduction electrons $N_e$ is given by

$$N_e = \sum_{\text{CB}} f_e(\epsilon), \tag{8.6}$$

where the sum is over all the conduction band states. The number of holes is

$$N_h = \sum_{\text{VB}} [1 - f_e(\epsilon)] = \sum_{\text{CB}} f_h(\epsilon), \tag{8.7}$$

where the sum is over all the valence band states and

$$f_h(\epsilon) \equiv 1 - f_e(\epsilon) = \frac{1}{\exp[(\mu - \epsilon)/kT] + 1} \tag{8.8}$$

gives the probability that a quantum state at energy $\epsilon$ is unoccupied (or equivalently, occupied by a hole).

Thus the concentrations $n_e = N_e/V$ and $n_h = N_h/V$ in (8.4) depend on $\mu$.

## 8.2 NONDEGENERATE CASE

For $f_e \ll 1$ and $f_h \ll 1$, the FD distribution function reduces to the classical distribution (Fig. 7.1). This is possible when $\mu$ lies inside the energy gap (Fig. 8.1), and

$$\epsilon_c - \mu \gg kT, \quad \mu - \epsilon_v \gg kT. \tag{8.9}$$

When $\epsilon_c - \mu$ and $\mu - \epsilon_v$ are positive and at least a few times greater than $kT$, the semiconductor is said to be *nondegenerate*. Then

$$f_e(\epsilon) \simeq \exp[-(\epsilon - \mu)/kT], \quad f_h \simeq \exp[-(\mu - \epsilon)/kT], \tag{8.10}$$

and the density of states is (Fig. 8.1)

$$\begin{aligned} g_e(\epsilon) &= 4\pi V (2m_e/h^2)^{3/2} (\epsilon - \epsilon_c)^{1/2} & \epsilon > \epsilon_c \\ g(\epsilon) &= 0 & \epsilon_v < \epsilon < \epsilon_c \\ g_h(\epsilon) &= 4\pi V (2m_h/h^2)^{3/2} (\epsilon_v - \epsilon)^{1/2} & \epsilon < \epsilon_v. \end{aligned} \tag{8.11}$$

The conduction electron concentration $n_e \equiv N_e/V$ is

$$\begin{aligned} n_e \equiv N_e/V &= \frac{1}{V} \int_{\epsilon_c}^{\infty} f_e(\epsilon)\, g_e(\epsilon)\, d\epsilon \\ &= 4\pi \left(\frac{2m_e}{h^2}\right)^{3/2} \exp(\mu/kT) \int_{\epsilon_c}^{\infty} \exp(-\epsilon/kT)(\epsilon - \epsilon_c)^{1/2}\, d\epsilon \\ &= n_c \exp[(\mu - \epsilon_c)/kT], \end{aligned} \tag{8.12}$$

where

$$n_c = 2\left(\frac{2\pi mkT}{h^2}\right)^{3/2} = \frac{2}{\lambda^3}. \tag{8.13}$$

The *quantum concentration* $n_c$ for conduction electrons for actual semiconductors varies as $T^{3/2}$, as in (8.13), but differs in magnitude by a proportionality factor. A formal way to accommodate this is to replace $m$ by an *effective mass* $m^*$ in (8.13). Thus

$\epsilon_{ionization} \quad 2\epsilon_s^2 \hbar^2 \quad (m\epsilon_s^2)$

where $\epsilon_s$ is the dielectric constant of Si. Using $\epsilon_s = 12$ for Si, and $m^*/m \sim 0.1$, we find $\epsilon_{ionization} \sim 0.01$ eV. Thus the energy level of the extra electron is just below the conduction band edge (Fig. 8.4). It is a *donor level* because the thermal energy can raise this extra electron to the conduction

$$n_c = 2\,(2\pi m_e^* kT/h^2)^{3/2} = 4.82\times 10^{15}\,(m_e^*/m)^{3/2}\,T^{3/2}. \qquad (8.14)$$

The $n_c$ represents the effective number of states per unit volume (*effective*



band. The substance is called an *n-type* impurity semiconductor because we have added an electron without creating a hole in the valence band.

If Ga atom (three valence electrons) is introduced as an impurity in Si lattice, we have one too few electrons to form the four covalent bonds with the neighbouring Si atoms. Consequently, a valence band electron can be easily thermally excited to fill this 'lack of an electron' level just above the valence band edge (Fig. 8.4). This will create a hole in the valence band. The energy level which now accepts the electron is called an *acceptor level* and the substance is a *p-type* impurity semiconductor.

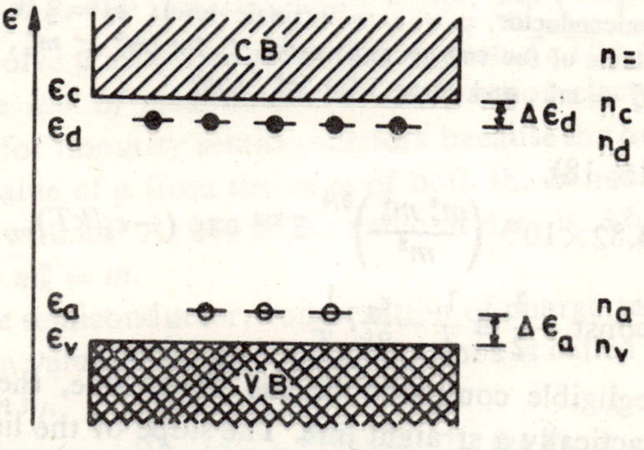


**Fig. 8.4 The donor (—●—) and acceptor (—o—) impurity levels located in the energy gap of a semiconductor.**

Let us assume that (i) each donor atom contributes one electron which can enter the conduction band or fill one hole in the valence band; (ii) each acceptor atom removes one electron either from the conduction band or from the valence band. This is called the *approximation of fully ionized impurities.* All impurities when ionized are either positively charged donors $D^+$ or negatively charged acceptors $A^-$.

From (8.4, 18), $\Delta n = n_e - n_h = n_d^+ - n_a^-$, $n_h = n_i^2/n_e$ and so

$$n_e^2 - n_e\,\Delta n = n_i^2. \qquad (8.22)$$

Solving this and remembering that $n_e > 0$.

$$n_e = +\frac{\Delta n}{2}\left\{\left[1+\frac{4n_i^2}{(\Delta n)^2}\right]^{1/2}+1\right\}, \qquad (8.23)$$

$$n_h = n_e - \Delta n = \frac{\Delta n}{2}\left\{\left[1+\frac{4n_i^2}{(\Delta n)^2}\right]^{1/2}-1\right\}. \qquad (8.24)$$

Usually the doping concentration is large compared to the intrinsic concentration. It means, either $n_e$ or $n_h$ is much large than $n_i$,

$$|\Delta_n| \gg n_i. \qquad (8.25)$$

This inequality defines an *extrinsic semiconductor*, and we can write

$$(\Delta n)\left[1+\frac{4n_i^2}{(\Delta n)^2}\right]^{1/2} \simeq |\Delta n| + 2n_i^2/|\Delta n|. \qquad (8{\cdot}26)$$

An example of the extrinsic semiconductor of the donor (or $n-$) type is $n_a = 0$ with a wide energy gap $\epsilon_g \gg \Delta\epsilon_d$ (Fig. 8.4).

In an $n$-type semiconductor $\Delta n > 0$ and so from (8.23, 24, 26)

$$n_e \simeq \Delta n + (n_i^2/\Delta n) \simeq \Delta n, \quad n_h \simeq n_i^2/\Delta n \ll n_i. \tag{8.27}$$

In a $p$-type semiconductor $\Delta n < 0$ and so

$$n_e \simeq n_i^2/|\Delta n| \ll n_i, \quad n_h \simeq |\Delta n| + (n_i^2/|\Delta n|) \simeq |\Delta n|. \tag{8.28}$$

Solving (8.12) or (8.15) for $\mu$,

$$\mu = \epsilon_c - kT \ln (n_c/n_e) = \epsilon_v + kT \ln (n_v/n_h). \tag{8.29}$$

Use of (8.23, 24) now gives the Fermi level in extrinsic semiconductor as a function of $T$ and doping concentration $\Delta n$ ($|\Delta n| \sim 10^{12}$ cm$^{-3}$). As an extrinsic semiconductor is cooled, the Fermi level approaches the conduction band edge for $n$-type ($\Delta n > 0$) and the valence band edge for $p$-type ($\Delta n < 0$), Fig. 8.5.

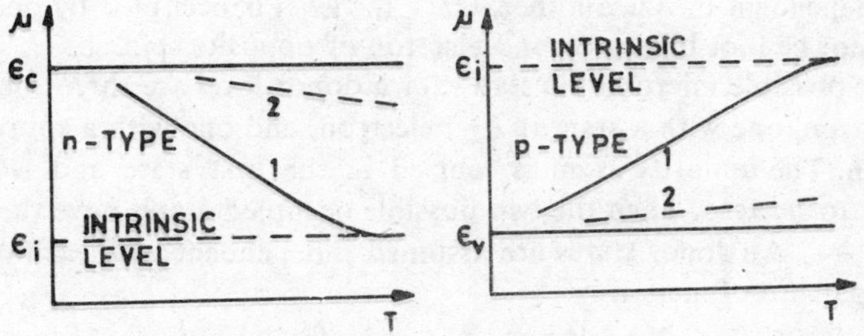


Fig. 8.5 The temperature dependence of the Fermi level in a semiconductor which is doped by a donor or an acceptor. The curves marked 1 are for $|\Delta n|_1$ and 2 for $|\Delta n|_2 > |\Delta n|_1$.

## 8.4 DEGENERATE SEMICONDUCTORS

For large carrier concentrations, approaching the quantum concentration, we cannot use the approximation (8.10) for that carrier. The method of Ch. 7 must now be used. Thus, (8.12) becomes

$$n_e = \frac{N}{V} = \int d\epsilon\, g(\epsilon) f_e(\epsilon)$$

$$= \frac{1}{2\pi^2}\left(\frac{8\pi^2 m_e^*}{h^2}\right)^{3/2} \int_{\epsilon_c}^{\infty} \frac{(\epsilon-\epsilon_c)^{1/2}\, d\epsilon}{1 + \exp[(\epsilon-\mu)/kT]}$$

$$= n_c \left[\frac{2}{\sqrt{\pi}} \int_0^{\infty} \frac{x^{1/2}\, dx}{\frac{1}{\eta_a} e^x + 1}\right] = n_c\, G_{3/2}(\eta_a), \tag{8.30}$$

where $n_c$ is given by (8.13), $x = (\epsilon - \epsilon_c)/kT$, $\eta_a = \exp[(\mu - \epsilon_c)/kT]$, and $G_{3/2}(\eta_a)$ is the FD-integral of (7.8).

When $\epsilon_c - \mu \gg kT$,

$$n_e/n_c \simeq \frac{2}{\sqrt{\pi}} \eta_a \int_0^\infty e^{-x} x^{1/2} \, dx = \frac{2}{\sqrt{\pi}} \Gamma(3/2) \eta_a = \exp[(\mu - \epsilon_c)/kT]. \tag{8.31}$$

This is the known result for the ideal gas.

The value of $\mu$ given by (8.30) can be written approximately as*

$$\ln \eta_a \simeq \ln r + (1/\sqrt{8}) r + \ldots, \quad r = n_e/n_c, \tag{8.32}$$

which is good for values of $r$ upto about 10. In semiconductors this is the usual range of values of $r$.

## 8.5 OCCUPATION OF DONOR LEVELS

We assume that one and only one electron with either spin up or spin down can occupy a donor level. We then have two different quantum states with the same energy. However, the occupations of two such quantum states are not independent of one another. Once the level is occupied by one electron, the donor cannot bind a second electron of opposite spin.

The possible microscopic states for a donor level are three: one without an electron, one with a spin up ($\uparrow$) electron, and one with a spin down ($\downarrow$) electron. The impurity atom is ionized in the first state and we take its energy to be zero. Then the two possible occupied states have the common energy $-\overset{0}{\epsilon}_d$. All donor states are assumed independent with each other. The grand partition function is

$$\mathscr{Z} = 1 + 2 \exp[(\mu + \overset{0}{\epsilon}_d)/kT]. \tag{8.33}$$

The probability $P(N, \epsilon)$ that impurity atom is ionized ($N = 0$) is

$$P(0, 0) = \frac{1}{\mathscr{Z}} = \frac{1}{1 + 2\exp[(\mu + \overset{0}{\epsilon}_d)/kT]}, \tag{8.34}$$

and that it is neutral (unionized) is

$$P(1\uparrow, -\overset{0}{\epsilon}_d) + P(1\downarrow, -\overset{0}{\epsilon}_d) = 1 - P(0, 0). \tag{8.35}$$

If we measure the energy of a singly occupied donor state relative to the origin of energy (Fig. 8.2), we must replace $-\overset{0}{\epsilon}_d$ by $\epsilon_d$ in (8.33, 34). From (8 34), the probability that the donor state is vacant (donor ionized), is then written as

$$f(D^+) = \frac{1}{1 + 2\exp[(\mu - \epsilon_d)/kT]}. \tag{8.36}$$

It gives the distribution for holes over donor states. From (8.35), the probability that the donor state is occupied by an electron (donor neutral, unionized), is

$$f(D) = \frac{1}{1 + \frac{1}{2}\exp[(\epsilon_d - \mu)/kT]}. \tag{8.37}$$

It gives the distribution of electrons over donor states.

---

* W.B. Joyce and R.W. Dixon, *Appl. Phys. Lett.* **31**, 354, (1977).

The case of acceptors is different. For an ionized acceptor $A^-$, the chemical bond with a neighbouring atom of the semiconductor involves a pair of electrons with antiparallel spins. Only one such state is possible. Therefore, $A^-$ contributes only one term, $\exp[(\mu-\epsilon_a)/kT]$, to $\mathscr{Z}$ for the acceptor. For neutral acceptor $A$, one electron is missing from the bond. This missing electron can have either spin up or spin down. Therefore, it contributes the term, 1, twice to the $\mathscr{Z}$. We can write the thermal average occupancy as

$$f(A^-) = \frac{\exp[(\mu-\epsilon_a)/kT]}{2+\exp[(\mu-\epsilon_a)/kT]} = \frac{1}{1+2\exp[(\epsilon_a-\mu)/kT]}. \tag{8.38}$$

It gives the distribution of electrons over acceptor states. For $A$ neutral (acceptor state unoccupied),

$$f(A) = \frac{2}{2+\exp[(\mu-\epsilon_a)/kT]} = \frac{1}{1+\frac{1}{2}\exp[(\mu-\epsilon_a)/kT]}. \tag{8.39}$$

It gives the distribution function for holes over acceptor states.

The concentrations of $D^+$ and $A^-$ are

$$n_d^+ = n_d f(D^+), \quad n_a^- = n_a f(A^-), \tag{8.40}$$

where $n_d$ and $n_a$ are concentrations of donor and acceptor impurities on doping. The neutrality condition (8.4), with $\Delta n = n_d^+ - n_a^-$, gives

$$n^- \equiv n_e + n_a^- = n_h + n_d^+ \equiv n^+. \tag{8.41}$$

It is useful to display these results on a logarithmic plot of $n^-$ and $n^+$ as function of $\mu$ (Fig. 8.6). The dashed lines represent $n_d^+$, $n_h$, $n_e$, $n_a^-$. The solid curves are for $n^+ = n_h + n_d^+$ = sum of all positive charges, and $n^- = n_e + n_a^-$ = sum of all negative charges. The actual Fermi level $\mu$ is given by the intersection of the $n^+$ and $n^-$ curves ($n^+ = n^-$).

The charge carriers are called *majority carriers* when their concentration exceeds that of intrinsic carriers $n_i$ at the given $T$. Thus, in donor-doped semiconductors the majority carriers are electrons. If either $n_h$ or $n_e$ is very small, the majority carrier concentration can be easily calculated. Suppose we have an $n$-type semiconductor with no acceptors. The neutrality point is then given by the intersection of the $n^+$ curve with the $n_e$ curve ($n^+ = n_e$). If the donor concentration is not too high, we can use (8.12) for the straight part of the $n_e$ curve,

$$\exp(\mu/kT) = (n_e/n_c)\exp(\epsilon_c/kT), \tag{8.42}$$

$$\exp[(\mu-\epsilon_d)/kT] = (n_e/n_c)\exp[(\epsilon_c-\epsilon_d)/kT] = n_e/n_e^*. \tag{8.43}$$

Here

$$n_e^* \equiv n_c \exp(-\Delta\epsilon_d/kT), \quad \Delta\epsilon_d = \epsilon_c - \epsilon_d, \tag{8.44}$$

is the electron concentration for the case $\mu = \epsilon_d$, and $\Delta\epsilon_d$ is the donor ionization energy.

From (8.36, 40, 43), with $n_e = n_d^+$,

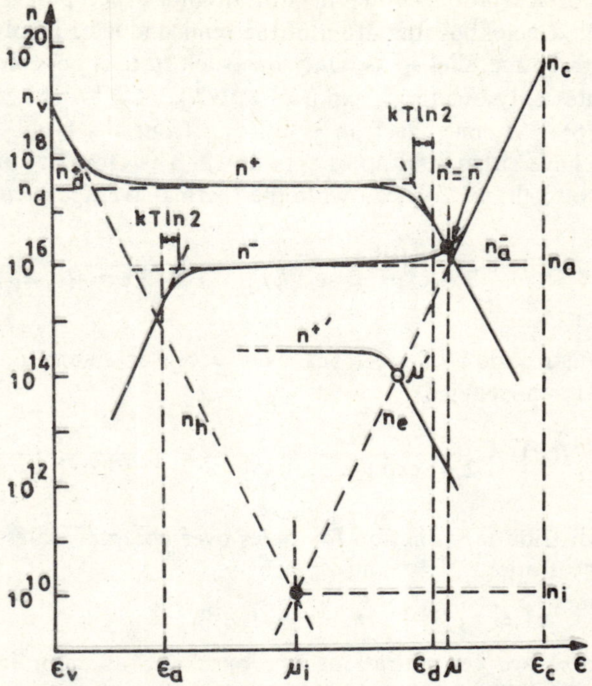


**Fig. 8.6 Determination of $\mu$ from the intersection of curves (solid) on a logarithmic plot for $n^+ = n_h + n_d^+$ and $n^- = n_e + n_a^-$ for an $n$-type semiconductor having both acceptors and donors. The dashed curves are for $n_d^+$, $n_h$, $n_e$, $n_a^-$. The value of $n_d/n_c \sim 10^{-2}$ and of $n_d'/n_c \sim 10^{-6}$. For $n_a = 0$, the actual Fermi level is at the intersection point of $n_e$ and $n^+$ (or $n^{+\prime}$) curves.**

$$n_e = \frac{n_d}{1 + 2\,(n_e/n_e^*)}, \tag{8.45}$$

$$n_e^2 + \frac{1}{2} n_e^* n_e - \frac{1}{2} n_d n_e^* = 0, \tag{8.46}$$

$$n_e = +\frac{1}{4} n_e^* \{[1 + (8n_d/n_e^*)]^{1/2} - 1\}$$

$$\simeq n_d - 2(n_d^2/n_e^*) = n_d\,[1 - 2(n_d/n_e)]. \tag{8.47}$$

This is a good approximation for the case of weak doping, $8n_d \ll n_e^*$. For complete ionization $n_e = n_d$. The term $-2n_d^2/n_e^*$ gives the first order departure from complete ionization.

## 8.6 ELECTROSTATIC PROPERTIES OF p-n JUNCTIONS

A *p-n* junction consists of a *p* and an *n* semiconductor in intimate contact (Fig. 8.7). On joining, the Fermi levels become equal because at equilibrium the chemical potential is the same for the two semiconductors. This equilibrium is achieved by a small transfer of electrons from *n*-type to *p*-type,

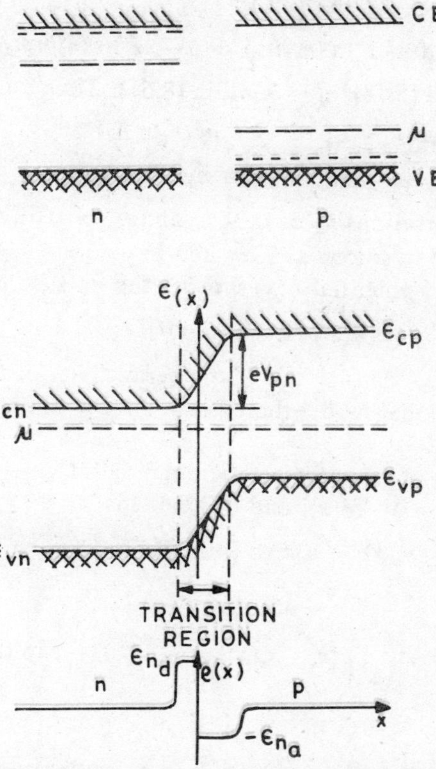


**Fig. 8.7** **A $p$ and an $n$ semiconductor when in intimate contact form a $p$-$n$ junction with the electrostatic potential $\epsilon(x)$ showing a transition region. The space charge $\rho(x)$ is shown at the bottom.**

where they recombine with free holes and produce a space charge (negative on the $p$ side and positive on the $n$ side) which inhibits further flow. This leaves a region with no free carriers near the junction (*depletion layer*). The ionized donors and acceptors in this layer make the $n$-type positively charged and the $p$-type negatively charged. The dipolar layer near the interface so formed produces a potential gradient in the transition region (Fig. 8.4). This built-in electrostatic potential step $V_{pn}$ exists even in the absence of an externally applied voltage. The potential step of height $eV_{pn}$ is needed to equalize the total chemical potential of two semiconductors when their intrinsic chemical potentials are unequal.

Let the two semiconductors be extrinsic and nondegenerate, that is,

$$n_i \ll n_d \ll n_c, \quad n_i \ll n_a \ll n_v. \tag{8.48}$$

For fully ionized donors on the $n$ side and fully ionized acceptors on the $p$ side,

$$n_e \simeq n_d, \quad n_h \simeq n_a, \tag{8.49}$$

where we have omitted the superscripts $\pm$ from $n_d$, $n_a$.

From (8.12), the conduction band energies on the $n$ and $p$ sides are

$$\epsilon_{cn} = \mu - kT \ln (n_d/n_c), \tag{8.50}$$

$$\epsilon_{cp} = \mu - kT \ln (n_{ep}/n_c) = \mu - kT \ln (n_i^2/n_a n_c), \tag{8.51}$$

where we have used (8.18), $n_i^2 = n_e n_h$, in (8.51). Thus

$$\begin{aligned} eV_{pn} &= \epsilon_{cp} - \epsilon_{cn} = -kT \ln (n_i^2/n_a n_d) \\ &= \epsilon_g - kT \ln (n_c n_v/n_a n_d). \end{aligned} \tag{8.52}$$

For doping concentrations $n_d \simeq 0.01\ n_c$ and $n_a \simeq 0.01\ n_v$, we get $eV_{np} \simeq \epsilon_g - 9.2\ kT \simeq 0.9\ eV$ in silicon at $T = 300$ K.

The electrostatic potential $\phi(x)$ satisfies the Poisson equation

$$d^2 \phi(x)/dx^2 = -\rho(x)/\mathcal{E} \tag{8.53}$$

where $\rho$ is the charge density and $\mathcal{E}$ the permittivity of the semiconductor. In this one-dimensional model there is an abrupt change from $n$ to $p$ at $x = 0$.

With the origin of the electrostatic potential at $x = -\infty$, we have $\phi(-x) = 0$, $\epsilon_c(x) = \epsilon_c(-\infty) - e\phi(x)$, and (8.12) reads

$$n_e(x) = n_d \exp [e\phi(x)/kT]. \tag{8.54}$$

From (8.53, 54),

$$\frac{d^2\phi}{dx^2} = -\frac{\rho}{\mathcal{E}} = -\left(\frac{e}{\mathcal{E}}\right) [n_d - n_e(x)] = -\frac{en_d}{\mathcal{E}} [1 - \exp (e\phi(x)/kT)]. \tag{8.55}$$

Rewrite it as

$$\left(2\frac{d\phi}{dx}\right)\frac{d^2\phi}{dx^2} \equiv \frac{d}{dx}\left(\frac{d\phi}{dx}\right)^2 = -\frac{2n_d}{\mathcal{E}}\frac{d}{dx} [e\phi - kT \exp (e\phi/kT)], \tag{8.56}$$

and now integrate with $\phi(-\infty) = 0$,

$$\left(\frac{d\phi}{dx}\right)^2 = -\frac{2n_d}{\mathcal{E}} [e\phi + kT - kT \exp (e\phi/kT)]. \tag{8.57}$$

At the interface, $x = 0$, let

$$-e\phi(0) = eV_n \gg kT, \tag{8.58}$$

where $V_n$ is the part of $V_{pn}$ on the $n$ side. Neglecting the exponential term in (8.57), the $x$ component of the electric field, $E = -d\phi/dx$, at the interface is

$$E = (2n_d/\mathcal{E})^{1/2} (eV_n - kT)^{1/2}. \tag{8.59}$$

If $V_p$ is the part of $V_{pn}$ on the $p$ side, we have similarly

$$E = (2n_a/\mathcal{E})^{1/2} (eV_p - kT)^{1/2}. \tag{8.60}$$

Matching these values of $E$ and using $V_{pn} = V_p + V_n$,

$$E = \left[\left(\frac{2e}{\mathcal{E}}\right)\frac{n_a\, n_d}{n_a + n_d}\left(V_{pn} - \frac{2kT}{e}\right)\right]^{1/2}. \tag{8.61}$$

The space charge is positive on the $n$ side (Fig. 8.4). The field $E$ is the same as if all the electrons on the $n$ side are depleted from $x = 0$ to a distance $w_n = \mathcal{E}E/en_d$. There is no depletion at $|x| > w_n$. The $w_n$ is the thickness of the space charge transition region on the $n$ side. On the $p$ side, $w_p = \mathcal{E}E/en_a$. Therefore, the total depletion width is

$$w = w_n + w_p = \frac{\mathcal{E}E}{e}\left(\frac{n_a + n_d}{n_a n_d}\right) \tag{8.62}$$

where $E$ is given by (8.61).

## *PROBLEMS*

8.1 In (8.30), $G_{3/2} = \frac{2}{\sqrt{\pi}} \int_0^\infty \frac{x^{1/2}\, ex}{\exp(x-\xi)+1}$ where $\xi = (\mu - \epsilon_c)/kT$. Show that the Fermi integral $G_{3/2}(\xi)$ can be approximated as

$$G_{3/2}(\xi) = \begin{cases} e\xi & -\infty < \xi < -1 \text{ nondegenerate,} \\ \dfrac{1}{(1/4) + e^{-\xi}} & -1 < \xi < 5, \\ \dfrac{4}{3\sqrt{4}}\,\xi^{3/2} & 5 < \xi < \infty \text{ completely degenerate} \end{cases}$$

8.2 If $n_e$ is not small compared to $n_c$, show that the law of mass action can be expressed as

$$n_e n_h = n_i^2 \exp\left[-n_e/\sqrt{8}\, n_c + \ldots\right].$$

8.3 Verify that the Boltzmann constant $k = 8.6167 \times 10^{-5}$ eV K$^{-1}$, and $1/k = 11605.4$ K.eV$^{-1}$, that is, 1 eV corresponds to 11605.4 K. Show that $(T, kT) = (1\text{K}, 8.6 \times 10^{-5}$ eV), (300 K, 0.0258 eV). Note that the energy of ionization of impurity atoms is of the order of hundredths of an eV.

8.4 Taking $\epsilon_g = 1.1$ eV and $m_e^*/m_h^* = 6$ in silicon, show that Fermi level in the intrinsic material is 0.55 eV at $T = 8$ K and 0.516 eV at $T = 300$ K. If $m_e^*/m = 0.8$, where $m$ is free-electron mass, show that the density of free electrons is $2.24 \times 10^{10}$ per cm$^3$ at $T = 300$ K.

8.5 In an $n$-type semiconductor the donor levels lie $\epsilon_d$ *below* the bottom of the conduction band. If $N_d$, $n_d$ and $n_c$ are the number of donors, the number of electrons in the donor levels, the number of conduction electrons per unit volume, respectively, then show that the free energy of electrons that are on the donor levels is

$$F = -n_d\epsilon_d - kT \ln\left[\frac{N_d!}{n_d!\,(N_d - n_d)!}\, 2^{n_d}\right].$$

Use $\mu = \partial F/\partial n_d$ to obtain

$$n_d = \frac{N_d}{\frac{1}{2}\exp[-(\epsilon_d + \mu)/kT] + 1}.$$

For the nondegenerate case show that

$$n_e = n_c \exp(\mu/kT), \quad n_c = 2\left(\frac{2\pi m_e^* kT}{h^2}\right)^{3/2}.$$

From these results obtain

$$\frac{n_e(N_d - n_d)}{n_d} = \tfrac{1}{2} n_c \exp(-\epsilon_d/kT).$$

8.6 Derive the result of Prob. 8.2 by using the grand partition function

$$\mathscr{Z}_d = [+ 2\eta_a \exp\{\epsilon_d/kT\}]^{N_d}, \quad \eta_a = \exp(\mu/kT).$$

[Hint: $n_d = \eta_a (d \ln \mathscr{Z}_d/\partial\eta_a)$].

8.7 The transition temperature from impurity to intrinsic conductivity, depends on the concentration of impurity in the specific semiconductor and on the band gap width $\epsilon_g = \epsilon_c - \epsilon_v$ for fixed impurity concentration. If this transition is defined symbolically by the condition $n_h = n_d$, or $n_e = 2n_d$, show that the temperature (depletion temperature $T_{\text{dep}}$) is given by

$$T_{\text{dep}} = \frac{\epsilon_g}{k \ln \dfrac{n_c(T_{\text{dep}})\, n_v(T_{\text{dep}})}{2n_d^2}}$$

for a donor-depend semiconductor. [Hint: $n_i^2 = n_e n_h = 2n_d^2$.].

8.8 At low temperatures the impurity concentration plays the leading part. Show that for a nondegenerate semiconductor ($n_d \neq 0$, $n_a = 0$, or $\Delta n = n_d$) $\mu$ is given by

$$\mu = \epsilon_d + kT \ln \left[\frac{1}{4}\left(1 + \frac{8n_d}{n_c} \exp[\Delta\epsilon_d/kT]\right)^{1/2} - 1\right], \; 0 < T < T_{\text{dep}},$$

$$\mu = \epsilon_c + kT \ln \left\{\frac{n_d}{2n_c}\left[1 + \left(1 + \frac{4n_i^2}{n_d^2}\right)^{1/2}\right]\right\}, \; T > T_{\text{dep}},$$

where $T_{\text{dep}}$ is given by Prob. 8.4. Interpret these equations and plot $\mu$ versus $T$ curves for various values of $n_d$.

8.9 Obtain expressions corresponding to Prob. 8.5 for an acceptor impurity.

8.10 For P in Si at room temperature $\Delta\epsilon_d/kT \simeq 1.74$. Calculate $n_e^*$ from (8.44). Use (8.47) to show that 11.4 pct of the donors remain unionized, if $n_d/n_c = 1/100$.

# 9
# NONEQUILIBRIUM STATES

## 9.1 BOLTZMANN TRANSPORT EQUATION

A complete description of a system in equilibrium is provided by the grand cononical distribution. In practice, often we have systems that are not in statistical equilibrium. The study of such systems is very complicated. We shall consider only the case of an ideal gas. Each particle of an ideal gas moves independently The total distribution function is determined by the one particle distribution function. Therefore, it is enough to establish a non-equilibrium distribution for a single particle.

Boltzmann proposed to find the distribution function from an equation similar in meaning to the Liouville equation. The state of a particle is given by its three coordinates $(x, y, z)$ and three momentum components $(p_x, p_y, p_z)$. We can work in the six-dimensional space of Cartesian coordinates $\mathbf{r}$ and velocity $\mathbf{v}$. The classical distribution function $f(\mathbf{r}, \mathbf{v})$ is defined as

$$f(\mathbf{r}, \mathbf{v}) = \text{number of particles in a volume element } d\mathbf{r}\, d\mathbf{v}. \tag{9.1}$$

By the Liouville theorem, as time lapses the volume element moves along a flowline in such a way that the distribution is conserved,

$$f(t + dt, \mathbf{r} + d\mathbf{r}, \mathbf{v} + d\mathbf{v}) = f(t, \mathbf{r}, \mathbf{v}), \tag{9.2}$$

in the absence of collisions. In the presence of collisions,

$$f(t + dt, \mathbf{r} + d\mathbf{r}, \mathbf{v} + d\mathbf{v}) - f(t, \mathbf{r}, \mathbf{v}) = dt\,(\partial f/\partial t)_{\text{collisions}}, \tag{9.3}$$

whence

$$dt\,(\partial f/\partial t) + d\mathbf{r}\cdot\nabla_{\mathbf{r}} f + d\mathbf{v}\cdot\nabla_{\mathbf{v}} f = dt\,(\partial f/\partial t)_{\text{collisions}}. \tag{9.4}$$

If $\mathbf{a}$ denotes the acceleration $d\mathbf{v}/dt$,

$$\partial f/dt + \mathbf{v}\cdot\nabla_{\mathbf{r}} f + \mathbf{a}\cdot\nabla_{\mathbf{v}} f = (\partial f/\partial t)_{\text{coll}}. \tag{9.5}$$

This is the *Boltzmann transport equation*. In the steady state

$$\partial f/\partial t = 0, \quad (\textit{equilibrium}). \tag{9.6}$$

The particle collisions restore equilibrium which was disturbed by the external forces. If $f$ does not vary greatly from its equilibrium value $f_0$, then we can write the collision term as

$$(\partial f/\partial t)_{\text{coll}} = -(f - f_0)/\tau, \tag{9.7}$$

where the *relaxation time* $\tau$ depends, in general, on $\mathbf{r}$ and $\mathbf{v}$. The Boltzmann equation in this $\tau$-approximation simplifies to

$$\frac{\partial f}{\partial t} + \mathbf{v}\cdot\nabla_{\mathbf{r}} f + \mathbf{a}\cdot\nabla_{\mathbf{v}} f = -\frac{f - f_0}{\tau}. \tag{9.8}$$

Let a nonequilibrium distribution of velocities result due to external forces which are suddenly removed. Using $\partial f_0/\partial t = 0$, (9.7) is

$$\frac{\partial\,(f - f_0)}{\partial t} = -\frac{f - f_0}{\tau}, \tag{9.9}$$

and its solution is

$$(f - f_0)_t = (f - f_0)_{t=0}\, e^{-t/\tau}. \tag{9.10}$$

## 9.2 PARTICLE DIFFUSION

We often have an isothermal system with a gradient in the particle concentration. The steady state Boltzmann transport equation in the $\tau$-approximation is

$$v_x\,(df/dx) = -(f - f_0)/\tau. \tag{9.11}$$

The nonequilibrium distribution function $f$ varies in the $x$ direction. To first order, with $\partial f/\partial x$ replaced by $\partial f_0/\partial x$, (9.11) is

$$f_1 \simeq f_0 - \tau v_x\,(df_0/dx). \tag{9.12}$$

By iteration, the second order solution is

$$\begin{aligned} f_2 &= f_0 - \tau v_x\,(df_1/dx) \\ &= f_0 - \tau v_x\,(df_0/dx) + \tau^2 v_x^2\,(d^2 f_0/dx^2). \end{aligned} \tag{9.13}$$

We need (9.13) for the study of nonlinear effects.

### Classical Limit

In the classical limit the FD distribution function is

$$f_0 = \exp\,[-(\epsilon - \mu)/kT]. \tag{9.14}$$

Because (9.8) is linear in $f$ and $f_0$, the normalization in (9.14) need not be the same as in (9.1). From (9.12, 14),

$$\begin{aligned} f &= f_0 - \tau v_x\,(df_0/d\mu)\,(d\mu/dx) \\ &= f_0 - \tau v_x\,(f_0/kT)\,(d\mu/dx). \end{aligned} \tag{9.15}$$

The particle flux density in the $x$ direction, $j_x$, is

$$j_x = \int v_x f g(\epsilon)\,d\epsilon = -(d\mu/dx)\int (\tau v_x^2 f_0/kT)\,g(\epsilon)\,d\epsilon, \tag{9.16}$$

where $g(\epsilon) = (4\pi^2)^{-1} (8\pi^2 m/h^2)^{3/2} \epsilon^{1/2}$ for a particle of zero spin. The left out part, $\int v_x f_0 g(\epsilon)\, d\epsilon$, vanishes because $f_0$ is an even function of $v_x$ while $v_x$ is an odd function. In general,

$$\tau = Av^s,\ A > 0,\ S = 0, -1, -2, \text{etc.}$$

If $\tau$ happens to be independent of velocity ($s = 0$),

$$\begin{aligned} j_x &= -(d\mu/dx)(\tau/kT) \int v_x^2 f_0 g(\epsilon)\, d\epsilon \\ &= -(d\mu/dx)(2\tau/mkT) \int \left(\frac{1}{2} mv_x^2\right) f_0\, g(\epsilon)\, d\epsilon. \end{aligned} \tag{9.17}$$

Recalling (4.47), the integral is $\frac{1}{2}nkT$ with $\int f_0\, g(\epsilon)\, d\epsilon = n$ as the concentration. Thus, with $\mu = kT \ln n +$ constant,

$$j_x = -(n\tau/m)(d\mu/dx) = -(\tau kT/m)(dn/dx). \tag{9.18}$$

The driving force of isothermal diffusion is given by the gradient of the particle concentration, that is by *Fick's law*,

$$\mathbf{j} = -D \text{ grad } n, \tag{9.19}$$

where $D$ is the particle diffusion constant (*diffusivity*). Clearly (9.18) is of the form (9.19) with

$$D = \tau (kT/m) = (1/3)\, \overline{v^2}\, \tau. \tag{9.20}$$

Another useful case is $s = -1$, $\tau \propto 1/v$, or $\tau = l/v$, where $l$ is the mean free path. Then

$$\begin{aligned} j_x &= -(d\mu/dx)(l/kT) \int (v_x^2/v) f_0\, g(\epsilon)\, d\epsilon \\ &= -(d\mu/dx)(l/kT)(\tfrac{1}{3} n\bar{c}) = -\tfrac{1}{3} l\bar{c}\,(dn/dx), \end{aligned} \tag{9.21}$$

where $\bar{c}$ is the average speed. In this case

$$D = \tfrac{1}{3} l\bar{c}, \qquad (\tau \propto 1/v). \tag{9.22}$$

**FD Distribution**

The Fermi-Dirac distribution is

$$f_0 = \frac{1}{\exp[(\epsilon-\mu)/kT] + 1}. \tag{9.23}$$

Following the discussion of (7.17), at low temperature, $df_0/d\mu = (-df_0/d\epsilon)$ is small everywhere except in the region $\epsilon \simeq \mu$ where it is very large. Thus $df_0/d\mu$ resembles a delta function, $df_0/d\mu \simeq \delta(\epsilon-\mu)$, or

$$df_0/dx = \delta(\epsilon-\mu)(d\mu/dx). \tag{9.24}$$

From (9.16, 24),

$$j_x = -(d\mu/dx)\,\tau(\mu) \int v_x^2\, \delta(\epsilon-\mu)\, g(\epsilon)\, d\epsilon, \tag{9.25}$$

where $\tau(\mu)$ is the relaxation time at $\epsilon = \mu$. Using $g(\epsilon) = g(\mu) = 3n/2\mu_0$, (7.30), at absolute zero, the integral has the value

$$\tfrac{1}{3} v_F^2 (3n/2\mu_0) = n/m, \tag{9.26}$$

where $\mu_0 \equiv \epsilon_F (T = 0) \equiv \frac{1}{2} m v_F^2$ gives the Fermi velocity $v_F$ on the Fermi surface. Using $\mu_0 = (h^2/2m)(3n/8\pi)^{2/3}$, (7.11),

$$d\mu/dx = \frac{2}{3}(h^2/2m)(3/8\pi)^{2/3} n^{-1/3}(dn/dx)$$

$$= \frac{2}{3}(\mu_0/n)(dn/dx). \tag{9.27}$$

From (9.25-27),

$$j_x = (-n\tau/m)(d\mu/dx) = -(2\tau/3m)\mu_0 (dn/dx)$$

$$= -\frac{1}{3}\tau v_F^2 (dn/dx) \equiv -D(dn/dx), \tag{9.28}$$

where

$$D = \frac{1}{3} v_F^2 \tau(\mu_0), \qquad \text{(FD)}. \tag{9.29}$$

It resembles (9.20) for the classical distribution of velocities.

## 9.3 ELECTRICAL CONDUCTIVITY

We wish to find the electric conductivity of a metal at temperature $T$ in the presence of an electric field $\mathbf{E}$. The electrons are assumed to be strongly degenerate. From (9.8),

$$\frac{\partial f}{\partial t} + \mathbf{v}\cdot\frac{\partial f}{\partial \mathbf{r}} - e\mathbf{E}\cdot\frac{\partial f}{\partial \mathbf{p}} = -\frac{f - f_0}{\tau}, \tag{9.30}$$

where $\mathbf{p} = m\mathbf{v}$. The equilibrium distribution function is

$$f_0 = \frac{1}{\exp\{[\epsilon(\mathbf{p}) - \mu]/kT\} + 1}, \qquad \epsilon(\mathbf{p}) = \frac{p^2}{2m}. \tag{9.31}$$

Assuming steadiness and uniformity, (9.30) becomes

$$-e\mathbf{E}\cdot\frac{\partial f}{\partial \mathbf{p}} = -\frac{f - f_0}{\tau}. \tag{9.32}$$

As $f_0$ is a function of $\epsilon$, we have to first order, (9.12),

$$f = f_0 + e\tau\frac{\partial f_0}{\partial \epsilon}\mathbf{v}\cdot\mathbf{E} \tag{9.33}$$

where $\mathbf{v} = \partial\epsilon/\partial\mathbf{p}$ is the velocity. To find the current density multiply (9.33) by $-e\mathbf{v}$ and integrate over all values of the momentum,

$$\mathbf{j} = -e\int \mathbf{v} f \frac{g_s\, d\mathbf{p}}{h^3} = e^2 \int \left(-\frac{\partial f_0}{\partial \epsilon}\right) \tau\, \mathbf{v}\mathbf{v}\cdot\mathbf{E}\, \frac{2d\mathbf{p}}{h^3}, \tag{9.34}$$

where $d\mathbf{p} = dp_x\, dp_y\, dp_z$. The electric conductivity $\tilde{\sigma}$ is defined by $\mathbf{j} = \tilde{\sigma}\,\mathbf{E}$.

Thus the components of the electric conductivity tensor are

$$\tilde{\sigma}_{\alpha\beta} = e^2 \int \tau\, v_\alpha\, v_\beta \left(-\frac{\partial f_0}{\partial \epsilon}\right)\frac{2d\mathbf{p}}{h^3} = e^2 \int \tau\, \overline{v_\alpha\, v_\beta} \left(-\frac{\partial f_0}{\partial \epsilon}\right) g(\epsilon)\, d\epsilon, \quad (9.35)$$

where $\alpha, \beta = x, y, z$; $\overline{v_\alpha v_\beta}$ is the average of $v_\alpha v_\beta$ taken over the energy surface $\epsilon$; and $g(\epsilon) = (4\pi^2)^{-1} (8\pi^2 m/h^2)^{3/2} \epsilon^{1/2}$ is the density of states of the conduction band.

At low temperatures, $-\partial f_0/\partial \epsilon$ has a sharp maximum at $\epsilon = \mu$, so that

$$\tilde{\sigma}_{\alpha\beta} = e^2\, \tau\, (\mu)\, (\overline{v_\alpha\, v_\beta})_F\, g(\mu), \quad (9.36)$$

where $(\overline{v_\alpha v_\beta})_F$ is the value of $\overline{v_\alpha v_\beta}$ on the Fermi surface $\epsilon = \mu$. For a spherical energy surface $\epsilon\, (\mathbf{p}) = p^2/2m$, we have $\mathbf{v} = \mathbf{p}/m$, $\overline{v_\alpha v_\beta} = \delta_{\alpha\beta}\ (2\,\epsilon/3\, m)$ and

$$\tilde{\sigma} = \frac{1}{3}\, e^2\, \tau\, (\mu) \frac{8\pi\, (2m\mu)^{3/2}}{mh^3} = \frac{ne^2}{m}\, \tau\, (\mu), \quad (9.37)$$

where $n = 8\pi\, (2m\, \mu)^{3/2}/3h^3$ is given by (7.10).

Note that the isothermal electric conductivity (9.37) also follows from the particle diffusivity result (9.18) if we multiply the particle flux density by $-e$ and replace $d\mu/dx$ by the gradient $-ed\phi/dx = eE_x$ of the external potential.

**Nondegenerate Case**

If (9.31) is approximated by the MB distribution (9.14) and $\tau = Av^s$, we can still use the arguments leading to (9.34). We have

$$-\frac{\partial f_0}{\partial E(\mathbf{p})} = \frac{1}{kT} f_0, \quad (f_0 = \exp\,[-(\epsilon-\mu)/kT]),$$

$$\tilde{\sigma} = \frac{e^2}{3kT} \int \tau\, \overline{\mathbf{v}^2}\, f_0\, (\epsilon)\, g\, (\epsilon)\, d\epsilon$$

$$= \frac{e^2}{3kT} \int_0^\infty A \left(\frac{2\epsilon}{m}\right)^{(1/2)s} \exp\,[(\mu-\epsilon)/kT]\, \frac{8\pi\, (2\epsilon m)^{3/2}}{mh^3}\, d\epsilon$$

$$= \frac{e^2}{m} A \left(\frac{2kT}{m}\right)^{(1/2)s} \frac{8\pi\, (2kTm)^{3/2}}{3h^3} \exp\,(\mu/kT)\, \Gamma\left(\frac{s+5}{2}\right).$$

Using

$$n = \int_0^\infty f_0\, (\epsilon)\, g\, (\epsilon)\, d\epsilon = \frac{8\pi^{3/2}\, (kTm)^{3/2}}{h^3} \exp\,(\mu/kT)$$

we get

$$\tilde{\sigma} = \frac{ne^2}{m} A \left(\frac{2kT}{m}\right)^{(1/2)s} \frac{2\sqrt{2}}{3\pi^{1/2}}\, \Gamma\left(\frac{s+5}{2}\right)$$

When $s = 0$ (or $\tau$ = constant),

$$\tilde{\sigma} = \frac{ne^2}{\sqrt{2}\, m}\, \tau. \tag{9.38}$$

## 9.4 THERMAL CONDUCTIVITY

We wish to calculate the heat conductivity of a metal in which the electric field vanishes and instead a uniform stationary temperature gradient $\partial T/\partial \mathbf{r}$ is introduced. The $f_0$ is given by (9.31) and (9.32) is replaced by

$$\mathbf{v}\cdot\frac{\partial f}{\partial \mathbf{r}} = -\frac{f-f_0}{\tau}. \tag{9.39}$$

The $f$ on the left side can be replaced by $f_0$, as in (9.33). Then

$$\mathbf{v}\cdot\frac{\partial f_0}{\partial \mathbf{r}} = \frac{\partial f_0}{\partial \epsilon} T\mathbf{v}\cdot\frac{\partial}{\partial \mathbf{r}}\left(\frac{\epsilon-\mu}{T}\right) = \frac{\partial f_0}{\partial \epsilon}\mathbf{v}\cdot\left[-\epsilon\,\frac{\partial \ln T}{\partial \mathbf{r}} - T\frac{\partial}{\partial \mathbf{r}}\left(\frac{\mu}{T}\right)\right],$$

$$f = f_0 - \frac{\partial f_0}{\partial \epsilon}\,\tau\,\mathbf{v}\cdot\left[-\epsilon\,\frac{\partial \ln T}{\partial \mathbf{r}} - T\frac{\partial}{\partial \mathbf{r}}\left(\frac{\mu}{T}\right)\right]. \tag{9.40}$$

The flow of electrons is given by

$$\mathbf{j} = \int \mathbf{v} f\,\frac{2d\mathbf{p}}{h^3} = \int\left(-\frac{\partial f_0}{\partial \epsilon}\right)\tau\,\overline{\mathbf{v}\,\mathbf{v}}\cdot\left[-\epsilon\,\frac{\partial \ln T}{\partial \mathbf{r}} - T\frac{\partial}{\partial \mathbf{r}}\left(\frac{\mu}{T}\right)\right] g(\epsilon)\, d\epsilon.$$

Taking

$$\epsilon(\mathbf{p}) = p^2/2m,\ \overline{\mathbf{v}_\alpha\,\mathbf{v}_\beta} = \delta_{\alpha\beta}\, 2\epsilon/3m, \text{ and } g(\epsilon) = (4\pi^2)^{-1}(8\pi^2 m/h^2)^{3/2}\,\epsilon^{1/2},$$

$$\mathbf{j} = R_1\left(-\frac{\partial \ln T}{\partial \mathbf{r}}\right) + R_0\left[-T\frac{\partial}{\partial \mathbf{r}}\left(\frac{\mu}{T}\right)\right], \tag{9.41}$$

where, with $\tau = A\, v^s$, $A > 0$, $s > -7$,

$$\begin{aligned} R_\nu &= \int\left(-\frac{\partial f_0}{\partial \epsilon}\right)\tau\,\frac{\overline{v^2}}{3}\,\epsilon^\nu\, g(\epsilon)\, d\epsilon \\ &= \frac{A}{m}\left(\frac{2}{m}\right)^{\frac{1}{2}s}\frac{8\pi\,(2m)^{3/2}}{3h^3}\int_0^\infty\left(-\frac{\partial f_0}{\partial \epsilon}\right)\epsilon^{\nu+\frac{1}{2}(s+3)}\, d\epsilon \\ &= \frac{n}{m}\,\mu_0^\nu\, A\left(\frac{2\mu_0}{m}\right)^{\frac{1}{2}s}\left\{1 + \frac{1}{6}\left[\nu + \frac{1}{2}(s+3)\right]\left(\nu + \frac{1}{2}s\right)\left(\frac{\pi kT}{\mu_0}\right)^2 + \ldots\right\}. \end{aligned} \tag{9.42}$$

For $\mathbf{j}$ to vanish, $\mathbf{j} = 0$, choose the gradient of the chemical potential as

$$-T\frac{\partial}{\partial \mathbf{r}}\left(\frac{\mu}{T}\right) = -\frac{R_1}{R_0}\left(-\frac{\partial \ln T}{\partial \mathbf{r}}\right). \tag{9.43}$$

The energy flow is given by

$$\mathbf{Q} = \int \epsilon \, \mathbf{v} f \frac{2d\mathbf{p}}{h^3} = \int \left( -\frac{\partial f_0}{\partial \epsilon} \right) \tau \, \epsilon \, \overline{\mathbf{v}\,\mathbf{v}} \cdot \left[ -\epsilon \, \frac{\partial \ln T}{\partial \mathbf{r}} - T \frac{\partial}{\partial \mathbf{r}} \left( \frac{\mu}{T} \right) \right] g(\epsilon) \, d\epsilon$$

$$= R_2 \left( -\frac{\partial \ln T}{\partial \mathbf{r}} \right) + R_1 \left[ -T \frac{\partial}{\partial \mathbf{r}} \left( \frac{\mu}{T} \right) \right]. \tag{9.44}$$

The energy flow in the case $\mathbf{j} = 0$ is just the thermal current. From (9.43,44),

$$\mathbf{Q} = -\frac{R_0 R_2 - R_1 R_1}{R_0 T} \cdot \frac{\partial T}{\partial \mathbf{r}} \equiv -K \frac{\partial T}{\partial \mathbf{r}}, \tag{9.45}$$

where $K = (R_0 \, R_2 - R_1^2)/(R_0 \, T)$ is the *thermal conductivity*. Clearly $K = 0$ if we use only the first term in (9.43). If we keep the second order term in (9.43),

$$K = \frac{n}{m} A \left( \frac{2\mu_0}{m} \right)^{\frac{1}{2}s} \frac{\mu_0^2}{3T} \left( \frac{\pi k T}{\mu_0} \right)^2$$

$$= \frac{n}{m} \tau(\mu_0) \, \tfrac{1}{3} (\pi k)^2 T. \tag{9.46}$$

Combining (9.37) and (9.46), we get the *Lorentz number*

$$L_{\tilde{\sigma}, K} = \frac{K}{\tilde{\sigma} \, T} = \frac{\pi^2}{3} \left( \frac{k}{e} \right) = 2.72 \times 10^{-13} \text{ e.s.u. deg}^{-2}$$

$$= 2.45 \times 10^{-8} \text{ watt-ohm. deg}^{-2}. \tag{9.47}$$

It does not involve $n$ and $m$. It does not involve $\tau$ if the relaxation times are same for the electrical and thermal processes. The value in (9.47) is close to the experiment. The result (9.47) states the empirical *Wiedemann-Franz law*.

The presence of $R_1$, both in (9.41) for $\mathbf{j}$ and in (9.44) for $\mathbf{Q}$, reflects the *Onsager reciprocal theorem*. According to this theorem there is reciprocity of interference effects between irreversible processes. In this case, it implies a close correlation between the flow of electrons due to a temperature gradient and the thermal flow generated by an electric field.

## 9.5 ISOTHERMAL HALL EFFECT

A metal or a semiconductor is placed in a magnetic field $\mathbf{H}$ and current density $\mathbf{j}$ passed through it. A transverse electric field $\mathbf{E}_H$ is set up given by

$$\mathbf{E}_H = R_H \, \mathbf{H} \times \mathbf{j}. \tag{9.48}$$

This equation defines the *Hall coefficient* $R_H$. The effect arises from the Lorentz force $\frac{e}{c} \, \mathbf{v} \times \mathbf{H}$ on the moving charge in $\mathbf{H}$.

Consider a conductor in a longitudinal electric field $E_x$ and a transverse magnetic field $H_z$ (Fig. 9.1). If $E_x$ produces a current $j_x$ in the $x$-direction, the $H_z$ deflects the electrons. They accumulate on the lower face of the plate.

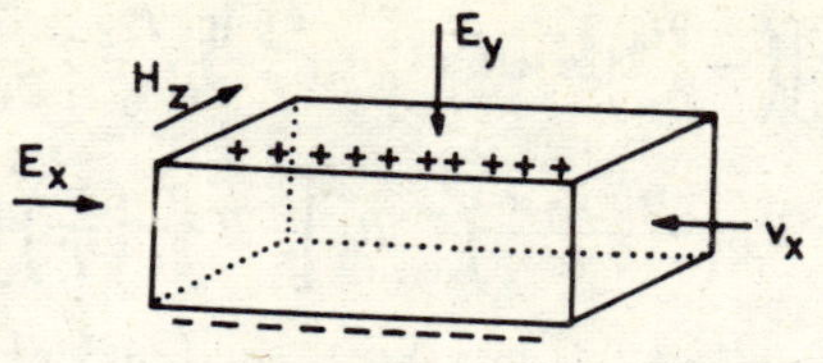


**Fig. 9.1 Geometry of the Hall effect.**

A positive ion excess is formed on the upper face until the *Hall voltage* is set up in the $y$-direction just enough to cancel the force due to the magnetic field.

In the theory of Drude, under $E_x$ the electrons acquire a drift velocity $v_x$ and so experience a force

$$F_y = -e\left(E_y - \frac{1}{c}\, v_x H_z\right) \tag{9.49}$$

It is zero if $E_y = v_x H_z / c$. As $j_x = -ne\, v_x$,

$$R_H = \frac{E_y}{j_x H_z} = -\frac{1}{nec}. \tag{9.50}$$

Thus $R_H \propto 1/n$ and its sign is same as that of the charge carriers. We now derive (9.50) for electrons obeying Fermi-Dirac distribution.

The Boltzmann equation for a metal, under electric fields in the $x$ and $y$ directions and a magnetic field in the $z$-direction, is, in the steady state,

$$v_x \frac{\partial f}{\partial x} + v_y \frac{\partial f}{\partial y} - \frac{e}{m}\left(E_x + \frac{v_y H_z}{c}\right)\frac{\partial f}{\partial v_x} - \frac{e}{m}\left(E_y - \frac{v_x H_z}{c}\right)\frac{\partial f}{\partial v_y} = -\frac{f - f_0}{\tau}. \tag{9.51}$$

Following (9.12), put

$$f = f_0 + v_x X + v_y Y, \tag{9.52}$$

where $X$ and $Y$ depend on $v^2 = v_x^2 + v_y^2 + v_z^2$. For weak $E_x$ and $E_y$, neglect the products of $X$ and $Y$ with $E_x$, $E_y$, $v_x$ and $v_y$. Then

$$v_x \frac{\partial f_0}{\partial x} + v_y \frac{\partial f_0}{\partial y} - \frac{e}{m}\left(E_x \frac{\partial f_0}{\partial v_x} + E_y \frac{\partial f_0}{\partial v_y}\right)$$

$$-\frac{e}{mc}\left(v_y H_z \frac{\partial f}{\partial v_x} - v_x H_z \frac{\partial f}{\partial v_y}\right) + \frac{v_x X + v_y Y}{\tau} = 0. \tag{9.53}$$

Using $\epsilon = \frac{1}{2} m v^2$,

$$\frac{\partial f_0}{\partial v_x} = \frac{\partial f_0}{\partial \epsilon}\frac{d\epsilon}{dv_x} = m v_x \frac{\partial f_0}{\partial \epsilon}, \qquad \frac{\partial X}{\partial v_x} = m v_x \frac{\partial X}{\partial \epsilon}, \text{ etc.,}$$

we can write

$$v_x \frac{\partial f_0}{\partial x} + v_y \frac{\partial f_0}{\partial y} - e\,(E_x v_x + E_y v_y)\frac{\partial f_0}{\partial \epsilon}$$

$$-\frac{e}{mc}\,(v_y H_z X - v_x H_z Y) + \frac{v_x X + v_y Y}{\tau} = 0. \tag{9.54}$$

The values of $v_x$ and $v_y$ being arbitrary,

$$\frac{\partial f_0}{\partial x} - eE_x \frac{\partial f_0}{\partial \epsilon} + \frac{e}{mc} H_z Y + \frac{X}{\tau} = 0, \tag{9.55}$$

$$\frac{\partial f_0}{\partial y} - eE_y \frac{\partial f_0}{\partial \epsilon} - \frac{e}{mc} H_z X + \frac{Y}{\tau} = 0, \tag{9.56}$$

or

$$X = \frac{-\tau}{1+a^2}\left[\left(\frac{\partial f_0}{\partial x} - eE_x \frac{\partial f_0}{\partial \epsilon}\right) - a\left(\frac{\partial f_0}{\partial y} - eE_y \frac{\partial f_0}{\partial \epsilon}\right)\right], \tag{9.57}$$

$$Y = \frac{-\tau}{1+a^2}\left[a\left(\frac{\partial f_0}{\partial x} - eE_x \frac{\partial f_0}{\partial \epsilon}\right) + \left(\frac{\partial f_0}{\partial y} - eE_y \frac{\partial f_0}{\partial \epsilon}\right)\right], \tag{9.58}$$

where $a = \tau\, eH_z/mc$.

The current density components $j_x$, $j_y$ are given by

$$j_x = -e \int_{-\infty}^{\infty} v_x^2 X \, dv_x \, dv_y \, dv_z, \tag{9.59}$$

$$j_y = -e \int_{-\infty}^{\infty} v_y^2 Y \, dv_x \, dv_y \, dv_z. \tag{9.60}$$

If the current cannot flow out of the conductor in the $y$ direction, we must have

$$j_y = 0. \tag{9.61}$$

If there is no thermal gradient in the $x$ or $y$ direction,

$$\frac{\partial f_0}{\partial x} = \frac{\partial f_0}{\partial y} = 0. \tag{9.62}$$

Substitute (9.57, 58, 61, 62) in (9.59, 60), and put $v_x^2 = v^2/3$, $dv_x \, dv_y \, dv_z = 4\pi v^2 dv$, to get*

$$j_x = -\frac{4\pi e^2}{3}[E_x I_1 - E_y I_2], \tag{9.63}$$

$$j_y = 0 = -\frac{4\pi e^2}{3}[E_x I_2 + E_y I_1], \tag{9.64}$$

where, using $\epsilon = \frac{1}{2}mv^2$,

$$I_1 = \int_0^{\infty} \tau \frac{v^4}{1+a^2} \frac{\partial f_0}{\partial \epsilon} \, dv = \frac{2}{m^2} \int_0^{\infty} l \frac{\epsilon}{1+a^2} \frac{\partial f_0}{\partial \epsilon} \, d\epsilon,$$

*Note that formally

$$\begin{pmatrix} j_x \\ j_y \end{pmatrix} = \begin{pmatrix} \sigma_{xx} & \sigma_{xy} \\ \sigma_{yx} & \sigma_{yy} \end{pmatrix} \begin{pmatrix} E_x \\ E_y \end{pmatrix}$$

$$I_2 = \int_0^\infty \tau \frac{av^4}{1+a^2}\frac{\partial f_0}{\partial \epsilon}\, dv = \frac{2}{m^2}\int_0^\infty l \frac{a\epsilon}{1+a^2}\frac{\partial f_0}{\partial \epsilon}\, d\epsilon,$$

with $l \equiv v\tau$.

For a degenerate electron gas, use

$$f_0 = \frac{2m^3}{h^3}\cdot \frac{1}{\exp[(\epsilon-\mu)/kT]+1} \tag{9.65}$$

where $2m^3/h^3$ is just a normalization factor for the velocity distribution function $f_0$, and the fact that $\partial f_0/\partial\epsilon$ has a sharp maximum at $\epsilon = \mu$, to get

$$j_x = \frac{16\pi me^2}{3h^3}\left(E_x \frac{l_F\mu}{1+a_F^2} - E_y a_F \frac{l_F\mu}{1+a_F^2}\right), \tag{9.66}$$

$$0 = \frac{16\pi me^2}{3h^3}\left(E_x a_F \frac{l_F\mu}{1+a_F^2} + E_y \frac{l_F\mu}{1+a_F^2}\right), \tag{9.67}$$

where $l_F = \tau(\mu)\, v_F$ = mean free path at the Fermi energy, and $a_F = (l_F/v_F)(eH_Z/mc)$. Thus, using (9.67) to eliminate $E_x$ or $E_y$,

$$E_x = \frac{3h^3}{16\pi me^2}\cdot\frac{1}{l_F\mu}\, j_x, \tag{9.68}$$

$$E_y = -\frac{3h^3}{16\pi me^2 c}\cdot\frac{H_z}{v_F\mu}\, j_x, \tag{9.69}$$

and using $v_F = (2\mu/m)^{1/2}$,

$$R_H = \frac{E_y}{j_x H_z} = -\frac{1}{nec}. \tag{9.70}$$

This is same as (9.50). For monovalent metals like Li, Na, K, etc., it gives satisfactory results. Our derivation is not restricted to metals or to semiconductors.

In Gaussian units the order of magnitude of $R_H$ in a simple metal such as K is

$$R_H \sim -\frac{1}{(10^{22})(5\times 10^{-10})(3\times 10^{10})} \sim -10^{-23} \text{ esu.}$$

To obtain it in v-cm/amp-gauss, multiply by $300\times(3\times 10^9) = 9\times 10^{11}$.

The sign of $R_H$ is same as the sign of carrier. For holes $R_H > 0$. The measurement of $R_H$ gives a direct measure of the number of carriers present if all carriers, are of the same sign.

**Semiconductors**

For an intrinsic semiconductor both the electrons in the conduction band and the holes in the valence band contribute to conduction. Therefore,

$$j = e(-n_e v_e + n_h v_h), \tag{9.71}$$

$$j_x = eE_x(n_e\tilde{\mu}_e + n_h\tilde{\mu}_h), \tag{9.72}$$

where $\tilde{\mu}_e$ and $\tilde{\mu}_h$ are called the *mobilities* of the conduction electrons and of the holes, respectively. The mobilities are defined as the velocities in the direction of the field for unit electric field.

The forces $F_e$ on the electron and $F_h$ on the hole, due to $H_z$, are

$$F_e = -\frac{e}{c}\,\tilde{\mu}_e E_x H_z, \quad F_h = -\frac{e}{c}\,\tilde{\mu}_h E_x H_z. \tag{9.73}$$

At equilibrium there is a field $E_y$ so that the net forces are

$$-eE_y + F_e, \quad eE_y + F_h. \tag{9.74}$$

The condition $j_y = 0$ at equilibrium gives

$$n_e\tilde{\mu}_e\left(-eE_y - \frac{e}{c}\,\tilde{\mu}_e E_x H_z\right) = n_h\tilde{\mu}_h\left(eE_y - \frac{e}{c}\,\tilde{\mu}_h E_x H_z\right). \tag{9.75}$$

From (9.72, 75),

$$R_H = \frac{E_y}{j_x H_z} = -\frac{1}{ec}\cdot\frac{n_e\tilde{\mu}_e^2 - n_h\tilde{\mu}_h^2}{n_e\tilde{\mu}_e + n_h\tilde{\mu}_h}. \tag{9.76}$$

This form is correct, like (9.70), if FD distribution is applicable.

## 9.6 NONEQUILIBRIUM SEMICONDUCTORS

Consider an $n$ type semiconductor. If light of quantum energy greater than the energy gap falls on it, an electron in the valence band can absorb a photon and make a transition to the conduction band (*photoconductivity*). The electron and hole concentrations so created are larger than their equilibrium concentrations. The excess electrons and holes finally recombine with each other. The recombination time can vary over a large range, $10^{-9}$s to $10^{-3}$s, depending upon the semiconductor. It is always much longer than the times, $\sim 10^{-12}$s, taken by the conduction electrons (holes) to reach thermal equilibrium with each other in the conduction band (valence band) at room temperature. It means that even though the total number of holes is not in equilibrium with the total number of electrons, the holes and electrons occupy the quantum states in a way that closely resembles the FD distributions in each band separately. The quasi-equilibrium state can be described by associating different *quasi-Fermi levels* $\mu_c$ and $\mu_v$ with two bands,

$$f_c(\epsilon, T) = \frac{1}{\exp[(\epsilon - \mu_c)/kT] + 1}, \quad f_v(\epsilon, T) = \frac{1}{\exp[(\epsilon - \mu_v)/kT] + 1}. \tag{9.77}$$

At a uniform temperature, any conduction electron flow in a semiconductor is described by the electric current density

$$\mathbf{j}_e = \tilde{\mu}_e n_e \operatorname{grad} \mu_c \tag{9.78}$$

if the gradient of the quasi-Fermi level $\mu_c$ is small. Here $n_e$ is the concentration of conduction electrons. The $j_e$ is zero if its driving force grad $\mu_c$ vanishes. The flow of electrons is from high to low chemical potential.

For electron concentration in the intrinsic but nondegenerate range, $n_i \ll n_e \ll n_c$, we can use (8.29) for the quasi-Fermi level $\mu_c$,

$$\mu_c = \epsilon_c - kT \ln (n_c/n_e). \tag{9.79}$$

Then

$$\mathbf{j}_e = \tilde{\mu}_e n_e \text{ grad } \epsilon_c + \tilde{\mu}_e kT \text{ grad } n_e.$$

The gradient at the bottom of the conduction band is produced by a gradient in the electrostatic potential $\phi$ and so from an electric field $\mathbf{E}$,

$$\text{grad } \epsilon_c = -e \text{ grad } \phi = e\mathbf{E}. \tag{9.80}$$

The electron diffusion coefficient $D_e$ is defined by the Einstein relation

$$D_e = \tilde{\mu}_e kT/e, \tag{9.81}$$

whenever conduction and diffusion take place by the same mechanism. Thus, finally

$$\mathbf{j}_e = e\tilde{\mu}_e n_e \mathbf{E} + eD_e \text{ grad } n_e. \tag{9.82}$$

We have two different contributions to $\mathbf{j}_e$. The first is due to $\mathbf{E}$ and the second due to grad $n_e$.

For holes

$$\mathbf{j}_h = \tilde{\mu}_h n_h \text{ grad } \mu_v = e\tilde{\mu}_h n_h \mathbf{E} - eD_h \text{ grad } n_h,$$

$$D_h = \tilde{\mu}_h kT/e. \tag{9.83}$$

## 9.7 ELECTRON-HOLE RECOMBINATION

If the number of excess carriers is small, we can define the rates of recombination in terms of lifetimes $\tau_e$ and $\tau_h$ of the electrons and holes,

$$\frac{dn_e}{dt} = -\frac{n_e - n_e^0}{\tau_e}, \quad \frac{dn_h}{dt} = -\frac{n_h - n_h^0}{\tau_h}, \tag{9.84}$$

where $n_e^0$ ($n_h^0$)) is the thermal equilibrium value of $n_e$($n_h$).

Particle flow must satisfy the requirement of continuity, this is, particle conservation. In a small volume element the rate of change of particle density is given by thermal or photo generation rate $\tilde{g}$ minus recombination rate $\tilde{r}$ and the divergence of the current,

$$dn_e/dt = \tilde{g} - \tilde{r} + (1/e)\,\nabla \cdot \mathbf{j}_e, \tag{9.85}$$

$$dn_h/dt = \tilde{g} - \tilde{r} - (1/e)\,\nabla \cdot \mathbf{j}_h. \tag{9.86}$$

The flow of electrons and holes is determined by (9.82, 83, 85, 86).

### Diffusion Flow

Consider an $n$-type semiconductor rod. Excess holes are generated at $x = 0$ by light or by some other method (say, forward bias). This creates a concentration gradient and diffusion in the $x$ direction. The holes will diffuse even if $\mathbf{E} = 0$. From (9.83, 86),

$$\frac{dn_h}{dt} = \tilde{g} - \frac{n_h - n_h^0}{\tau_h} + D_h \frac{\partial^2 n_h}{\partial x^2}. \tag{9.87}$$

In the steady state condition, $\tilde{g} = 0$, $dn_h/dt = 0$, we get

$$\frac{d^2 n_h}{dx^2} = \frac{n_h - n_h^0}{L_h^2}, \quad L_h = \sqrt{D_h \tau_h}, \tag{9.88}$$

where $L_h$ is the *diffusion length* of holes. For a long $n$-type rod, (length of rod)$/L_h \to \infty$, (9.88) has a solution

$$n_h - n_h^0 = n_h^0 \exp(-x/L_h). \tag{9.89}$$

Then, from (9.83, 89),

$$j_h = -e n_h^0 (D_h/L_h) \exp(-x/L_h). \tag{9.90}$$

For silicon: $\tilde{\mu}_h = 1350\ \text{cm}^2\ \text{v}^{-1}\ \text{s}^{-1}$, $D_h = 12\ \text{cm}^2\ \text{s}^{-1}$, $\tau_h \sim 10^{-6}$ to $10^{-4}$ s.

**Diode Action of p-n Junction**

Assume that the transition region is small compared with the diffusion lengths of the electrons or the holes. Let a potential $V$ be applied across the junction, with the $p$ side positive (forward bias). A positive potential lowers electron energy. The flow of electrons from $p$ to $n$ is unaltered. At a given energy level, the concentration on the $n$ side is a factor $\exp(eV/kT)$ greater than on the $p$ side. So the flow of electrons from $n$ to $p$ is greater. The opposite happens for the holes (Fig. 9.2a). The situation for the reverse bias ($p$ side negative) is shown in Fig. 9.2b.

As the potential drop occurs in the transition region where $E = 0$, we can use (9.88) in steady state conditions, $\tilde{g} = 0$, $dn_e/dt = 0$,

$$-\frac{d^2}{dx^2} \Delta n_e = \frac{\Delta n_e}{L_e^2}, \quad \Delta n_e = n_e - n_e^0, \quad L_e = \sqrt{D_e \tau_e}. \tag{9.91}$$

Its solution is

$$\Delta n_e = A \exp(-x/L_e) + B \exp(x/L_e). \tag{9.92}$$

The second term diverges for $x \to \infty$ and so $B = 0$. At $x = 0$, according to Boltzmann statistics,

$$n_e = n_e^0 + \Delta n_e = n_e^0 \exp(eV/kT),$$

$$\Delta n_e = n_e^0 [\exp(eV/kT) - 1] = A. \tag{9.93}$$

Therefore, (9.22) becomes

$$n_e = n_e^0 + n_e^0 [\exp(eV/kT) - 1] \exp(-x/L_e). \tag{9.94}$$

The electron current at $x = 0$ is

$$j_e(0) = -e D_e \nabla n_e \Big|_{x=0} = e \frac{D_e}{L_e} n_e^0 [\exp(eV/kT) - 1]. \tag{9.95}$$

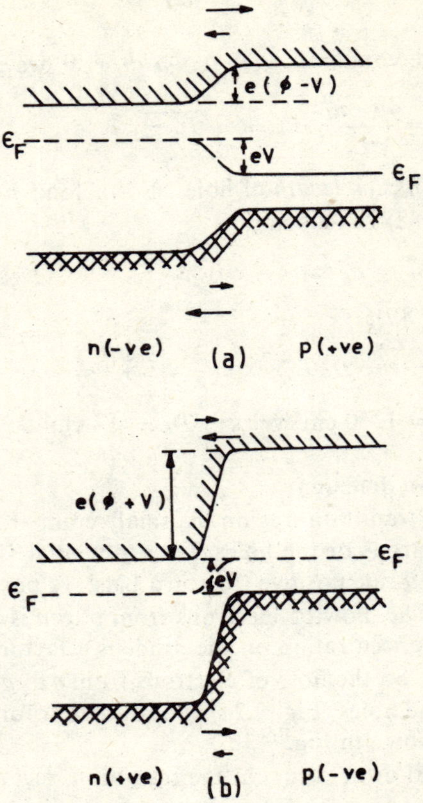


**Fig. 9.2** **Effect of bias on the energy bands of a *p-n* junction: (a) Forward bias, showing a large net current from *p* to *n*. (b) Reverse bias showing small net current from *n* to *p*.**

Similarly, for holes,

$$j_h(0) = e\,\frac{D_h}{L_h}\,n_h^0\,[\exp\,(eV/kT)-1]. \tag{9.96}$$

The total current is

$$\begin{aligned} j &= j_e(0) + j_h(0) \\ &= e\left(\frac{D_h n_h^0}{\tau_h} + \frac{D_e n_e^0}{\tau_e}\right)[\exp\,(eV/kT)-1]. \end{aligned} \tag{9.97}$$

The current is small for $V < 0$. Therefore, *pn* junction acts as a rectifier (Fig. 9.3).

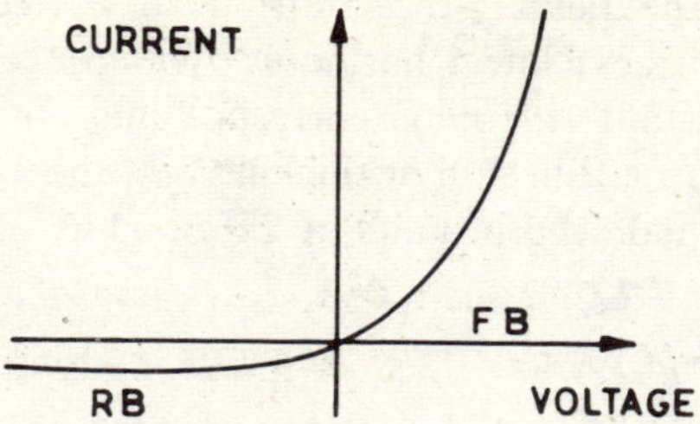


**Fig. 9.3 Current as a function of bias voltage for the regions of reverse bias (RB) and forward bias (FB).**

## 9.8 QUANTUM HALL EFFECT

In a magnetic field **H** an electron experiences the Lorentz force perpendicular to the field. In Gaussian units

$$m\,(d^2\mathbf{r}/dt^2) = (e/c)\,(d\mathbf{r}/dt)\times\mathbf{H}. \tag{9.98}$$

Let $\mathbf{H} = (0, 0, H)$, so that $[\mathbf{v}\times\mathbf{H}]_x = v_yH$, $[\mathbf{v}\times\mathbf{H}]_y = -v_xH$, $[\mathbf{v}\times H]_z = 0$, and

$$md^2x/dt^2 = (e/c)v_yH,\ md^2y/dt^2 = -(e/c)v_xH,\ md^2z/dt^2 = 0. \tag{9.99}$$

In the $(xy)$-plane the inward directed magnetic force $(e/c)vH$ gives rise to motion in a circle of radius $r$ with 'cyclotron' frequency $\omega$,

$$mv^2/r = (e/c)vH,\ \omega = 2\pi/T = 2\pi/(2\pi r/v) = eH/mc. \tag{9.100}$$

For very large $H$, $r$ is small, and in the correspondence principle limit of very small orbits the associated energy is *quantized*

$$E_\perp = (n + \tfrac{1}{2})\hbar\omega,\ \omega = eH/mc, \tag{9.101}$$

where $n$ is an integer. In the SI units $\omega = (e/m)H$.

The Schrödinger equation for a free electron in a magnetic field is

$$\frac{1}{2m}\left(-i\hbar\nabla - \frac{e}{c}\mathbf{A}\right)^2\psi = E\psi \tag{9.102}$$

where $\nabla\times\mathbf{A} = \mathbf{H}$. The choice

$$\mathbf{A} = (0, xH, 0) \tag{9.103}$$

gives $H$ in the $z$-direction. From (9.102, 103),

$$-\frac{\hbar^2}{2m}\left(\frac{\partial^2\psi}{\partial x^2} + \frac{\partial^2\psi}{\partial z^2}\right) - \frac{\hbar^2}{2m}\left(\frac{\partial}{\partial y} - \frac{ieH}{\hbar c}x\right)^2\psi = E\psi. \tag{9.104}$$

Try the solution

$$\psi(x, y, z) = u(x)\exp\,[i(k_yy + k_zz)]. \tag{9.105}$$

Then

$$-\frac{\hbar^2}{2m}\frac{\partial^2u}{\partial x^2} + \frac{\hbar^2}{2m}\left(k_y - \frac{eH}{\hbar c}x\right)^2u = E_\perp u, \tag{9.106}$$

$$E_\perp = E - (\hbar^2/2m)\,k_z^2. \tag{9.107}$$

The motion along the field ($z$-direction) is of a free particle. In the ($xy$)-plane, (9.106) describes a linear harmonic oscillator centred at $x_0 = \hbar c k_y/(eH)$, with spring constant $(eH/mc)^2$ corresponding to the cyclotron frequency $\omega = (e/mc)\,H$ In a thin film of thickness $d_z$, the $k_z$ (perpendicular to the surface) also gets quantized in units of $2\pi/d_z$. Thus

$$E_\perp = (n + \tfrac{1}{2})\,\hbar\omega,\ n = 0, 1, 2, \ldots, (xy)\text{-plane},$$
$$(k_z)_l = (2\pi/d_z)\,l,\ l = 0, \pm 1, \pm 2, \ldots, \text{ along } z\text{-axis},$$
$$E_{nl} = E_\perp + E_z = (n + \tfrac{1}{2})\,\hbar\omega + (\hbar^2/2m)\,k_z)_l^2. \qquad (9.108)$$

For small $d_z$, the levels are separated enough to stratify the electron states at low temperatures ($T \to 0$). Then it is enough to consider the $k_z = 0$ state. The quantized states $E_{nl}$ are called *Landau levels* (Fig. 9.4). The discrete levels $E_\perp$ separated by $\hbar\omega$ correspond to the circular orbits in the ($xy$)-plane.

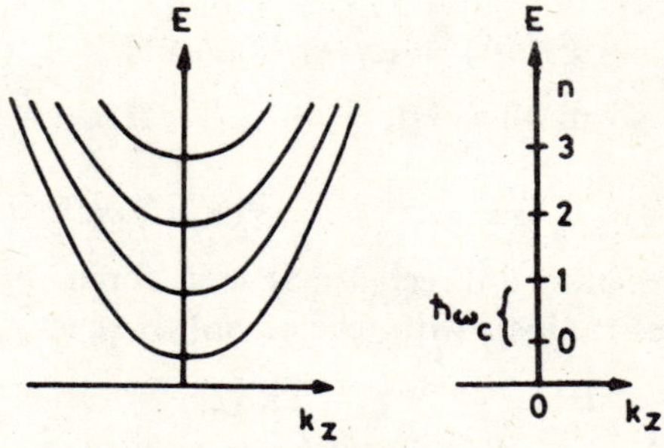


**Fig. 9 4 Landau levels.**

Compare the continuously distributed classical orbits $(\hbar^2/2m)\,(k_x^2 + k_y^2)$ and the discrete levels $(e\hbar/mc)\,H\,(n + \tfrac{1}{2})$, for a given $k_z$ (or $l$). Equal area concentric annular rings in the ($xy$)-plane will contain equal number of states. When $H$ is applied, the equally spaced levels of $E_\perp$ result when all states initially in an annular ring collapse to one discrete level (Fig. 9.5).

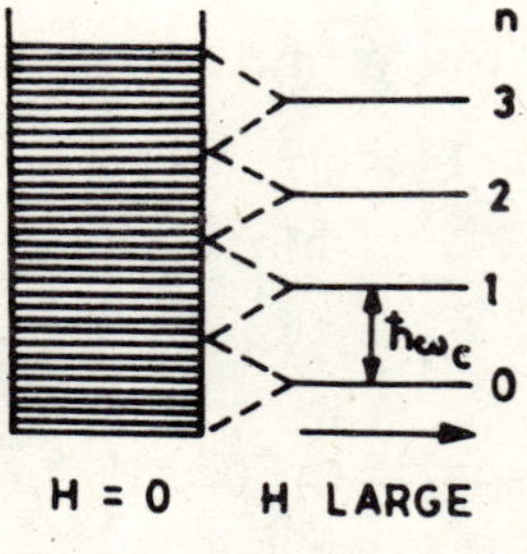


**Fig. 9.5 The grouping of the quasi-continuum of states given by $\hbar^2 k_z^2/2m$, all with the same $k_z$, to form Landau levels in the presence of a strong magnetic field.**

The degeneracy $g_L$ the Landau level $E_{nl}$, for a given $l$, is same for all $n$, and

$$g_L = \text{no. of states } (\hbar^2/2m)\,(k_x^2 + k_y^2) < (e\hbar/mc)\,H. \tag{9.109}$$

If $p^2/2m = (\hbar^2/2m)\,k^2 = (e\hbar/mc)\,H$ defines $p$, then the number of states within each Landau level is (area)$\times \pi p^2/\hbar^2$, or per unit area,

$$g_L = (e/hc)\,H, \qquad \text{neglecting spin.} \tag{9.110}$$

Using $hc/e = 4.14 \times 10^{-7}$ G-cm$^2$, $H = 1$ kilogauss, area $= 1$ cm$^2$, one gets $g_L \sim 10^{10}$. It is a complete degenerate gas at $T \to 0$.

A single electron state is now labelled $(n, l, \alpha)$, $n = 0, 1, 2, \ldots; l = 0, \pm 1, \pm 2, \ldots; \alpha = 1, 2, \ldots, g_L$. The partition function is

$$Z = \sum_{\{n_\lambda\}}{}' \exp\left(-\beta \sum_\lambda E_\lambda n_\lambda\right), \qquad \lambda = \{n, l, \alpha\} \tag{9.111}$$

where $\Sigma'$ implies the restrictions $n_\lambda = 0, 1$; $\sum_\lambda n_\lambda = N$; in the sum.

**Quantum Hall Effect**

A two-dimensional electron system (9.110) exists in a metal oxide semiconductor field effect transistor (MOSFET) (Fig. 9.6). In an $n$-channel MOSFET the Fermi level $E_F$ can be made to enter the conduction band by

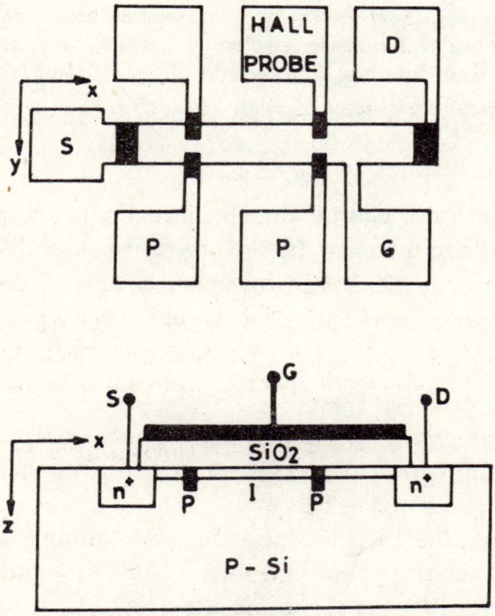


Fig. 9.6 A top (above) and a sectional view (bottom) of the MOSFET. G: gate, H: strong magnetic field, S: source, D: drain, I: $n$-type induced channel between the insulator $SiO_2$ and the p-type silicon, PP: potential probes. For quantum Hall effect Klitzing et al. kept the length of the device 400 $\mu$m, width 50 $\mu$m, and the distance between the potential probes (PP) 130 $\mu$m, at $T = 1.5$ K.

applying a large positive gate voltage $V_g$ (Fig. 9.7). This creates an inversion layer which forms a two-dimensional electron gas. The carriers in it are electrons (majority carriers are holes in the rest of the material).

Let a current $j_x$ flow along the $x$-direction and a magnetic field $H$ be applied along the $z$-direction (Fig. 9.6). A measurement of Hall effect gives, (9.50),

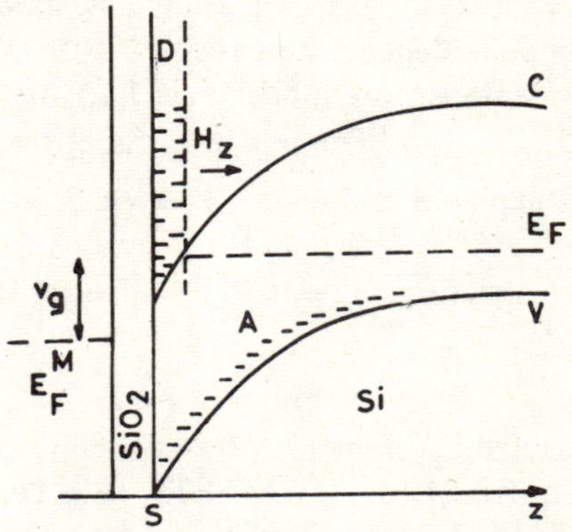


**Fig. 9.7** **Formation of the inversion layer I ($n$-type induced channel) in $p$-silicon at the interface with the insulator $SiO_2$ in a MOSFET. A: acceptor states (negatively charged), C: conduction band. D: depletion layer, S: surface of the semiconductor, V: valence band, E: energy, z: distance from interface into the crystal, $E_F$: semiconductor Fermi level, $E_F^M$: metal Fermi level, $V_g$: gate voltage.**

$$j_x = (n_e ec/H)\, E_y = \sigma_{xy}\, E_y, \tag{9.112}$$

where $n_e$ is the density of charge carriers and $\sigma_{xy} = (n_e\, ec)/H$, relating $j_x$ and $E_y$, is the Hall conductivity. Its reciprocal is the Hall resistivity $\rho_{xy}$ and $R_H = \rho_{xy}/H = 1/(n_e ec)$ is the Hall coefficient.

Writing the carrier concentration as $n_e = n_L = g_L \nu$, $\nu = 1, 2, 3, \ldots$, one gets from (9.110, 112),

$$\sigma_{xy} = j_x/E_y = \nu\,(e^2/h), \quad \nu = 1, 2, 3, \ldots \tag{9.113}$$

Thus the Hall conductivity of the two-dimensional system (inversion layer) in the quantum limit (small orbits) is quantized in integral multiples of $e^2/h$. Here $\nu$ narrow levels are lying below $E_F$.

Klitzing received the 1985 Nobel prize for observing the quantum Hall effect* (9.113) by measuring $\rho_{xy}$ in a silicon MOSFET and so $e^2/h$ (or $e^2/\hbar c$) accurately (1 part in $10^5$). As $\rho_{xy} \propto 1/n_e$, the sign and value of $n_e$ is found by measuring $\rho_{xy}$.

When a potential difference is applied across the $SiO_2$ crystal in a MOSFET, an inversion layer (25-50 Å thick) is formed at the $SiO_2$-Si interface. The density of electrons in the layer depends on the potential diffe-

*K.V. Klitzing, G. Dorda and M. Pepper, *Phys. Rev. Letts.* **45**, 494 (1980).

rence. The inversion layer region is a rectangular box of thickness $d_z$, (9.108). The lowest excitation energy $(\hbar^2/2m)(k_z)_1^2$ for the motion of the electron along $z$-axis is about 20 milli-eV. Below 10 K temperature, the typical energies are only about 1 milli-eV. Thus the motion perpendicular to the layer is frozen out and it behaves as a two-dimensional layer.

Klitzing *et al.* varied $V_g$ (that is, $n_e$) with a constant source-drain current $j_x$ and measured the Hall voltage, which gave $\rho_{xy}$. For small $H$, the expected smooth monotonic decrease of $\rho_{xy}$ with $n_e$, $\rho_{xy} \propto 1/n_e$, was found (Fig. 9.8). To have $\hbar\omega > kT$ in (9.101), or $1.6\, H \times 10^{-20} > 1.38\ 10^{-16}\, T$, one

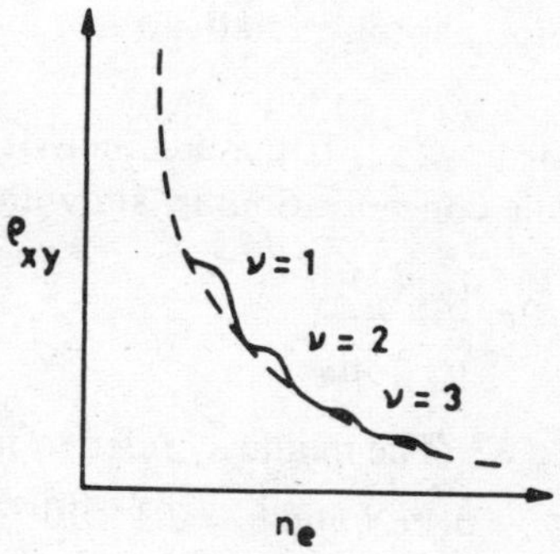


**Fig. 9.8** **Plot of $\rho_{xy}$ versus $n_e$ at small H according to $\rho_{xy} \propto 1/n_e$ (dashed), and large H (quantum Hall effect) showing steps according to $\rho_{xy} = h/(\nu e^2)$, $\nu = 1, 2, 3, \ldots$ (solid).**

gets $H > 42$ kilogauss for $T = 4.2$ K. In a run a $T = 1.5$ K, $H = 180$ kilogauss, they found that over certain ranges of $n_e$ (surface charge density), the $\rho_{xy}$ remains constant (*Hall steps*) (Fig. 9.8). These steps are manifestations of quantization and occur at $\rho_{xy} = h/(\nu e^2)$, (9.113). The value of $e^2/\hbar c$ is found to be 1/137.0353. It is also found that, unlike the ordinary Hall effect, the normal resistivity measured by potential probes vanishes, as in a superconductor, in the regions of Hall steps. If $E_F$ for some $V_g$ is between two Landau levels, the density of states at $E_F$ is zero and an inversion layer carrier cannot be scattered. Then the flow of current is lossless. The normal resistivity vanishes at Hall steps.

Use of $n_e = n_L = g_L\nu$ in (9.113) implies that $\nu$ Landau levels are completely occupied. This is most unlikely because of the very large degeneracy of each Landau level ($g_L \sim 10^{10}$). Therefore, (9.113) cannot be taken as an explanation of the *integral* quantum Hall effect. Tsui *et al** have also observed *fractional* quantum Hall effect with $\nu = 2/3, 4/3. 5/3, 2/5$, etc.

The results are insensitive to the location or type of impurities, and also to the type of host material. This suggests that some fundamental principle

*D. Tsui et al. *Phys, Rev. Letts.* **48**, 1559 (1982).

is at play. It is suspected that the quantum Hall effect is due to the long-range phase rigidity characteristic of a supercurrent*.

To conclude, the Hall conductivity shows plateaus as a function of the variables (for example, the magnetic field) which determine the number of electrons participating in the effect. In a MOSFET the change in the carrier density is produced by varying the gate voltage. At very low temperatures the plateaus are surprisingly quite wide, implying a corresponding large deviation from the integer filling of the Landau levels**.

## *PROBLEMS*

9.1 Use the Liouville theorem to derive the Boltzmann transport equation.

9.2 Calculate the rate of photon-radiative recombination of electrons and holes.

9.3 In a semiconductor $\tilde{\mu}_h < \tilde{\mu}_e$. If the conductivity $\tilde{\sigma}$ is considered as a function of the hole concentration $n_h$, show that

$$\tilde{\sigma}_{\text{minimum}} = 2\sigma_i \frac{(\tilde{\mu}_e \tilde{\mu}_h)^{1/2}}{\tilde{\mu}_e + \tilde{\mu}_h},$$

when $n_h = n_i (\tilde{\mu}_e / \tilde{\mu}_h)^{1/2}$. The $n_i$ and $\tilde{\sigma}_i$ refer to intrinsic material.

9.4 For Na we have $\mu = 3.1$ eV and $\tilde{\sigma} = 2.1 \times 10^{17}$ esu at 0K. Calculate the relaxation time from (9.37). [Ans: $3.3 \times 10^{-14}$ s.]

9.5 Show that the diffusion equation (9.88),

$$\frac{d^2 \Delta n_h}{dx^2} = \frac{\Delta n_h}{L_h^2}$$

has a solution, for an $n$-type rod of length $l$ terminated at $x = l$ by a boundary of recombination velocity $S_v$, given by

$$\frac{\Delta n_h}{\Delta n_h(0)} = \cosh X - \sinh X \left[\frac{\sinh W + (S_v L_h / D_h) \cosh W}{\cosh W + (S_v L_h / D_h) \sinh W}\right],$$

$$X = x/L_h, \; W = l/L_h,$$

for the boundary conditions

$$\Delta n_h = \Delta n_h(0) \text{ at } x = 0,$$

$$-D_h \left(\frac{d\Delta n_h}{dx}\right) = S_v \Delta n_h \text{ at } x = l.$$

Discuss the cases (i) $W \rightarrow \infty$, (ii) $S_v$ large ($\gg D_h/L_h$), and (iii) $S_v$ small ($\ll D_h/L_h$).

9.6 Show that if $\Delta n_h(0)$, the excess density at $x = 0$, is a sinusoidal function of time,

$$\Delta n_h(0) = \Delta n_h'(0)\,[1 + a e^{i\omega t}],$$

*R.B. Laughlin, *Phys. Rev.* B 23, 5635 (1981); P, Streda, *J. Phys.* C **15**, L 717 (1982); H, Aoki, *Phys. Rev. Letts.* **55**, 1136 (1985).

**S, Trugman, *Phys. Rev.* B 27 7539 (1983).

where $a$ is a constant less than unity, then the results of Prob. 9.4 remain same if $L_h$ is replaced by $L_h' = L_h(1 - i\omega t)^{-1/2}$.

9.7 A magnetic field of flux density 1 tesla is applied perpendicular to the largest faces of a semiconductor crystal of size $12\text{mm} \times 5\text{mm} \times 1\text{mm}$. For a 20 mA current along the length of the crystal, the voltage across the width is found to be 7.4 mV. Calculate the Hall Coefficient.
[Ans.: $3.7 \times 10^{-4}$ m$^3$ coulomb$^{-1}$].

# 10
# FLUCTUATIONS

## 10.1 INTRODUCTION

With the passage of time the properties of a system vary about the mean of equilibrium values. Similar behaviour will apply to the elements of an ensemble. So far we have assumed that these fluctuations are quite small. It is useful to study the conditions for this to be true.

The equivalence of the three ensembles studied also depends on the fluctuations being very small. For example, if energy $E$ fluctuations in the systems of a canonical ensemble are small, it is equivalent to a microcanonical ensemble. If both $N$ and $E$ of the systems in a grand canonical ensemble fluctuate negligibly, all the three ensembles are equivalent.

## 10.2 MEAN-SQUARE DEVIATION

Consider a quantity $n$. Its average value is $\bar{n}$. The deviation $\delta n$ is defined by

$$\delta n \equiv n-\bar{n}. \tag{10.1}$$

Clearly

$$\overline{\delta n} = \bar{n}-\bar{n} \equiv 0. \tag{10.2}$$

A first rough measure of the fluctuation is provided by the *mean-square deviation*

$$\begin{aligned}\overline{(\delta n)^2} = \overline{(n-\bar{n})^2} &= \overline{n^2}-2\bar{n}\bar{n}+(\bar{n})^2 \\ &\equiv \overline{n^2}-(\bar{n})^2. \end{aligned}\tag{10.3}$$

$\overline{n^2}$ is called the *second moment* of the distribution. The *standard deviation* $\Delta n$, that is, the root mean-square deviation from the mean, is defined by

$$\Delta n = [\overline{(n-\bar{n})^2}]^{1/2} \tag{10.4}$$

In general, if $P_i$ is the probability of finding a system in the state $i$, and if $f_i$ is the value of a physical quantity $f$ when the system is in the state $i$, then the average value of $f$ is defined by

$$\bar{f} = \sum_i P_i f_i, \tag{10.5}$$

with

$$\sum_i P_i = 1. \tag{10.6}$$

Then

$$\begin{aligned} \overline{f-\bar{f}} &= \Sigma P_i(f_i - \bar{f}) \\ &= \Sigma P_i f_i - \bar{f}\, \Sigma P_i = \bar{f} - \bar{f} = 0, \end{aligned} \tag{10.7}$$

$$\begin{aligned} \overline{(f-\bar{f})^2} &= \Sigma P_i(f_i-\bar{f})^2 = \Sigma P_i f_i^2 - 2\bar{f}\, \Sigma P_i f_i + \bar{f}^2\, \Sigma P_i \\ &= \overline{f^2} - 2\bar{f}^2 + \bar{f}^2 = \overline{f^2} - \bar{f}^2, \end{aligned} \tag{10.8}$$

$$\Delta f = [\overline{f^2} - \bar{f}^2]^{1/2}. \tag{10.9}$$

## 10.3 FLUCTUATIONS IN ENSEMBLES

*Canonical Ensemble*: Fluctuations occur in energy because the system is in thermal equilibrium with the reservoir. For a canonical ensemble,

$$Z = \sum_i \exp(-\beta E_i), \tag{10.10}$$

$$\bar{E} = \sum_i P_i E_i = \frac{\sum_i E_i \exp(-\beta E_i)}{\sum_i \exp(-\beta E_i)} = -\frac{\partial Z/\partial \beta}{Z}, \tag{10.11}$$

$$\overline{E^2} = \frac{\sum_i E_i^2 \exp(-\beta E_i)}{\sum_i \exp(-\beta E_i)} = \frac{\partial^2 Z/\partial \beta^2}{Z}, \tag{10.12}$$

$$\begin{aligned} -\frac{\partial \bar{E}}{\partial \beta} &= \frac{1}{Z}\left(\frac{\partial^2 Z}{\partial \beta^2}\right) - \frac{1}{Z^2}\left(\frac{\partial Z}{\partial \beta}\right)^2 \\ &= \overline{E^2} - \bar{E}^2 = \overline{(\delta E)^2}, \end{aligned} \tag{10.13}$$

$$\begin{aligned} C_V &= \left(\frac{\partial \bar{E}}{\partial T}\right)_V = \left(\frac{\partial \bar{E}}{\partial \beta}\right)_V \frac{d\beta}{dT} = \left(\frac{\partial \bar{E}}{\partial \beta}\right)_V (-k\beta^2) \\ &= k\beta^2\, \overline{(\delta E^2)}. \end{aligned} \tag{10.14}$$

A measure of the energy fluctuation is the ratio

$$\frac{\Delta E}{\bar{E}} = \frac{(\overline{(\delta E)^2})^{1/2}}{\bar{E}} = \frac{(kT^2 C_V)^{1/2}}{\bar{E}}. \tag{10.15}$$

For large $T$ the right hand side of (10.13) is of order $1/N^{1/2}$ because the extensive quantities $C_V$ and $\bar{E}$ are proportional to the number of molecules $N$ (for an ideal gas, $\bar{E} \simeq NkT$, $C_V \simeq Nk$). For a macroscopic system $N \simeq 10^{23}$, $\Delta E/\bar{E} \simeq 10^{-11}$, that is, the fluctuations are very small. In fact in such a canonical ensemble the distribution of energies is so peaked about the ensemble average energy that the ensemble is practically a microcanonical ensemble.

For a solid at low temperatures the Debye law (5.75, 76) gives

$$\bar{E} = U \simeq NkT/\Theta_D)^3,\ C_V \simeq Nk(T/\Theta_D)^3, \text{ for } T \ll \Theta_D,$$

and so $\Delta E/E = [N^{-1}(\Theta_D/T)^3]^{1/2}$. For $\Theta_D = 160$ K, $T = 10^{-2}$ K, and $N \simeq 10^{17}$ for a piece 0.01 cm on a side, we find $\Delta E/E \simeq 0.006$, which is small but not negligible. Thus statistical thermodynamics is unreliable for a tiny piece of matter.

*Grand Canonical Ensemble*: The energy fluctuations are calculated as in the case of canonical ensemble. We study here the possibility of *concentration fluctuations*. In this case (4.58, 67) give

$$\mathscr{Z} = \sum_{N,i} \exp\,[(N\mu - E_{Ni})/\theta], \tag{10.16}$$

$$\bar{N} = -\left(\frac{\partial \Omega_g}{\partial \mu}\right)_{V,T} = \theta \frac{\partial}{\partial \mu} \ln \mathscr{Z} = \frac{\theta}{\mathscr{Z}} \frac{\partial \mathscr{Z}}{\partial \mu},$$

$$\overline{N^2} = \frac{\sum\limits_{N,i} N^2 \exp\,[(N\mu - E_{Ni})/\theta]}{\sum\limits_{N,i} \exp\,[(N\mu - E_{Ni})/\theta]} = \frac{\theta^2}{\mathscr{Z}} \frac{\partial^2 \mathscr{Z}}{\partial \mu^2},$$

$$\overline{(\delta N)^2} = \overline{N^2} - (\bar{N})^2 = \theta^2 \left[\frac{1}{\mathscr{Z}} \frac{\partial^2 \mathscr{Z}}{\partial \mu^2} - \frac{1}{\mathscr{Z}^2}\left(\frac{\partial \mathscr{Z}}{\partial \mu}\right)^2\right]$$

$$= \theta \frac{\partial \bar{N}}{\partial \mu}. \tag{10.17}$$

For an ideal classical gas (4.75) gives

$$\bar{N} = e^{\mu/\theta} \frac{(2\pi m\theta)^{3/2}}{h^3} V, \tag{10.18}$$

$$\partial \bar{N}/\partial \mu = \bar{N}/\theta, \tag{10.19}$$

$$\overline{(\delta N)^2} = \bar{N} = PV/kT, \tag{10.20}$$

$$\Delta N/N = [\overline{(\delta N)^2}/(\bar{N})^2]^{1/2} = 1/(\bar{N})^{1/2}, \qquad \text{(MB)}. \tag{10.21}$$

The smaller the volume of the gas studied (the smaller the value of $\bar{N}$) the greater is the fractional fluctuation, $\Delta N/N$, of the number of particles.

## 10.4 CONCENTRATION FLUCTUATIONS IN QUANTUM STATISTICS

Consider $\bar{n}_i$, the average number of particles in the single particle quantum state $i$. We have on using upper (lower) sign for the FD(BE) statistics,

$$\frac{\partial \bar{n}_i}{\partial \mu} = \frac{\partial}{\partial \mu}\left[\frac{1}{\exp\,[(\epsilon_i - \mu)/\theta] \pm 1}\right] = \frac{1}{\theta} \frac{\exp\,[(\epsilon_i - \mu)/\theta]}{\{\exp\,[(\epsilon_i - \mu)/\theta] \pm 1\}^2}$$

$$= \frac{1}{\theta}\left[\frac{\{\exp\,[(\epsilon_i - \mu)/\theta] \pm 1\} \mp 1}{\{\exp\,[(\epsilon_i - \mu)/\theta] \pm 1\}^2}\right]$$

$$= \frac{1}{\theta}(\bar{n}_i \mp \bar{n}_i^2) = \frac{1}{\theta}\,\bar{n}_i(1 \mp \bar{n}_i). \tag{10.22}$$

From (10.17),

$$\overline{(\delta n_i)^2} = \theta(\partial \bar{n}_i/\partial \mu) = \bar{n}_i(1 \mp \bar{n}_i), \tag{10.23}$$

$$(\Delta n_i/n_i) = [\overline{(\delta n_i)^2}/\bar{n}_i^2\,]^{1/2} = (\bar{n}_i^{-1} \mp 1)^{1/2}, \tag{10.24}$$

or

$$(\Delta n_i/n_i) = \begin{cases} (n_i^{-1} - 1)^{1/2} & \text{(FD)},\\ (n_i^{-1} + 1)^{1/2} & \text{(BE)},\\ (n_i^{-1}\,)^{1/2} & \text{(MB)}. \end{cases} \tag{10.25}$$

Thus the fractional fluctuation in concentration is smaller than the MB case (10.21) for FD statistics and larger for the BE statistics. The fractional fluctuation in all cases is greatest for the least occupied state ($n_i \ll 1$).

*FD Case*: The Pauli exclusion principle allows the maximum value $\bar{n}_i = 1$. Then the fluctuation vanishes for the FD case. In fact, (10.23) shows that fluctuations vanish for $\bar{n}_i = 0$ and 1, that is, for empty high energy states and for fully occupied states deep below the Fermi level. The fluctuation is large for high energy occupied state.

*BE case*: For the BE gas as $T \to 0$, a large number of particles condense in the ground state, $\epsilon_{(gr)} = 0$, $\bar{n}_{(gr)} \simeq N$. Then from (10.3)

$$\overline{(\delta n_0)^2} = \overline{n^2} - (\bar{n})^2 \simeq \overline{n^2} - N^2, \tag{10.26}$$

and from (10.23)

$$\overline{(\delta n_0)^2} = \bar{n}_0(1 + \bar{n}_0) \simeq N + N^2 \simeq N^2, \tag{10.27}$$

so that

$$\overline{n^2} \simeq 2N^2. \tag{10.28}$$

Instead of considering the single-particle state $n_i$, we can consider a group of $g$ neighbouring states, all having the same mean occupation number $\bar{n}$. We can sum (10.23) over such a group of $g$ neighbouring states containing $\bar{N} = g\bar{n}$ particles. The statistical independence of the probability distribution of the different single-particle states allows us to write

$$\overline{(\delta N)^2} = g\,\overline{(\delta n)^2} = g\bar{n}(1 \mp \bar{n}) = \bar{N}\left(1 \mp \frac{1}{g}\bar{N}\right). \tag{10.29}$$

The relation (10.23) is applicable to photons as well, even though (10.22) cannot be used, since $\mu = 0$ for photons.

We can use (10.29) for photons that obey BE statistics, $n(\epsilon) = (e^{\epsilon/\theta} - 1)^{-1}$. The number of quantum states of the photons with frequencies between $\nu$ and $\nu + \Delta\nu$ is given by (4.118), $g = 8\pi V(\nu^2/c^3)\,\Delta\nu$. The total energy of the quanta in the frequency range is $E_{(\Delta\nu)} = Nh\nu$.

If we multiply (10.29) by $(h\nu)^2$,

$$\overline{(\delta E^{\mathrm{ph}}_{(\Delta\nu)})^2} = h\nu\,E_{(\Delta\nu)} + \frac{c^3(E_{(\Delta\nu)})^2}{8\pi V\,\nu^2 d\nu}. \tag{10.30}$$

This result was derived by Einstein. The first term on the right involving $h$ is typical of the corpuscular nature of radiation. The second term, not involving $h$, reqresents the classical result for the energy fluctuations of black-body radiation. The result (10.30) implies that photons like to travel in bunches. Large photon density fiuctuations have been experimentally observed*.

## 10.5 ONE DIMENSIONAL RANDOM WALK

A drunk sailor, who has lost the sense of direction, takes a *random walk* in one dimension. Suppose he takes $N$ steps of equal length $l$, each step being random (say) to the east or to the west. Each step has a probability $\frac{1}{2}$ of being in either direction. Let us find the probability that he is at a distance $x$ from the starting point after such a walk.

Denote by $P(m, N)$ the probability that the sailor is at a point $m$ steps away after $N$ steps. The probability of any given sequence of $N$ steps is $(\frac{1}{2})^N$, because each step has a probability of $\frac{1}{2}$. Hence

$$P(m, N) = (\text{number of distinct sequences that reach } m \text{ after } N \text{ steps}) \times (\tfrac{1}{2})^N$$

To arrive at the point $m$, some set of $n_1 = \frac{1}{2}(N+m)$ steps out of $N$ must be positive, and the remaining $n_2 = \frac{1}{2}(N-m)$ steps must be negative. Therefore, the number of distinct sequences that reach $m$ is

$$W(m) = \frac{N!}{[\frac{1}{2}(N+m)]!\,[\frac{1}{2}(N-m)]!}, \tag{10.31}$$

and

$$P(m, N) = (\tfrac{1}{2})^N W(m). \tag{10.32}$$

For large $N$ use the Stirling approximation in its more exact form (Appendix II), $N! = (2\pi N)^{1/2} N^N e^{-N}$, or

$$\begin{aligned} \ln N! &= N \ln N - N + \tfrac{1}{2}\ln(2\pi N) \\ &= (N + \tfrac{1}{2}) \ln N - N + \tfrac{1}{2} \ln 2\pi. \end{aligned} \tag{10.33}$$

Then

$$\begin{aligned} \ln P(m, N) = (N + \tfrac{1}{2}) \ln N &- \tfrac{1}{2}(N + m + 1) \ln \tfrac{1}{2}(N+m) \\ &- \tfrac{1}{2}(N - m + 1) \ln \tfrac{1}{2}(N-m) - \tfrac{1}{2} \ln 2\pi - N \ln 2. \end{aligned} \tag{10.34}$$

Since $m \ll N$, expand

$$\ln\left(1 \pm \frac{m}{N}\right) = \pm\frac{m}{N} - \frac{m^2}{2N^2} \pm \cdots \tag{10.35}$$

so that, using $\ln \frac{1}{2}(N \pm m) = \ln \frac{1}{2} N + \ln[1 \pm (m/N)]$,

* R.H. Brown and R.Q. Twiss, *Nature* 177, 27(1956); E.M. Purcell, *Nature*, 178, 1449 (1956).

$$\ln P(m, N) \simeq (N + \tfrac{1}{2}) \ln N - \tfrac{1}{2} \ln 2\pi - N \ln 2$$
$$- \tfrac{1}{2} (N + m + 1) \left( \ln N - \ln 2 + \frac{m}{N} - \frac{m^2}{2N^2} \right)$$
$$- \tfrac{1}{2} (N - m + 1) \left( \ln N - \ln 2 - \frac{m}{N} - \frac{m^2}{2N^2} \right)$$
$$\simeq - \tfrac{1}{2} \ln N + \ln 2 - \tfrac{1}{2} \ln 2\pi - (m^2/2N), \qquad (10.36)$$

or

$$P(m,N) \simeq (2/\pi\ N)^{1/2} \exp(-m^2/2N). \qquad (10.37)$$

As $x = ml$ and $m = n_1 - n_2 = n_1 - (N - n_1) = 2n_1 - N$, the probability that the sailor is between $x$ and $x + dx$ after $N$ steps is

$$P(x, N)\, dx = P(m, N)\, dm = P(m, N) \frac{dx}{2l}. \qquad (10.38)$$

We write $dx = 2l\, dm$, because $m$ can take only integral values separated by an amount $\Delta m = 2$.

From (10.37, 38)

$$P(x, N)\, dx = (2\pi l^2 N)^{-1/2} \exp(-x^2/2Nl^2)\, dx. \qquad (10.39)$$

This is the *normal* or *Gaussian distribution*, which is usually written as

$$P(x) = (2\pi)^{-1/2} \gamma^{-1} \exp(-x^2/2\gamma^2),$$
$$\int_{-\infty}^{+\infty} P(x)\, dx = 1. \qquad (10.40)$$

It has a symmetrical peak situated at $x = 0$. The width of the peak increases with $\gamma$ (Fig. 3.8).

To introduce time, we assume that the sailor takes $N = nt$ steps in time $t$. Then the probability of the sailor being in the interval $dx$ at $x$ after time $t$ is

$$P(x)\, dx = (2\pi l^2\, nt)^{-1/2} \exp(-x^2/2l^2 nt)\, dx. \qquad (10.41)$$

The mean square distance travelled is given by the mean square fluctuation

$$\overline{(\delta x)^2} = \overline{x^2} = \int_{-\infty}^{+\infty} x^2\, P(x)\, dx = l^2\, nt = \gamma^2. \qquad (10.42)$$

where we have used $\int_{-\infty}^{+\infty} x^2 \exp(-ax^2)\, dx = \tfrac{1}{2} (\pi/a^3)^{1/2}$.

If $\tau$ is the time taken for each step, then $t = \tau N$ and $1/\tau = v$ is the velocity. We can write the *conditional probability*, that the sailor will be within $dx$ at $x$ at time $t$ if he was at $x = 0$ at $t = 0$, as

$$P(0, 0; x, t) = (4\pi Dt)^{-1/2} \exp(-x^2/4Dt)\, dx,\ D = \tfrac{1}{2} v^2 \tau. \qquad (10.43)$$

Note that $n\tau = 1$. The spread of the distribution increases with $t$, and

$$\overline{x^2} = (v\tau)^2\, N = 2Dt. \qquad (10.44)$$

$D$ is the particle diffusion constant.

The problem of $N$ particles, each having a magnetic moment $\mu$ which may be either parallel or antiparallel to a magnetic field $H$ was discussed in Sec. 3.8. The calculation of the probability distribution of the total magnetic moment $M$ for $H = 0$ is identical with that in the random walk problem [compare (10.32, 37) with (3.94)]. If we write $M = m\mu_H$, then (10.37) gives for the entropy

$$\sigma = \ln P(m,N) \simeq \text{constant} - (m^2/2N). \tag{10.45}$$

In the presence of the magnetic field $H$,

$$E = -m\mu_H H,\ F = E - kt\sigma \simeq -m\mu_H H + (m^2kt/2N) + \text{constant}.$$

If $F$ is minimum, $\partial F/\partial m = 0$ gives

$$m/N = \mu_H H/kT, \tag{10.46}$$

$$M = m\mu_H \simeq N\mu_H^2 H/kT. \tag{10.47}$$

Apart from numerical factors and replacement of $kT$ by $\epsilon_F(0)$, (10.47) agrees with (7.45).

## 10.6 RANDOM WALK* AND BROWNIAN MOTION

A very small particle immersed in a liquid exhibits a random type of motion. It is called *Brownian motion*. It is produced by the thermal fluctuation of pressure on the particle. Because of the fluctuations, the forces do not always cancel and the particle is knocked about in a random way.

The Brownian motion in one dimension is like a random walk along a line. At the end of each period of time $\tau$ the particle has either moved a distance $l = v\tau$ to the right or a distance $l$ to the left. If the direction of each successive step is a *random variable*, then the probability that during $+N$ periods the particle has made $s$ positive and $N-s$ negative steps, resulting in net displacement $x_s = [s-(N-s)]\, l = (2s-N)l$, is

$$P_s(N) = \frac{N!}{s!\,(N-s)!} P^s Q^{N-s}. \tag{10.48}$$

It is called the binomial distribution (Appendix I) and reduces to (10.32) for $P = 1-Q = \frac{1}{2}$ and $m = 2s-N$. By definition

$$\bar{x}_s = \sum_{s=0}^{N} x_s P_s(N) = \sum_{s=0}^{N} (2_s - N)\, l P_s(N), \tag{10.49}$$

$$\overline{(x_s-\bar{x}_s)^2} = \sum_{s=0}^{N} (x_s-\bar{x}_s)^2 P_s(N) = \sum_{s=0}^{N} [(2s-N)\, l - \bar{x}_s]^2 P_s(N). \tag{10.50}$$

The binomial expansion gives

$$\sum_{s=0}^{N} P(s, N) = \sum_{s=0}^{N} \frac{N!}{s!\,(N-s)!} P^s Q^{N-s} = (P+Q)^N = 1^N = 1, \tag{10.51}$$

---

*For a detailed survey see, for example, S. Chandrasekhar, *Rev. Mod. Phys.* **15**, 1 (1943),

$$\bar{s} = \sum_{s=0}^{N} sP_s(N) = NP \sum_{s-1=0}^{N} \frac{(N-1)!\, P^{s-1}\, Q^{N-1-s+1}}{(s-1)!\,(N-1-s+1)!}$$

$$= NP(P+Q)^{N-1} = NP, \tag{10.52}$$

$$\overline{(\delta s)^2} = \overline{(s-\bar{s})^2} = \overline{s^2} - (\bar{s})^2 = \sum_{s=0}^{N} s^2 P_s(N) - (\bar{s}^2)$$

$$= \overline{s(s-1)} + \bar{s} - N^2P^2$$

$$= NP(N-1)P + NP - N^2P^2 = NP(1-P). \tag{10.53}$$

The *fluctuation formula* (10.53) reduces to $\overline{(\delta s)^2} = NP = \bar{s}$ when $P$ is a very small fraction.

From (10.49-53)

$$\bar{x}_s = (2NP-N)\,l = (2P-1)\,Nl = 0, \text{ for } P = \tfrac{1}{2}. \tag{10.54}$$

$$\overline{(x_s-\bar{x}_s)^2} = \Sigma\,[(2s-N)\,l-(2P-1)\,Nl]^2\, P_s$$

$$= \Sigma\,[2s-2PN)\,l]^2\, P_s$$

$$= 4l^2\, \Sigma\,(s^2-2sPN + P^2N^2)\, P_s$$

$$= 4l^2\, \{[NP(1-P) + P^2N^2] - 2NPPN + P^2N^2\}$$

$$= 4l^2\, NP(1-P)$$

$$= l^2N = (v\tau)^2\, N, \qquad \text{for } P = \tfrac{1}{2}. \tag{10.55}$$

For the Brownian motion considered, we have $P=\frac{1}{2}$. Therefore, $\bar{x}_s = 0$ and the tendency to stay away from the origin is measured by

$$\overline{x_s^2} = (v\tau)^2\, N = v^2\tau t = 2Dt, \tag{10.56}$$

in agreement with (10.44) which was obtained by integration.

If $N$ particles are concentrated at $x = 0$ when $t = 0$, then the particle concentration $n_c\,(x, t)$ at $x$ after time $t$ is given by (10.43)

$$\frac{n_c(x, t)}{N} = P(x, t) = (4\pi Dt)^{-1/2} \exp(-x^2/4Dt), \tag{10.57}$$

$$D = \overline{x^2}/2t. \tag{10.58}$$

Following Einstein. we can estimate $D$ in a simple way. Imagine a cylinder of unit cross-section with its axis along the $x$ axis and faces separated by the distance $l$. Let the molecular concentration be $n_a$ at the end $A$ and $n_b$ at the end $B$. The concentration gradient $-dn/dx = (n_a-n_b)/l$ gives rise to the diffusion of molecules. From the gas law the osmotic pressure at $A$ is $p_a = n_akT$ and at $B$ is $p_b = n_bkT$. The resulting force pushing the cylinder in the positive $x$ direction is $(p_a-p_b) = (n_a-n_b)/kT$. If $n$ is the mean concentration of the molecules, the force acting on a single particle is

$$f = \frac{(n_a-n_b)\,kT}{nl} = -\frac{1}{n}\frac{dn}{dx}\,kT$$

$$= 6\pi\eta_v rv, \tag{10.59}$$

where the last step comes from Stokes' law for a spherical particle of radius $r$ moving with speed $v = 1/\tau$ in a medium of viscosity $\eta_v$.

By the definition of $D$, we can equate $-D(dn/dx)$ to the number of molecules $nv$ moving to the right per second per area,

$$-D\frac{dn}{dx} = nv = -\left(\frac{kT}{6\pi\eta_v r}\right)\frac{dn}{dx}$$

or

$$D = kT/(6\pi\eta_v r), \tag{10.60}$$

as given originally by Einstein. Every quantity can now be measured in the result

$$\overline{x^2} = 2Dt = (kT/3\pi\eta_v r)\, t = (R/N_a)\,(T/3\pi\eta_v r)\, t. \tag{10.61}$$

Perrin (1910) verified this relation by recording under a microscope the position of a particle at intervals of $t = 30$ s. The components $x$ of the observed displacements in 30 s intervals gave the mean value $\overline{x^2}$. Substituting it he found $N_a = 6.68 \times 10^{23}$ mol$^{-1}$.

## 10.7 FOURIER ANALYSIS OF A RANDOM FUNCTION

Many processes proceed by random walk. A particle continues to move in one direction until it comes across an obstacle. After some delay, it moves off in a new direction, and so on.

A *random process* or *stochastic process*, is a process $x(t)$ such that the variable $x$ does not depend in a well defined way on the independent variable (say) time $t$. Measurements on the various but similar systems of an ensemble give different functions $x(t)$. As we cannot determine $x(t)$ for the various systems, we try to study relevant probability distributions. It is convenient to resolve the variables into components according to a harmonic law.

As an example, consider a quantity $x(t) = F(t)$ varying spontaneously according to a random law, say, as a result of heat fluctuations. It may be the readings of a galvanometer in a closed circuit in the absence of external emf. Heat fluctuation can produce an irregular emf and so current of any sign in the circuit. The temperature and concentration of current carriers fluctuate in various parts of the circuit thereby producing a variable emf.

We can form an ensemble by cutting a long oscillogram record obtained in a long period of time into equal length strips. Each strip is an element of the ensemble. We can arrange them one below the other as shown in Fig. 10.1. The ensemble averages are taken in a vertical direction. The time averages are taken in the horizontal direction.

We can find, for example,

$p_1(x, t)\, dx$ = probability of finding $x$ between $x$ and $x + dx$ at time $t$,

$p_2\,(x_1 t_1, x_2 t_2)\, dx_1\, dx_2$ = probability of finding $x$ between $x_1$ and $x_1 + dx_1$ at time $t_1$ and between $x_2$ and $x_2 + dx_2$ at time $t_2$,

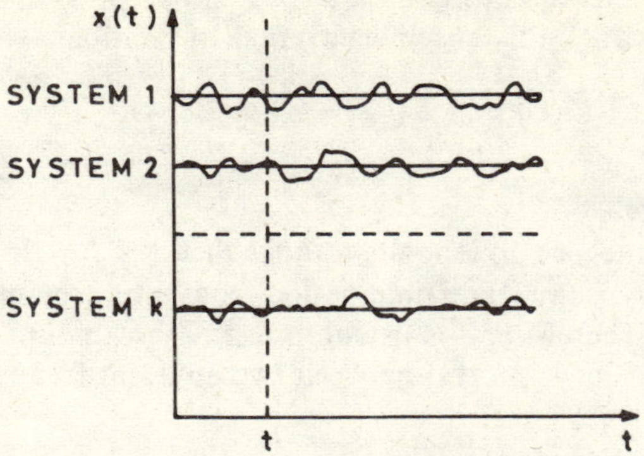


**Fig. 10.1** Plot of some physical quantity $x(t)$ versus time $t$ for various systems of an ensemble.

and similarly $p_3, p_4, \ldots, p_\infty$. Usually, it is enough to find $p_2$. When it is so, the random process is called a *Markoff process.* If the joint probability distributions $p_l$ are invariant under a transformation which shifts the origin of time, we get a *stationary* random process. We would be interested only in the stationary Markoff processes.

Consider a randomly fluctuating quantity $x(t) \equiv F(t)$ which varies with time. We study its behaviour for a particular system in the ensemble from $t = 0$ to $t = T$. This set of irregularly varying readings can be expressed as a Fourier series,

$$F(t) = \sum_{n=1}^{\infty} a_n \cos \frac{2\pi nt}{T} + \sum_{n=1}^{\infty} b_n \sin \frac{2\pi nt}{T}$$

$$= \sum_{n=1}^{\infty} (a_n \cos \omega_n t + b_n \sin \omega_n t), \quad \omega_n = 2\pi n/T. \tag{10.62}$$

Let { } denote the time average. We assume that $\{F(t)\} = 0$. Therefore, we have not taken any constant term in (10.62). The Fourier coefficients vary from one record of duration $T$ to another.

If $x(t)$ is (say) an electric current in a unit resistance, the instantaneous power dissipation is $F^2(t)$. The power in the $n$th component is

$$w_n = (a_n \cos \omega_n t + b_n \sin \omega_n t)^2. \tag{10.63}$$

The time average is

$$\{w_n\} = \tfrac{1}{2}(a_n^2 + b_n^2), \tag{10.64}$$

because $\{\cos^2 \omega_n t\} = \{\sin^2 \omega_n t\} = \frac{1}{2}$ and $\{\cos \omega_n t \sin \omega_n t\} = 0$. By definition

$$\{F^2(t)\} = \frac{1}{T}\int_0^T \sum_n (a_n \cos \omega_n t + b_n \sin \omega_n t)^2 \, dt$$

$$= \frac{1}{2}\sum_n (a_n^2 + b_n^2) = \sum_n \{w_n\}. \tag{10.65}$$

If we make measurements on a large number of systems (independent strips of the record), the ensemble average of the time average is

$$\overline{\{F^2(t)\}} = \tfrac{1}{2}\sum_n (\overline{a_n^2} + \overline{b_n^2}) = \sum_n \{\overline{w_n}\}. \tag{10.66}$$

**Spectral Density**

Each positive integer corresponds to a frequency $\omega_n = 2\pi n/T$. The power at this frequency is $\{\overline{w_n}\}$. The time period from 0 to $T$ being arbitrary, $\omega_n$ is also arbitrary. Therefore, it is useful to define the amount of power $G(\omega_n)$ $\Delta\omega_n$ in the frequency interval between two adjacent frequencies.

$$\Delta\omega_n = \omega_{n+1} - \omega_n = 2\pi/T. \tag{10.67}$$

Then

$$G(\omega_n)\,\Delta\omega_n = \tfrac{1}{2}\,(\overline{a_n^2} + \overline{b_n^2}) = \{\overline{\omega_n}\}, \tag{10.68}$$

$$\overline{\{F^2(t)\}} = \sum_n G(\omega_n)\,\Delta\omega_n = \int_0^\infty G(\omega)\,d\omega. \tag{10.69}$$

The $G(\omega)$ is the *spectral density* or *power spectrum* of the randomly fluctuating quantity $F^2(t)$. Its integral over all frequencies gives the ensemble averaged total power, which is independent of time.

*Correlation Function*: Suppose $x(t)$ is some displacement, so that it denotes the difference between the quantity and its mean value $\{x\} = 0$. There exists a correlation between the values of $x(t)$ at different instants. The value of $x$ at $t$ affects the probabilities of its various possible values at a later instant $t + \tau$. We characterize this time correlation by the mean value of the product, called the correlation function,

$$C(\tau) = \{x(t)\,x(t+\tau)\}, \tag{10.70}$$

where the average is over time $t$. It is a function of $\tau$ only. As $\tau$ increases, the correlation tends to zero, and so also $C(\tau) \rightarrow 0$.

With no change in the result, we can take an ensemble average of (10.70),

$$C(\tau) = \overline{\{x(t)\,x(t+\tau)\}}$$

$$= \overline{\frac{1}{T}\int_0^T \sum_{n,m} [a_n \cos\omega_n t + b_n \sin\omega_n t]\,[a_m \cos\omega_m (t+\tau) + b_m \sin\omega_m (t+\tau)]}$$

$$= \tfrac{1}{2}\sum_n \overline{(a_n^2 + b_n^2)}\cos\omega_n\tau, \tag{10.71}$$

where we have expanded $\sin\omega_m(t+\tau)$ and $\cos\omega_m(t+\tau)$. From (10.68, 71),

$$C(\tau) = \int_0^\infty G(\omega) \cos \omega\tau \, d\omega, \tag{10.72}$$

where the sum has been replaced by an integral. Thus $C(\tau)$ is the Fourier cosine transform of $G(\omega)$. The inverse transform gives

$$G(\omega) = 4 \int_0^\infty C(\tau) \cos \omega\tau \, d\tau. \tag{10.73}$$

These relations state the *Wiener-Khintchine theorem.*

As an example, take

$$C(\tau) = C(0) \exp(-\tau/\tau_0) = \overline{\{F^2(t)\}} \exp(-\tau/\tau_0). \tag{10.74}$$

The $\tau_0$ is a measure of the average time the system persists without changing its state by more than $e^{-1}$. Thus, $\tau_0$ is a persistence time or *correlation time* for the random process. From (10.73, 74).

$$G(\omega) = 4 \int_0^\infty \overline{\{F^2(t)\}} \cdot \exp(-\tau/\tau_0) \cos \omega\tau \, d\tau = \frac{4\tau_0 \overline{\{F^2(t)\}}}{1 + (\omega\tau_0)^2}. \tag{10.75}$$

This function is plotted in Fig. 10.2. The power spectrum is nearly flat up to $\omega \approx 1/\tau_0$, and then decreases rapidly as $1/\omega^2$.

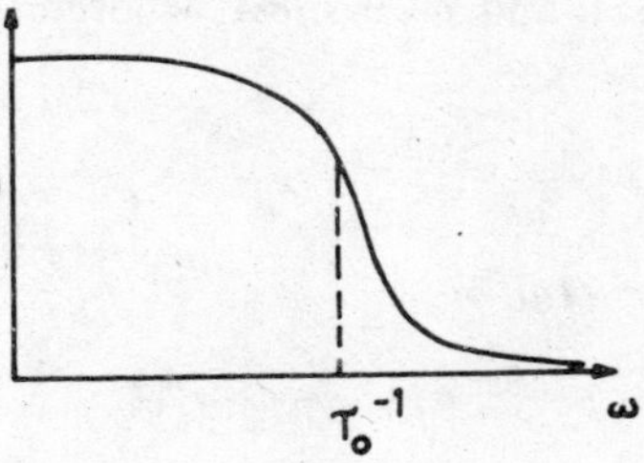


**Fig. 10.2 Plot of $G(\omega)$ versus $\omega$.**

## 10.8 ELECTRICAL NOISE (NYQUIST THEOREM)

Consider a resistor of area $A$, length $L$ and resistance $R$. Suppose it has $n$ electrons per unit volume and the time between collisions is $\tau_e$. By Ohm's law, the voltage $V$ is given by

$$V = RI = RAne\bar{u}, \tag{10.76}$$

where $I$ is the current, $e$ the electrical charge and $\bar{u}$ the average velocity components of the electrons along the length of the resistor. Because $nAL$ is the total number of electrons,

$$nAL\bar{u} = \sum_i u_i, \tag{10.77}$$

where the sum is over all electrons. From (10.76, 77),

$$V = (Re/L) \sum_i u_i = \sum_i V_i, \quad V_i = Re\, u_i/L. \tag{10.78}$$

The random variable is $u_i$, or $V_i$.

We can regard $\tau_e$ as the relaxation time, or mean time of flight of the conduction electrons. Then the correlation function may be given the form

$$C(\tau) = \overline{V_i(t)\, V_i(t+\tau)} = \overline{V_i^2} \exp(-\tau/\tau_e). \tag{10.79}$$

By the Wiener-Khintchine theorem, the spectral density function is

$$G(\omega) = 4\left(\frac{R_e}{L}\right)^2 \overline{u^2} \int_0^\infty \exp(-\tau/\tau_e) \cos \omega\tau \, d\tau$$

$$= 4\left(\frac{R_e}{L}\right)^2 \overline{u^2} \frac{\tau_e}{1 + (\omega\tau_e)^2}. \tag{10.80}$$

In metals at 300 K, $\tau_e \sim 10^{-12}$ s, or $\omega\tau_e \ll 1$ for the usual circuit analysis frequencies (dc to microwave). Then $1 + (\omega\tau_e)^2 \simeq 1$, and using $\frac{1}{2} m\overline{u^2} = \frac{1}{2} kT$,

$$\overline{V^2} = nAL\overline{V_i^2} = nAL\, G(\omega)\, \Delta\omega$$

$$= nAL\, 4\left(\frac{kT}{m}\right)\left(\frac{R_e}{L}\right)^2 \tau_e\, \Delta\omega. \tag{10.81}$$

As $\sigma = ne^2 \tau_e/m$, (9.37), and $R = L/\sigma A$, where $\sigma$ is the electrical conductivity, we get

$$\overline{V^2} = 4kTR\, \Delta\omega. \tag{10.82}$$

This is the *Nyquist theorem.*

## *PROBLEMS*

10.1 What is the probability that $n$ particles are found in a volume $v < V$ of an ideal classical gas?

[Ans. Poisson distribution $P_n = \dfrac{\bar{n}^n \exp(-\bar{n})}{n!}$, $\bar{n}$ = mean number of particles in the volume $v$]. [Hint: use 4.58, 59, 70, 71, 75].

10.2 Consider a microcanonical ensemble of $N$ elements and allow a very weak interaction among the systems (interaction energy $\delta E$). Let $w_{sr}$ be the probability per unit time that any system due to the interaction makes a transition from a state $r$ to a state $s$. By the *principle of detailed balance* $w_{sr} = w_{rs}$. If $N_r$ is the number of subsystems in the state $r$, we can write

$dN_r/dt$ = rate of entering $r$ − rate of leaving $r$

$= \Sigma' w_{rs} N_s - N_r \Sigma' w_{sr}$; (*master equation*).

Use these results to show that in equilibrium all states of the entire system have the same occupation numbers.

10.3 Boltzmann defined a quantity which is minus the entropy $\sigma$, $H \equiv -\sigma$. Use the definitions $\sigma = -\sum_r p_r \ln p_r$, $p_r = N_r/N$ and the master equation given in Prob. 10.2 to prove the *Boltzmann H theorem* $dH/dt \leqslant 0$. [Hint: $\sigma = -\sum (N_r/N) \ln (N_r/N) = N^{-1} (N \ln N - \sum N_r \ln N_r$, $d\sigma/dt = -N^{-1} \sum \dot{N}_r \ln N_r = (2N)^{-1} \sum' w_{rs} (N_r - N_s) (\ln N_r - \ln N_s) \geqslant 0$ as even term is positive or zero].

10.4 Consider a two-level system in contact with a heat reservoir. Let $E_2 - E_1 = \epsilon$. Show that $w_{21}/w_{12} = \exp(-\epsilon/kT)$, where $w_{21}$ is the probability per unit time that a particle in state 2 makes a transition to state 1.

10.5 A substance has nuclei of spin $\frac{1}{2}$ and magnetic moment $\mu_n$, and unpaired electrons ($\mu_e < 0$). It is placed in a magnetic field $H$ pointing in the $z$ direction. There is a hyperfine interaction due to the magnetic field produced by the electron at the position of the nucleus. Show that

$$\frac{n_+ N_-}{n_- N_+} = \exp [2\beta (\mu_n - \mu_e) H]$$

where $n_+ (N_+)$ is the mean number of nuclear (electron) up spins. [Hint: For the combined system $(n + e)$ in thermal contact with the lattice heat reservoir, use the principle of detailed balance $n_+ N_- w_{ne} (+\ominus \rightarrow -\oplus) = n_- N_+ w_{ne} (-\oplus \rightarrow +\ominus)$, where $+(\oplus)$ indicates the up orientations of nucleus (electron)].

10.6 Use the result of Prob. (10.5) to discuss the *Overhauser effect* whereby the nuclear polarization in a magnetic field is enhanced above the thermal equilibrium value. (Phys. Rev. **92**, 411, 1953).

10.7 Consider two types of quantum states 1 and 2. In a canonical ensemble show that

$$p_1 = \exp [-\beta (F_1 - F)],$$

where $p_1$ is the probability that the system is in state 1, $F$ is the free energy of the total system, and $F_1$ of the subsystem 1.

10 8 A container has exactly saturated vapour. What is the probability of finding a droplet of radius $R$?

10.9 In problem 10.7, if we have a microcanonical ensemble, $p_1 = n_1/n$, where $n_1$ is the number of quantum states of type 1 and $n$ is the total number, then show that $p_1 = \exp(\sigma_1 - \sigma) = \exp(\Delta\sigma)$. If entropy $\sigma$ is a function of some parameter $x$, show that the probability of a fluctuation giving $x$ is

$$p(x) \propto \exp \left[ \frac{1}{2} (x - x_0^2) \left( \frac{\partial^2 \sigma}{\partial x^2} \right)_{x_0} \right]$$

to second order. If $\Delta\sigma_t$ is the change in entropy in the fluctuation given by $\Delta\sigma_t = -R_{min}/T$, where $R_{min}$ is the minimum work for the reversible change in the variables, $R_{\min} = \Delta\epsilon - T_0\,\Delta\sigma + p_0 dV$, then show that

$$p(x) \sim \exp\left(\frac{\Delta p\,\Delta V - \Delta T\,\Delta\sigma}{2kT}\right),$$

$$\overline{(\Delta T)^2} = kT^2/C_V \text{ and } \overline{(\Delta V)^2} = -kT\,(\partial V/\partial p)_T.$$

10.10 Use the Fourier integral

$$x(t) = \int_{-\infty}^{\infty} A(\omega)\, e^{i\omega t}\, d\omega$$

and the reality of $x(t)$ to derive (10.72).

10.11 For a Gaussian random process

$$\omega(x) = \frac{1}{\rho(2\pi)^{1/2}} \exp\left(-x^2/2\rho^2\right).$$

Calculate $\overline{x^2(t)\, x^2(t+\tau)}$ for it.

10.12 Use the drift velocity equation

$$m\left(\frac{d}{dt} + \frac{1}{\tau_0}\right)\bar{u} = e\mathbf{E}$$

to derive $\sigma = Ne^2\,\tau_0/m$, as used in obtaining (10.82).

# 11
# COOPERATIVE PHENOMENA: ISING MODEL

## 11.1 PHASE TRANSITIONS OF THE SECOND KIND

Consider ferromagnetic substances, like iron and nickel. Some of the spins of the atoms become spontaneously (without any external field) polarized in the same direction, below the Curie temperature $T_c$. This creates a macroscopic magnetic field. As temperature is raised, the thermal energy makes it possible for some of the aligned spins to flip over. This tends to destory the initial ordered state. For $T > T_c$, the spins get oriented at random. The spontaneous magnetization vanishes. As $T_c$ is approached from both sides, the heat capacity of the metal approaches $\infty$. The transition from the nonferromagnetic state to the ferromagnetic state is called a *phase transition of the second kind*. It is associated with some kind of *change in symmetry* of the lattice. For example, in ferromagnetism the symmetry of spins is involved. The energy levels of the system are given by

$$E = -\sum_{ii'} \epsilon_{ii'}\, \sigma_i \sigma_{i'} - \mu_B H \sum_i \sigma_i, \tag{11.1}$$

where, on each lattice site $i$, the spin quantum number $\sigma_i$ is $+1$ or $-1$, $\epsilon_{ii'}$, is the interaction energy, and $\mu_B H$ is the interaction energy associated with the external magnetic field $H$. For spontaneous configuration $H = 0$.

The change of symmetry can also occur due to the change in the *ordering* of the crystal. For example, in an alloy $AB$ the atoms may be substituted for one another on a set of given lattice sites. Then we can say that $\sigma_i = +1$ for an atom $A$ on the site $i$, and $\sigma_i = -1$ for an atom $B$ on that site. At low temperatures the alloy $AB$ is ordered. Above a transition temperature it becomes disordered.

The difference between the nonferromagnetic-ferromagnetic transition and the order-disorder transition is that in the former case 'up' and 'down' spins can be transformed freely into one another, while in the latter case the total number of $A$ type and $B$ type atoms is fixed. However similar theoretical results hold in both the cases.

These transitions come under a large group of phenomena called *cooperative phenomena.* Certain subsystems, like spins or atoms, cooperate due to *exchange interactions* to form units below a certain critical point.

Note that a phase transition of the second kind, in contrast to ordinary phase transitions (of the first kind), is continuous in the sense that the state of the body changes continuously. Although the symmetry changes discontinuously at the transition point, at each instant the body belongs to one of the two phases. At a phase transition point of the first kind, the bodies in two different states are in equilibrium, while at a phase transition point of the second kind the states of the two phases are the same.

## 11.2 ISING MODEL

The theory of cooperative phenomena is very complicated, specially when all interactions are included and three-dimensional systems are considered.

We assume that in (11.1) the $\epsilon_{ii'}$, acts only between nearest neighbours in the lattice. This is the basic assumption of the *Ising model.* Then (11.1) is written as

$$E\{\sigma_i\} = -\epsilon \sum_{\langle i, j\rangle} \sigma_i\sigma_j - \mu_B H \sum_i \sigma_i, \tag{11.2}$$

where $\langle i, j\rangle$ means that the sum is over *pairs of nearest neighbours*, and the interactions are isotropic, that is, all $\epsilon_{ij}$ have the same value $\epsilon$. For $\epsilon > 0$, the neighbouring spins tend to be parallel and ferromagnetism is possible. *The spontaneous configuration of least energy is the completely polarized (ordered) configuration in which all the Ising spins are oriented in the same direction*, (11.2). This configuration is attained at $T = 0$. For $\epsilon < 0$, the neighbouring spins tend to be antiparallel and *antiferromagnetism* results. We will assume that $\epsilon > 0$.

In (11.2) no distinction is made between $\langle ij\rangle$ and $\langle ji\rangle$. The sum over $\langle ij\rangle$ has $\tilde{z}N/2$ terms, where $\tilde{z}$ is the number of nearest neighbours of a site (*coordination number* of the lattice) and $N$ the number of spins.

The thermodynamic quantities require the evaluation of the partition function

$$\begin{aligned} Z &= \sum_{\{\sigma_i\}} \exp(-\beta E\{\sigma_i\}) \\ &= \sum_{\sigma_1} \sum_{\sigma_2} \ldots \sum_{\sigma_N} \exp(-\beta E\{\sigma_i\}), \end{aligned} \tag{11.3}$$

where $\beta = 1/kT$ and the sum is taken over all the $2^N$ possible combinations of the $N$ spins.

It is extremely difficult to calculate (11.3). Several approximate methods have been developed for this. The Bragg-Williams (BW) approximation is the simplest.

## 11.3 BRAGG-WILLIAMS APPROXIMATION

Bragg and Williams assume that the distribution of spins is random. Let $N_+(N_-)$ be the number of spins for which $\sigma_i$ is $+1\,(-1)$. Then $N_+/N\ (N_-/N)$ is the probability of finding a spin $+1\,(-1)$, or up (down), on a given lattice site.

From (11.2), assuming random arrangement of spins over the whole lattice,

$$E = -\frac{1}{2}\tilde{z}N\epsilon\left[\left(\frac{N_+}{N}\right)^2 + \left(\frac{N_-}{N}\right)^2 - \frac{2N_+N_-}{N^2}\right] - \mu_B H(N_+ - N_-). \tag{11.4}$$

where $\tilde{z}$ is the number of nearest neighbours of a site, $N = N_+ + N_-$ is the number of spins, and we have taken $N_+ > N_-$ in the last term. The number $N_+/N$ is a measure of the *long-range order*, as it requires no correlation between nearest neighbours. It only requires that in the entire lattice a fraction $N_+/N$ of all the spins are up If $N_+/N$ is known in the neighbourhood of a given spin, then the same average value is likely to occur everywhere on the entire lattice.

As $\mu_B$ is the magnetic moment associated with the spin, the total magnetic moment is

$$M = \mu_B\,(N_+ - N_-). \tag{11.5}$$

Using $N = N_+ + N_-$,

$$\frac{N_\pm}{N} = \tfrac{1}{2}\,(1 \pm m), \qquad m \equiv \frac{M}{N\mu_B}, \tag{11.6}$$

$$E = -\tfrac{1}{2}\,\tilde{z}\,\epsilon N m^2 - \mu_B\, NmH. \tag{11.7}$$

The $m$ is called *long-range order parameter*, $m = (N_+ - N_-)/N = (2N_+/N) - 1$, $-1 \leqslant m \leqslant +1$. The order parameter $m$ may be magnetization in a ferromagnetic system, the dielectric polarization in a ferroelectric system, the fraction of neighbour $A-B$ bonds to total bonds in an alloy $AB$, or the fraction of superconducting electrons in a superconductor. In transitions, where the atoms are displaced from their positions in the symmetrical phase, $m$ can be taken as the amount of this displacement.

The number of arrangements of spins over the $N$ sites is given by the number of ways we can pick $N_+$ things out of $N$,

$$W_{BW} = \frac{N!}{N_+!\,(N-N_+)!}. \tag{11.8}$$

From (3.91),

$$S = k \ln W_{BW} = -Nk\left(\frac{N_-}{N}\ln\frac{N_-}{N} + \frac{N_+}{N}\ln\frac{N_+}{N}\right). \tag{11.9}$$

The Helmholtz free energy $F = E - TS$ is

$$F = -\tfrac{1}{2}\,\tilde{z}\,\epsilon\, Nm^2 - \mu_B NmH - NkT\,[-\ln 2 + \tfrac{1}{2}\,(1-m)\ln(1-m) + \tfrac{1}{2}\,(1+m)\ln(1+m)]. \tag{11.10}$$

The equilibrium value of $m$ (or $N_+ - N_-$) is determined by $\partial F/\partial m = 0$,

$$-\tilde{z}\epsilon Nm - \mu_B NH - NkT\left[-\tfrac{1}{2}\ln(1-m) - \tfrac{1}{2} + \tfrac{1}{2}\ln(1+m) + \tfrac{1}{2}\right] = 0,$$

or

$$\ln\frac{1+m}{1-m} = 2x, \qquad x \equiv \frac{\tilde{z}\epsilon m + \mu_B H}{kT}. \tag{11.11}$$

It gives the well-known result of the Weiss theory,

$$m \equiv \frac{M}{N\mu_B} = \frac{e^{2x}-1}{e^{2x}+1} = \tanh x. \tag{11.12}$$

For $H = 0$, the spontaneous magnetic moment is

$$M_s = N\mu_B \tanh\frac{\tilde{z}\epsilon M_s}{N\mu_B kT},$$

$$m_s = \tanh\frac{T_c m_s}{T}, \quad T_c \equiv \frac{\tilde{z}\epsilon}{k}, \quad m_s \equiv \frac{M_s}{N\mu_B}, \tag{11.13}$$

where $T_c$ is the Curie temperature*. We can solve (11.13) graphically (Fig. 11.1a) to obtain $m_s$ as a function of $T$ in the BW approximation (Fig. 11.1b). For this, plot the right and left sides separately as functions of $m_s$. The

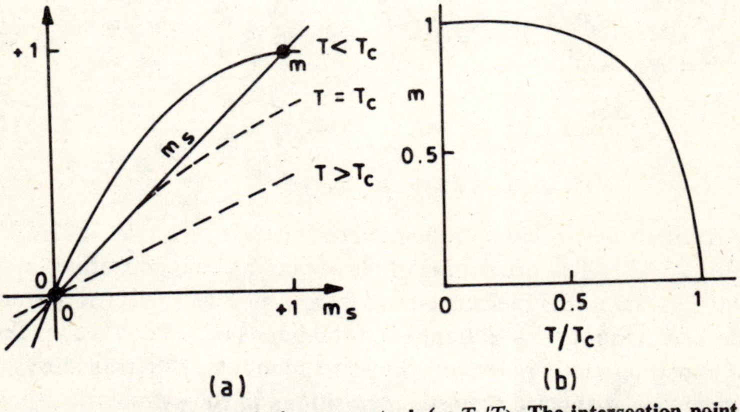


**Fig. 11.1** (a) Graphical solution $m_s = \tanh(m_s T_c/T)$. The intersection point $m$ moves up to $m_s = 1$ as $T \to 0$. All the magnetic moments are lined up at $T = 0$.

(b) The spontaneous magnetic moment in the Bragg-Williams approximation. The order parameter $m$ varies smoothly in a second-order phase transition.

intercepts of the two curves give the value of $m$ at the temperature of interest. Clearly the solution is such that $m_s = 0$ for $T_c/T < 1$ and $m_s = m$, $0$, $-m$ for $T_c/T > 1$. In the latter case the root $m_s = 0$ is not acceptable because it corresponds to a maximum of $F$, instead of minimum. Thus, $m_s = 0$ for $T > T_c$ and $\pm m$ for $T < T_c$, where $m$ is the root of (11.13) that

*Curie points are phase transition points of the second kind.

is greater than zero. The degeneracy $m_s = \pm m$ occurs because for $H = 0$ there is no intrinsic difference between 'up' and 'down'. This degeneracy does not affect $F$ as it is and even function of $m$. In general, $m$ is obtained numerically to yield Fig. 11.1b.

## 11.4 FOWLER-GUGGENHEIM APPROXIMATION

In (11.4) the energy of a spin depends on the distribution of spins over the entire lattice and not on its neighbours. Fowler and Guggenheim (FG) have handled the spins more carefully.

Let $N_{++}$ be the number of $(++)$ pairs, $N_{--}$ of $(--)$ pairs, and $N_{+-}$ of $(+-)$ pairs. There are only $\tilde{z}N_+$ *links* possible which end on $+$ spins (Fig. 11.2). Each of these links is counted twice in $N_{++}$ and once in $N_{+-}$. Thus

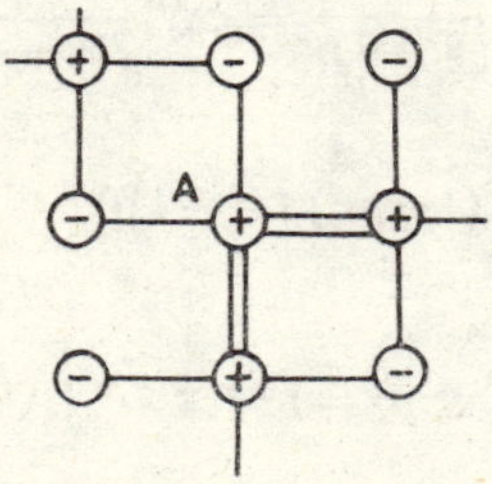


**Fig. 11.2** **A possible two-dimensional lattice ($\tilde{z} = 4$) for the result $\tilde{z}N_+ = 2N_{++} + N_{+-}$. First $\tilde{z}$ lines are drawn to connect $A$ to all the nearest neighbours. This is repeated for another $+$ site and continued for all $+$ sites. In all $\tilde{z}N_+$ lines are drawn.**

$2N_{++} + N_{+-} = \tilde{z}N_+$, and similarly $2N_{--} + N_{+-} = \tilde{z}N_-$. Clearly $N_{++}/(\frac{1}{2}\tilde{z}N)$ is a measure of the *short-range order* (local correlation). For a given spin, it gives the fraction of its nearest neighbours with spin up.

In BW approximation the possibility of local correlation between spins was ignored. We took $N_{++}/(\frac{1}{2}\tilde{z}N) = (N_+/N)^2$. For short-range order,

$$E = -(N_{++} + N_{--} - N_{+-}) - \mu_B H (N_+ - N_-). \tag{11.14}$$

In the *quasi-chemical approximation* of FG, a $(++)$ pair and a $(--)$ pair combine to form 2 $(+-)$ pairs according to the 'reaction'

$$(++) + (--) \rightarrow 2\,(+-).$$

Then, for chemical equilibrium

$$\frac{N_{++}N_{--}}{(N_{+-})^2} = \frac{1}{2^2}\frac{\exp(-\epsilon_{++}/kT)\exp(-\epsilon_{--}/kT)}{[\exp(-\epsilon_{+-}/kT)]^2} = \frac{1}{4}\exp(4\epsilon/kT) \equiv \frac{y^2}{4} \tag{11.15}$$

where $\epsilon_{++}$ is the potential energy of $(++)$ pair, etc.,

$$\epsilon_{++} = \epsilon_{--} = -\epsilon,\ \epsilon_{+-} = +\epsilon,\ (\epsilon > 0),$$

and the factor $2^2$ comes from the presence of the unsymmetrical bond $(+-)$. In (11.15), the pairs are treated as independent chemical bonds, even though they are not due to restrictions $2N_{++} + N_{+-} = \tilde{z}N_+$, etc.

We can write (11.15) as

$$\frac{(\tilde{z}N_+ - N_{+-})(\tilde{z}N_- - N_{+-})}{(N_{+-})^2} = y^2. \tag{11.16}$$

From (11.6, 16),

$$(y^2-1)\,N_{+-}^2 + (\tilde{z}N)N_{+-} - \tfrac{1}{4}\,\tilde{z}^2N^2\,(1-m^2) = 0,$$

or

$$N_{+-} = \tilde{z}N\,\frac{-1+\alpha}{2(y^2-1)} = \tilde{z}N\,\frac{1-m^2}{2(\alpha+1)}, \tag{11.17}$$

where

$$\alpha = [1 + (1-m^2)\,(y^2-1)]^{1/2}. \tag{11.18}$$

Thus

$$N_{\pm\pm} = \tfrac{1}{2}\,(\tilde{z}N_\pm - N_{+-}) = \tfrac{1}{4}\,\tilde{z}N\left[(1 \pm m) - \frac{1-m^2}{1+\alpha}\right]. \tag{11.19}$$

For spontaneous magnetization $(H = 0)$,

$$E = -\epsilon\,(N_{++} + N_{--} - N_{+-}) = -\epsilon\,\tfrac{1}{2}\,\tilde{z}N\left(1 - 2\,\frac{1-m^2}{1+\alpha}\right). \tag{11.20}$$

The calculation of $S$ in $F = E - TS = -kT \ln Z$ is tedious. To simplify it, define a quantity $E'$ as

$$F = E' - kT \ln W_{BW} = -kT \ln Z, \tag{11.21}$$

where $W_{BW}$ is given by (11.8). Then

$$E = -k\,\frac{\partial \ln Z}{\partial(T^{-1})} = \frac{\partial(E'/T)}{\partial(1/T)},\ E' = T\int_0^{1/T} E\,d\left(\frac{1}{T}\right), \tag{11.22}$$

where $E' = 0$ for $1/T = 0$.

From (11.15, 18),

$$\frac{4\epsilon}{kT} = \ln y^2 = \ln\left(1 + \frac{\alpha^2-1}{1-m^2}\right) = \ln\frac{\alpha^2 - m^2}{1-m^2}, \tag{11.23}$$

$$d\left(\frac{1}{T}\right) = \frac{k}{2\epsilon}\,\frac{\alpha\,d\alpha}{\alpha^2 - m^2}, \tag{11.24}$$

$$E' = -\tfrac{1}{4}\,\tilde{z}NkT\int_1^\infty\left(1 - 2\,\frac{1-m^2}{\alpha+1}\right)\frac{\alpha d\alpha}{\alpha^2 - m^2}$$

$$= -\tfrac{1}{4}\tilde{z}NkT \int_1^{\alpha} \left( \frac{2}{\alpha+1} + \frac{m-\frac{1}{2}}{\alpha - m} - \frac{m+\frac{1}{2}}{\alpha+m} \right) d\alpha$$

$$= -\tfrac{1}{4}\tilde{z}NkT \left[ 2 \ln \frac{\alpha+1}{2} + (m-\tfrac{1}{2}) \ln \frac{\alpha - m}{1-m} - (m+\tfrac{1}{2}) \ln \frac{\alpha+m}{1+m} \right]$$

$$= -\tfrac{1}{4}\tilde{z}NkT \left[ 2 \ln \frac{\alpha+1}{2} + (m-1) \ln \frac{m-\alpha}{m-1} - (m+1) \ln \frac{m+\alpha}{m+1} + \frac{2\epsilon}{kT} \right]. \tag{11.25}$$

Using (11.9, 25),

$$F = E' - kT \ln W_{BW}$$

$$= \frac{1}{2} NkT \left\{ (1+m) \ln (1+m) + (1-m) \ln (1-m) - 2 \ln 2 + \tfrac{1}{2}\tilde{z} \left[ (1+m) \ln \frac{\alpha+m}{1+m} + (1-m) \ln \frac{\alpha - m}{1-m} - 2 \ln \frac{\alpha+1}{2} - \frac{2\epsilon}{kT} \right] \right\} \tag{11.26}$$

The equilibrium condition, $\partial F/\partial m = 0$, gives

$$\ln \frac{1+m}{1-m} + \frac{1}{2}\tilde{z} \left[ \ln \frac{\alpha+m}{\alpha-m} - \ln \frac{1+m}{1-m} + \frac{1+m}{\alpha+m} \left( 1 + \frac{d\alpha}{dm} \right) - \frac{1-m}{\alpha-m} \left( 1 - \frac{d\alpha}{dm} \right) - \frac{2}{\alpha+1} \frac{d\alpha}{dm} \right] = 0. \tag{11.27}$$

Using (11.18),

$$\frac{d\alpha}{dm} = \frac{1-y^2}{\alpha} m = \frac{m}{\alpha} \left( \frac{1-\alpha^2}{1-m^2} \right),$$

$$\frac{\partial F}{dm} = (1 - \tfrac{1}{2}\tilde{z}) \ln \frac{1+m}{1-m} + \tfrac{1}{2}\tilde{z} \ln \frac{\alpha+m}{\alpha-m} = 0. \tag{11.28}$$

One root of (11.28) is always $m = 0$. A root $m \neq 0$ exists for low temperatures. At the Curie temperature the two roots merge to give the single root $m = 0$. It means the Curie temperature is given by the conditions

$$m = 0, \qquad \partial F/\partial m = 0, \qquad \partial^2 F/\partial m^2 = 0, \tag{11.29}$$

$$\frac{\partial^2 F}{\partial m^2} = (1 - \tfrac{1}{2}\tilde{z}) \left( \frac{1}{1+m} + \frac{1}{1-m} \right) + \tfrac{1}{2}\tilde{z} \left( \frac{1}{\alpha+m} + \frac{1}{\alpha-m} \right) = 0. \tag{11.30}$$

For $m = 0$, we have $\alpha = y$ and (11.30) gives

$$\frac{\tilde{z}}{\tilde{z}-2} = \alpha = y = \exp(2\epsilon/kT_c), \quad \frac{2\epsilon}{kT_c} = \ln \frac{\tilde{z}}{\tilde{z}-2}. \tag{11.31}$$

For $\tilde{z} = 2$, we have $T_c = 0$. Therefore, *a one-dimensional linear lattice cannot be ferromagnetic.* The BW approximation does not yield this result. Bethe used a different approach to obtain (11.31).

The values of $\tilde{z}$ and $\ln[\tilde{z}/(\tilde{z}-2)]$ for the common lattice types are as follows:

| *Lattice type* | $\tilde{z}$ | $\ln[\tilde{z}/(\tilde{z}-2)]$ |
|---|---|---|
| Simple cubic | 6 | 0.41 |
| Body-centered cubic | 8 | 0.26 |
| Face-centered cubic | 12 | 0.18 |
| Hexagonal close packed | 12 | 0.18 |
| Two-dimensional square net | 4 | 0.69 |

## 11.5 KIRKWOOD METHOD

For $H = 0$, (11.3) is

$$Z = \sum_{\{\sigma_i\}} e^w, \quad w = \frac{\epsilon}{kT} \sum_{\langle i,j \rangle} \sigma_i \sigma_j. \tag{11.32}$$

From (11.21, 32),

$$\begin{aligned}\frac{E'}{kT} &= -\ln\left(\frac{1}{W_{BW}} \sum_{\{\sigma_i\}} e^w\right) \\ &= -\ln\left[\frac{1}{W_{BW}} \sum_{\{\sigma_i\}} \left(1 + w + \frac{1}{2!} w^2 + \dots\right)\right] \\ &= -\ln\left(1 + \frac{\langle w \rangle}{1!} + \frac{\langle w^2 \rangle}{2!} + \dots\right) \\ &= -\left[\left(\frac{\langle w \rangle}{1!} + \frac{\langle w^2 \rangle}{2!} + \dots\right) - \frac{1}{2}\left(\frac{\langle w \rangle}{1!} + \frac{\langle w^2 \rangle}{2!} + \dots\right)^2 + \dots\right] \\ &= -\langle w \rangle - \frac{1}{2!}[\langle w^2 \rangle - \langle w \rangle^2] - \frac{1}{3!}[\langle w^3 \rangle - 3\langle w^2 \rangle\langle w \rangle + 2\langle w \rangle^3] - \dots,\end{aligned} \tag{11.33}$$

where

$$\langle w^n \rangle = \frac{1}{W_{BW}} \sum_{\{\sigma_i\}} w^n. \tag{11.34}$$

We have already calculated $W_{BW}$. To evaluate the averages $\langle w^n \rangle$, it is useful to let $i, k, \dots$ belong to one sublattice and $j, l, \dots$ to another. Then $\sigma_i$ and $\sigma_j$ can vary independently. Using (11.7),

$$\langle w \rangle = \frac{\epsilon}{kT} \sum_{\langle i,j \rangle} \langle \sigma_i \sigma_j \rangle = \frac{\epsilon}{kT} \sum_{i,j} \langle \sigma_i \rangle \langle \sigma_j \rangle$$

$$= -\tfrac{1}{2} \tilde{z} \frac{\varepsilon}{kT} Nm^2. \qquad (11.35)$$

The term

$$\langle w^2 \rangle = \frac{\epsilon}{kT} \sum_{(ij)(kl)} \langle \sigma_i \sigma_j \sigma_k \sigma_l \rangle \qquad (11.36)$$

has three cases (Fig. 11.3): (a) $i = k$, $j = l$, (b) $i \neq k$, $j = l$ or $i = k$, $j \neq l$, and (c) $i \neq k$, $j \neq l$. The number of terms in each case is (a)

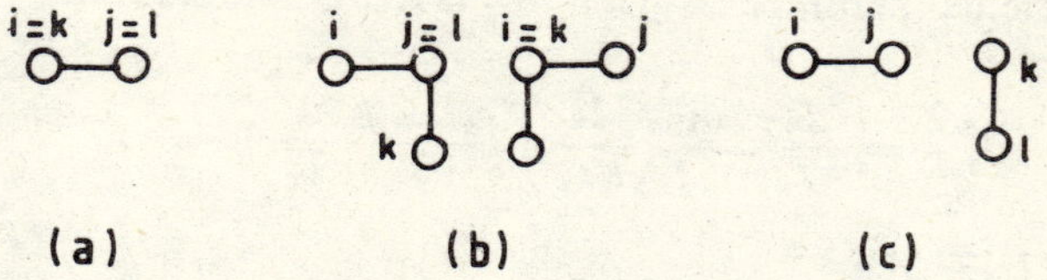


**Fig. 11.3 Construction for the three cases of (11.36).**

$q \equiv \frac{1}{2} N\tilde{z}$, (b) $q(\tilde{z}-1)$, and (c) $q^2 - q - 2q(\tilde{z}-1)$. The corresponding average values are: (a) 1 as $\sigma_i^2 = 1$, (b) for the pairs $(i, k)$ which are on the same sublattice, using (11.6),

$$\langle \sigma_i \sigma_k \rangle = \frac{\frac{1}{2}N_+(\frac{1}{2}N_+ - 1)}{\frac{1}{2}N(\frac{1}{2}N - 1)} + \frac{\frac{1}{2}N_-(\frac{1}{2}N_- - 1)}{\frac{1}{2}N(\frac{1}{2}N - 1)} - 2\frac{\frac{1}{2}N_+ \frac{1}{2}N_-}{\frac{1}{2}N(\frac{1}{2}N - 1)}$$

$$= m^2 - \frac{1-m^2}{\frac{1}{2}N - 1} \simeq m^2 - \frac{2}{N}(1-m^2) \equiv m^2 - \Delta = \langle \sigma_j \sigma_l \rangle,$$

and (c) $\langle \sigma_i \sigma_j \sigma_k \sigma_l \rangle = \langle \sigma_i \sigma_k \rangle \langle \sigma_j \sigma_l \rangle$ implies, to order $O(N)$,

$$\sum_{(ij)(kl)} \langle \sigma_i \sigma_j \sigma_k \sigma_l \rangle = q + 2q(\tilde{z}-1)(m^2 - \Delta) + q(q - 2\tilde{z} + 1)(m^2 - \Delta)^2$$

$$= q + 2q(\tilde{z}-1)m^2 + q^2 m^4 - 2q^2 m^2 \Delta - q(2\tilde{z}-1)m^4 + O(1)$$

$$= q^2 m^4 + q[1 + 2(\tilde{z}-1)m^2 - 2\tilde{z}m^2(1-m^2)$$
$$-(2\tilde{z}-1)m^4] + O(1)$$
$$= q^2 m^4 + q(1-m^2)^2. \qquad (11.37)$$

The final result has the form

$$-\frac{E'}{NkT} = \tfrac{1}{2}\tilde{z}\left[\frac{\epsilon}{kT} m^2 + \tfrac{1}{2}\left(\frac{\epsilon}{kT}\right)^2 (1-m^2)^2 + \tfrac{2}{3}\left(\frac{\epsilon}{kT}\right)^3 m^2(1-m^2) + \dots\right] \qquad (11.38)$$

$$F = \frac{1}{2} NkT \Big\{ (1+m)\ln(1+m) + (1-m)\ln(1-m) - 2\ln 2$$
$$- \tilde{z}\Big[\Big(\frac{\epsilon}{kT}\Big) m^2 + \tfrac{1}{2}\Big(\frac{\epsilon}{kT}\Big)^2 (1-m^2)^2 + \tfrac{2}{3}\Big(\frac{\epsilon}{kT}\Big)^3 m^2(1-m^2) + \ldots\Big]\Big\}. \tag{11.39}$$

The equilibrium value of $m$ is obtained by $\partial F/\partial m = 0$,

$$\ln\frac{1+m}{1-m} = \tilde{z}m\Big[\frac{2\epsilon}{kT} - \frac{1}{2!}\Big(\frac{2\epsilon}{kT}\Big)^2 (1-m^2)$$
$$+ \frac{1}{3!}\Big(\frac{2\epsilon}{kT}\Big)^3 (1-m^2)(1-3m^2) - \ldots\Big]. \tag{11.40}$$

The Curie temperature is given by (11.29). Differentiate (11.40) and put $m = 0$,

$$\frac{2}{\tilde{z}} = \frac{2\epsilon}{kT_c} - \frac{1}{2!}\Big(\frac{2\epsilon}{kT_c}\Big)^2 + \frac{1}{3!}\Big(\frac{2\epsilon}{kT_c}\Big)^3 - \ldots. \tag{11.41}$$

From (11.13, 31),

$$\text{BW:}\quad \frac{2}{\tilde{z}} = \frac{2\epsilon}{kT_c}, \tag{11.42}$$

$$\text{FG:}\quad \frac{2}{\tilde{z}} = 1 - \exp(-2\epsilon/kT_c) = \frac{2\epsilon}{kT_c} - \frac{1}{2!}\Big(\frac{2\epsilon}{kT_c}\Big)^2 + \frac{1}{3!}\Big(\frac{2\epsilon}{kT_c}\Big)^3 - \ldots. \tag{11.43}$$

Thus Bragg-Williams approximation agrees with Kirkwood's approximation (11.41) only in the first term, while quasi-chemical approximation (11.43) of Fowler and Guggenheim agrees up to the first three terms.

The heat capacity curves according to the three approximations are compared in Fig. 11.4. The discontinuity in the specific heat curve charac-

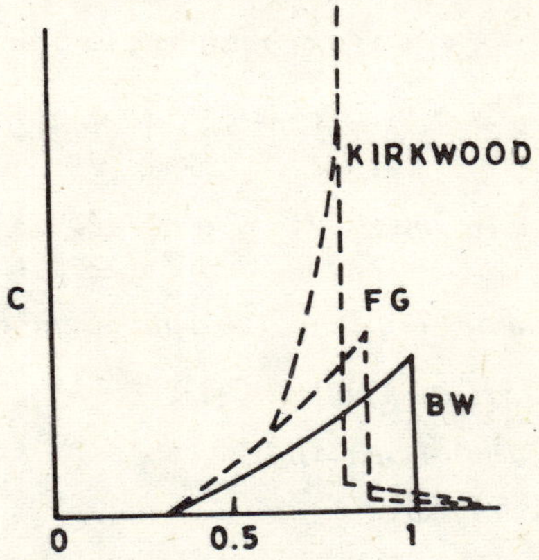


**Fig. 11.4** Magnetic heat capacity of a simple cubic lattice for $H = 0$ in the various approximations. The $(T/T_c)_{BW}$ scale is for $T/T_c = \frac{1}{3}(kT/2\epsilon)$.

terizes a second-order transition as the discontinuity is in the second derivative of $F$,

$$C_V = \frac{\partial}{\partial T} T^2 \frac{\partial}{\partial T}\frac{F}{T}.$$

For a simple cubic lattice, we have $\tilde{z} = 6$, and $2/\tilde{z} = 1/3 = 2\epsilon/kT_c$ in the BW approximation. Energy is required to produce disorder. This excess energy corresponds to a heat capacity in addition to what is given by the Dulong-Petit law for ordinary thermal motion. The rate of disordering increases from zero to a maximum value just below $T_c$ (Fig. 11.4). The contribution above $T_c$ comes from the local order which continues to need energy for its decrease at higher temperatures.

## 11.6 ONE-DIMENSIONAL ISING MODEL

The one-dimensional Ising model consists of a chain of $N$ spins, each spin interacting only with its two nearest neighbours. The energy for the configuration specified by $\{\sigma_1, \sigma_2, \ldots, \sigma_N\}$ is

$$E_I = -\epsilon \sum_{i=1}^{N} \sigma_i\sigma_{i+1}. \tag{11.44}$$

It is convenient to arrange the chain in a ring (Fig. 11.5) so that

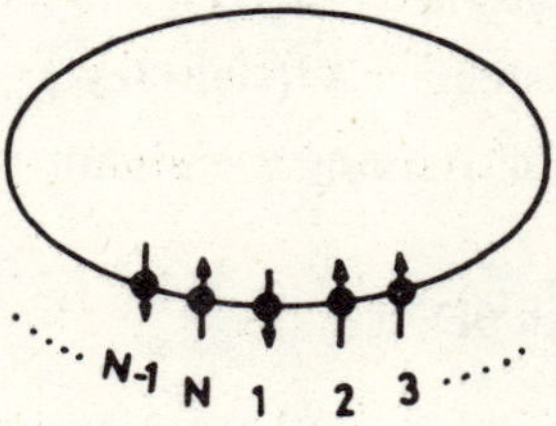


Fig. 11.5 N Ising spins arranged in a ring.

$$\sigma_{N+1} = \sigma_1. \tag{11.45}$$

The partition function is

$$Z = \sum_{\sigma_1=\pm1} \ldots \sum_{\sigma_N=\pm1} \exp\left[\beta\varepsilon \sum_{i=1}^{N} \sigma_i\sigma_{i+1}\right]$$

$$= \sum_{\sigma_1=\pm1} \ldots \sum_{\sigma_N=\pm1} \Pi\,(\cosh \beta\epsilon + \sigma_i\sigma_{i+1} \sinh \beta\epsilon), \tag{11.46}$$

where we have used

$$\exp(c\sigma\sigma') = \begin{Bmatrix} e^c & (\sigma\sigma' = 1) \\ e^{-c} & (\sigma\sigma' = -1) \end{Bmatrix} = \cosh c + \sigma\sigma' \sinh c,$$

which holds because $\sigma\sigma'$ can only be $+1$ or $-1$.

The expansion of products in (11.46) gives a sum of terms, each of which is a product of the form

$$(\cosh \beta\epsilon)^{N-s} (\sinh \beta\epsilon)^s (\sigma_i\sigma_{i+1} \dots \sigma_j\sigma_{j+1}). \tag{11.47}$$

We can display these terms graphically by thick and thin links forming the ring (Fig. 11.6). The thick link corresponds to (say) the factor $\sigma\sigma' \sinh \beta\epsilon$

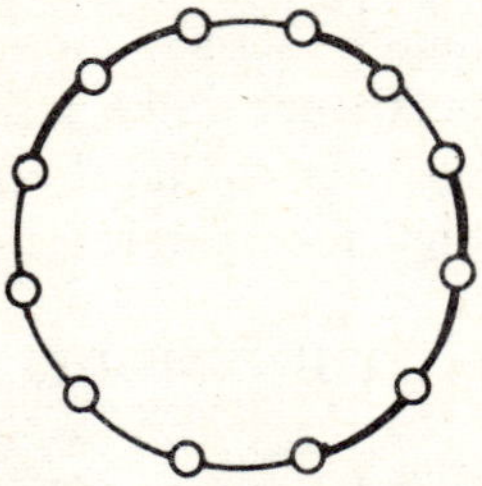

**Fig. 11.6** Construction for the terms (11.47).

and the thin link to $\cosh \beta\epsilon$. If there is a site at which thick and thin links join, then its spin occurs only once and the sum of the two values $\pm 1$ makes the product zero. At a site where two thick links join its spin occurs squared and gives a value 1, because $\sigma^2 = 1$. A non-zero contribution comes only when the chain of thick links, if present, has no ends. Thus the only nonzero terms are the first term $(\cosh \beta\epsilon)^N$, and the last term $(\sinh \beta\epsilon)^N$, giving

$$Z = 2^N [(\cosh \beta\epsilon)^N + (\sinh \beta\epsilon)^N], \tag{11.48}$$

For $\beta\epsilon = \epsilon/kT \neq \infty$ $(T \neq 0)$, $\cosh \beta\epsilon > \sinh \beta\epsilon$. Therefore, for $N \gg 1$, $(\cosh \beta\epsilon)^N \gg (\sinh \beta\epsilon)^N$,

$$Z = 2^N (\cosh \beta\epsilon)^N, \tag{11.49}$$

$$F = -kT \ln Z = -NkT \ln\left(2 \cosh \frac{\epsilon}{kT}\right). \tag{11.50}$$

The energy and heat capacity are

$$E = \frac{\partial}{\partial(1/T)} \frac{F}{T} = -N\epsilon \tanh \frac{\epsilon}{kT}, \tag{11.51}$$

$$C = \frac{\partial E}{\partial T} = \frac{N\epsilon^2}{kT^2} \left(\cosh^2 \frac{\epsilon}{kT}\right)^{-1}. \tag{11.52}$$

They are of the form (3.99, 100). Unlike Fig. 11.4, there is no transition temperature (Fig. 3.10). Therefore, *the one-dimensional Ising model cannot be ferromagnetic*. This result was first obtained by Ising in 1925.

A two-dimensional Ising model is a plane square lattice of $N$ points, at each of which is a spin (or dipole) pointing perpendicular to the lattice plane. The total number of possible configurations is $2^N$. For interaction between adjoining spins only,

$$E = -\epsilon \sum_{k,\, l=1}^{L} (\sigma_{kl}\, \sigma_{k,\, l+1} + \sigma_{kl}\, \sigma_{k+1,\, l}), \tag{11.53}$$

where $L$ is the number of points in a lattice line, $N = L^2$. The exact evaluation of partition function using (11.53) is extremely complicated but possible as shown by Onsager in 1944. The system is ferromagnetic with the Curie temperature

$$kT_c = \frac{2\epsilon}{\ln(1+\sqrt{2})}. \tag{11.54}$$

For three-dimensional Ising model only approximate series solutions exist.

A linear Ising chain is not ferromagnetic because it can be easily broken at any point. A single break destroys the long-range order and increases the energy by $2\epsilon$ (Fig. 11.7). This break can occur at any one of $N$ sites.

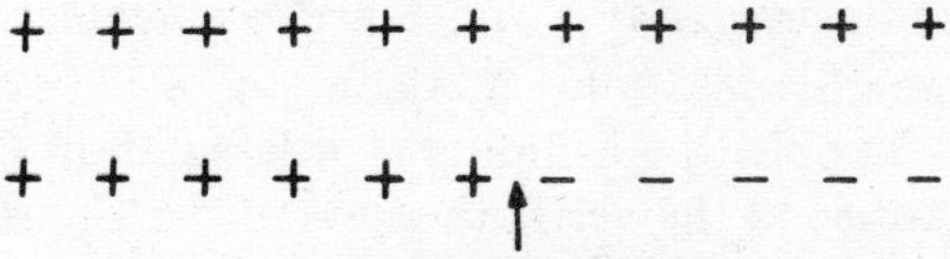


**Fig. 11.7** **Long-range order in a linear chain (above); A single break destroys the long-range order (below).**

Therefore, the gain in entropy is $k \ln N$. The change in free energy is

$$F = 2\epsilon - kT \ln N. \tag{11.55}$$

It can always be made negative even for every low $T$ if $N$ is sufficiently large. Therefore, an infinite linear chain has no singularities, (11.52).

A simple argument shows that a two-dimensional Ising model is ferromagnetic (11.54) Consider the connected region of reversed (−) spins shown in Fig. 11.8. It is bounded by a polygon of length $l$. There are $l$

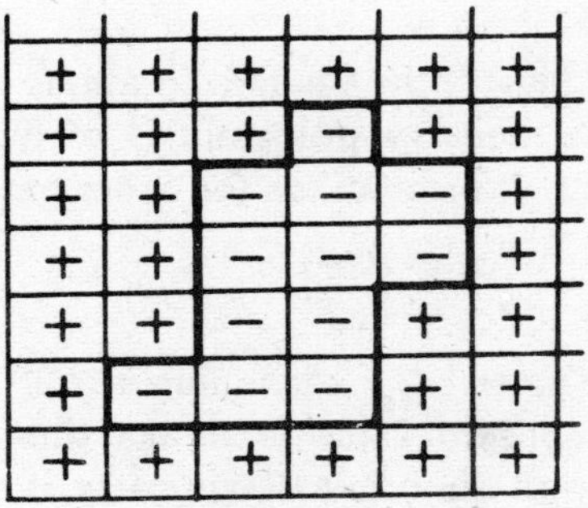

**Fig. 11.8** **A connected region of reversed spins.**

links of type (+ −) at the border of the two regions. The energy of the system is thus increased by $2l\epsilon$. At each node of the boundary there are three choices of direction, in general. We can say that there are $3^l$ ways of laying down the boundary. This is an over-estimate because the polygon must close somewhere. For large $l$ this is a small error. Thus

$$F \approx 2l\epsilon - kT \ln 3^l = l\,(2\epsilon - kT \ln 3). \tag{11.56}$$

These contributions are positive if $kT < 2\epsilon/\ln 3$. Therefore, the ordered state is stable below the Curie temperature,

$$kT_c \approx \frac{2\epsilon}{\ln 3}. \tag{11.57}$$

The correct result is (11.54).

The Ising model can simulate systems other than a ferromagnet.

## 11.7 ORDER-DISORDER IN ALLOYS

A binary alloy $AB$ is *ordered* if $A$ and $B$ are in a regular periodic arrangement. In a common ordered alloy all the atoms $A$ have atoms $B$ as nearest neighbours. For example, the completely ordered alloy CuZn has a cubic lattice with Zn atoms at the vertices, say, and the Cu atoms at the centers of the cubic cells (body-centred cubic structure or *bcc* structure) (Fig. 11.9a).

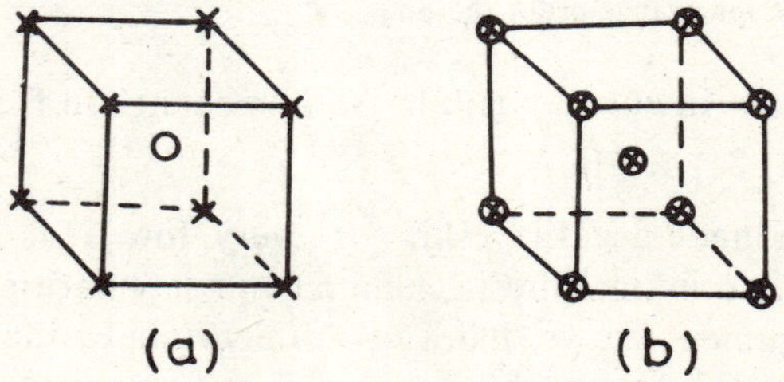


**Fig. 11.9** **(a) The CuZn alloy bcc structure with Cu (o) at the centre and Zn (x) at the vertices. (b) The disordered state for m = 0 showing greater symmetry as all sites are equivalent.**

As the temperature is raised, the Cu and Zn atoms change places creating disorder. We then have nonzero probabilities of finding atoms of either kind at every lattice site. We can difine the order parameter as

$$m = \frac{P_{\text{Cu}} - P_{\text{Zn}}}{P_{\text{Cu}} + P_{\text{Zn}}},$$

where $P_{\text{Cu}}$ ($P_{\text{Zn}}$) is the probability of finding a Cu (Zn) atom at a given lattice site. For any nonzero value of $m$ the sites (vertices and centres) remain nonequivalent and the original symmetry is unchanged. However, the symmetry of the system is *changed* when $m = 0$. For $P_{\text{Cu}} = P_{\text{Zn}}$, all sites become equivalent and *the symmetry of the crystal is increased* (Fig. 11.9b).

The bcc structure is made up of two interpenetrating simple cubic lattices, $a$ and $b$. The nearest neighbours of an atom on (say) lattice $a$ lie on the lattice $b$. If there are $N$ atoms $A$ and $N$ atoms $B$ in the alloy, the long-range order parameter $m$ is defined so that the number of $A$ atoms on the lattice $a$ is $\frac{1}{2}(1 + m)N$, and on the lattice $b$ is $\frac{1}{2}(1-m)N$. When $m = \pm 1$, the order is perfect (each lattice contains only one type of atom). When $m = 0$, each lattice contains equal number of $A$ and $B$ atoms (Fig. 11.9b) and there is no long-range order of the original kind but an increase in symmetry (X-ray powder photograph shows smaller number of lines because lattice points behave as if occupied by only one type of atom).

Note that any departure, however small, from the value $m = 0$ results in a lowering of the symmetry. Thus, in general, the symmetry of one phase (completely disordered phase here, $m = 0$) is higher than that of the other. Usually the more symmetrical case corresponds to higher temperatures. In particular, in a transition of second kind from an ordered (Fig. 11.9a) to a disordered state (Fig. 11.9b) the temperature increases. Exceptions to this rule are possible (for example, Rochelle salt is orthorhombic below the 'lower Curie point' and monoclinic above it).

Define $\sigma_i$ to be $+1$ for a site occupied by an atom $A$ and $-1$ for a site occupied by an atom $B$. Then

$$E = \epsilon_{AA}N_{AA} + \epsilon_{BB}N_{BB} + \epsilon_{AB}N_{AB}, \tag{11.58}$$

where $N_{ij}$ is the number of nearest neighbour $ij$ bonds, and $\epsilon_{ij}$ is the energy of an $ij$ bond. Define

$$2\epsilon = -\tfrac{1}{2}(\epsilon_{AA} + \epsilon_{BB}) - \epsilon_{AB}. \tag{11.59}$$

If $\epsilon$ is positive, lower energies are obtained by creating unlike pairs of atoms at the expense of like pairs. That is why in ordered structures like atoms tend to keep apart. A negative value of $\epsilon$ will make the like atoms to segregate into pure metals at lower temperatures.

If $n$ is the total number of pairs,

$$\sum_{\langle i,j\rangle} \sigma_i\sigma_j = N_{AA} + N_{BB} - N_{AB} = n - 2N_{AB}, \tag{11.60}$$

$$\begin{aligned} E &= \tfrac{1}{2}n(\epsilon_{AA} + \epsilon_{BB}) - 2\epsilon N_{AB} + \text{constant} \\ &= \epsilon \sum_{\langle i,j\rangle} \sigma_i\sigma_j + \text{constant}. \end{aligned} \tag{11.61}$$

As (11.61) is of the same form as (11.2), the order-disorder transformation can be described in the Ising model. We get the long-range order $m$ versus $T/T_c$ curve as shown in Fig. 11.1, when $m$ now refers to an $AB$ alloy. Thus passage through a phase transition point of the second kind has a continuous change of $m$ to zero. As a result the thermodynamic functions of the state of the body (volume, energy, entropy, etc) vary continuously as the transition point is passed. Therefore a phase transition of the second kind, unlike one of the first kind, does not involve latent heat. It is characterized by the derivatives of these thermodynamic quantities becoming discontinuous at a transition point (Fig. 11.4).

## 11.8 STRUCTURAL PHASE CHANGE

Many crystals change from one structure to another as the temperature is raised, or under external forces. For example, at high $T$ the ferroelectric crystal $BaTiO_3$ has a cubic (perovskite) strcuture (Fig. 11.10). The Ba atoms

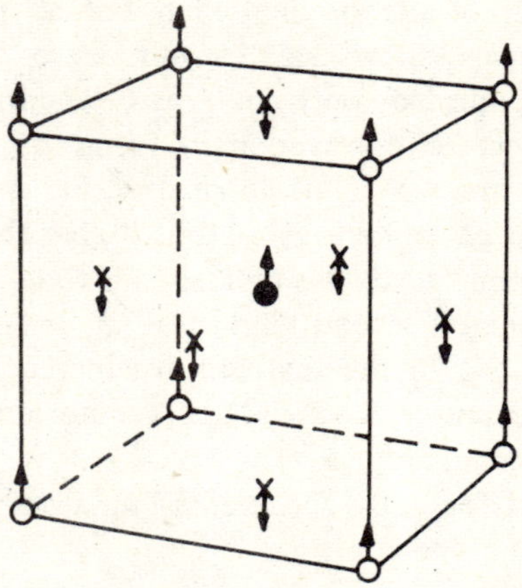

**Fig. 11.10 Unit cell of $BaTiO_3$ showing main distortion responsible for ferroelectricity: Ba(o), Ti(●) and O(x).**

are at the vertices, the O atoms at the centres of the faces, and the Ti atom at the centre of the unit cell. As $T$ is decreased below $T_c$, the positive ions begin to move relative to the negative ions, parallel to an edge of the cube. Thus in the low $T$ distorted phase the crystal acquires a macroscopic dipole moment. The electric polarizability shows a dramatic increase as $T \to T_c$. Below $T_c$ we get a spontaneous polarization.

The distortion due to the displacement of atoms results in a change of symmetry. In $BaTiO_3$ it changes from cubic to tetragonal. It is a phase transition of the second kind. The changes take place continuously in the configuration of atoms and reflect the properties of interatomic force. Even a small displacement of the atoms from their original position is enough to change the value of order parameter $m$. Here the $m$ may be taken as the amount of this displacement.

If $x$ represents a displacement in one particular mode of motion, the free energy $F$ is a function of even powers of $x$,

$$F = Ax^2 + Bx^4 + Cx^6 + \ldots. \tag{11.62}$$

The coefficients $A$, $B$, $C$, ... are functions of all other displacements and so of $T$. The $A$ is related to the characterstic frequency $\nu_l$ of the mode. If $A$ is large and positive, the $F$ has a minimum value near $x = 0$. An interesting case is when $A$ increase with $T$ such that

$$A \propto \nu_l^2(q) = c(T - T_c). \tag{11.63}$$

We get an instability as $T$ is decreased below $T_c$ because then $F$ has a local *maximum* at $x = 0$ (Fig. 11.11). It means the crystal (like $BaTiO_3$) tends to distort at low $T$.

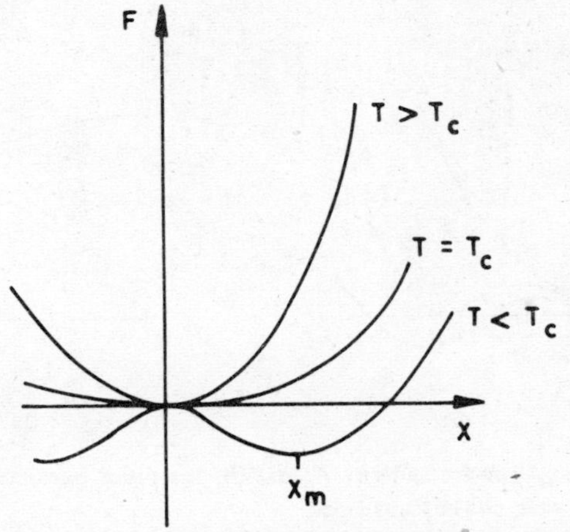


**Fig. 11.11** **The free energy $F$ as a function of displacement for a second-order phase transition.**

If $B > 0$, the minimum value of $F$, given by $\partial F/\partial x = 0$, shifts to the position $x_m$ given by

$$x_m^2 = c(T_c - T)/2B \quad \text{for} \quad T < T_c. \tag{11.64}$$

The distortion sets in below $T_c$ and is proportional to $(T_c - T)^{1/2}$ (Fig. 11.1b) with $m = x_m$. The $F$ and $x_m$ are continuous at the transition but the slope $\partial x_m/\partial T$ is not. The minimum of $F$ at thermal equilibrium is

$$\begin{aligned} F &= -c^2(T_c - T)^2/4B \qquad T/T_c < 1 \\ &= 0 \qquad\qquad\qquad\qquad\; T/T_c > 1. \end{aligned} \tag{11.65}$$

The specific heat, $C_v = c^2T/2B$, falls discontinuously to zero at $T = T_c$.

From (11.63), the phase transition of the second kind occurs when $\nu_l(q) \to 0$. This decrease in the mode frequency is called *softening*. It implies that the harmonic restoring forces have become so weak that finally a large displacement can occur which is limited only by the anharmonic forces. We can also say that for $\nu_l \to 0$ excitations require no energy. They are created spontaneously in a large number to produce the distortion.

On the other hand, a first-order transition occurs if $B < 0$ with stability provided by $C > 0$ in (11.62). A second minimum is found in $F$ besides the one at $x = 0$ (Fig. 11.12a). For $T < T_c$ this (absolute) minimum is at $F < 0$, and $x_m$ changes discontinuously (Fig. 11.12b). As this can occur away from $A = 0$, the softening of the mode is avoided. In such a case *latent heat is involved.*

The arguments given here provide basis for the Landau theory which is formulated in the language of magnetic systems (see next chapter for a formal treatment).

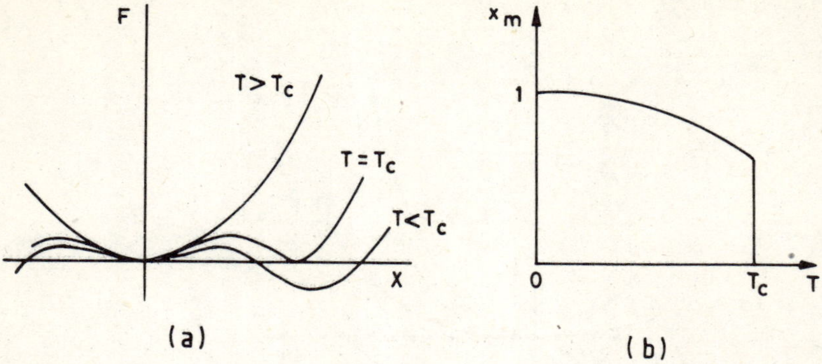


**Fig. 11.12** (a) The free energy $F$, and (b) the order parameter for a first-order phase transition.

## 11.9 LATTICE GAS

The Ising model can be applied to a *lattice gas*. It consists of atoms on (say) *a*-sites and vacancies on *b*-sites of an atom-vacancy lattice. Then it is formally equivalent to a binary alloy. The difference is that, as the atoms are free to move about, an enumeration of the occupied sites gives $N_{0c}!$ configurations instead of unity.

### *PROBLEMS*

11.1 Define a long-range order parameter $m$ and a short-range order parameter $s$ by

$$N_+/N \equiv (m + 1)/2 \quad (-1 \leqslant m \leqslant 1)$$

$$N_{++}/(\tfrac{1}{2}\tilde{z}N) \equiv (s + 1)/2 \quad (-1 \leqslant s \leqslant 1)$$

where $\tilde{z}$ is the number of nearest neighbours of a site. Show that

$$\sum_{\langle i, j\rangle} \sigma_i\sigma_j = \tfrac{1}{2}\tilde{z}N(2s-2m+1),$$

$$\sum_{i=1}^{N} \sigma_i = Nm,$$

$$\frac{1}{N}E(m, s) = -\frac{1}{2}\epsilon\tilde{z}(2s-2m+1)-\mu_B Hm.$$

11.2 Show that in the Bragg-Williams theory we can write

$$s \approx \tfrac{1}{2}(m+1)^2-1,$$

$$\frac{1}{N}E(m) \approx -\frac{1}{2}\epsilon\tilde{z}m^2-\mu_B Hm.$$

Now derive (11.13) in terms of $m$.

11.3 For $H = 0$, (11.13) can be expressed as

$$m = \tanh \frac{\tilde{z}\epsilon m}{kT},$$

where $m$ is defined in Prob. (11.1). Show that

$$m \approx 1-2\exp(-2T_c/T), \quad (T_c/T \ll 1),$$

$$m \approx \left[3\left(1-\frac{T}{T_c}\right)\right]^{1/2}. \quad \left(0 < 1-\frac{T}{T_c} \ll 1\right),$$

where $\tilde{z}\epsilon = kT_c$. Plot $m$ (Fig. 11.1b).

11.4 If $N$ = total number of lattice sites, $N_a$ = total number of atoms, and $N_{aa}$ = total number of nearest-neighbour pairs of atoms, write down the grand partition function for a lattice gas and obtain the equation of state. $\left[\text{Hint: } E = -\epsilon_0 N_{aa},\ Z = (N_a!)^{-1} \sum_{(a)} \exp(\beta\epsilon_0 N_{aa}),\right.$

$$\left.\mathscr{Z} = \sum_{N_a=0}^{\infty} z_a^{N_a} Z\right].$$

11.5 Relate the lattice gas with the Ising model as follows: (Lattice gas, Ising Model) = $(N_+, N_a)$, $(N_-, N-N_a)$, $(4\epsilon, \epsilon_0)$,

$$\left(-\frac{F}{N} + \frac{1}{2}\tilde{z}\epsilon - \mu_B H,\ P\right),\ \left(\frac{M}{2N} + \frac{1}{2}, \frac{1}{V}\right),$$

where $M$ is spontaneous magnetization.

11.6 For $H = 0$, (11.7) gives $E = -\frac{1}{2}\tilde{z}\epsilon N m^2$. Show that

$$C_V = \frac{Nkx^2}{\cosh^2 x - \dfrac{x}{\tanh x}}, \quad \frac{T}{T_c} = \frac{\tanh x}{x},$$

where $x = mT_c/T$. Expand tanh in the neighbourhood of $T_c$ to obtain

$$m^2 = 3\left(1-\frac{T}{T_c}\right) \text{ for } 0 < \left(1-\frac{T}{T_c}\right) \ll 1.$$

11.7 Calculate $C_V$ in Prob. 11.6 for $T \ll T_c$.

11.8 The probability that an atom $A$ on lattice a has an $AA$ bond is equal to the probability that an $A$ occupies a given nearest-neighbour site on $b$, times the number of nearest-neighbour sites (8 for bcc structure) in an alloy $AB$. Show that

$$N_{AA} = 2(1-m^2)N, \quad N_{BB} = 2(1-m^2)N$$

$$N_{AB} = 4(1+m^2)N,$$

where $m$ is the order parameter. If $E_0 = 2N(\epsilon_{AA} + \epsilon_{BB} + 2\epsilon_{AB})$ and $\epsilon = 2\epsilon_{AB} - (\epsilon_{AA} + \epsilon_{BB})$, show that (11.58) can be expressed as

$$E = E_0 + 2Nm^2\epsilon.$$

Show that the number of arrangements of these atom is

$$\left(\frac{N!}{[\frac{1}{2}(1+m)N]!\,[\frac{1}{2}(1-m)N]!}\right)^2,$$

and the condition for free energy $F$ to be minimum with respect to the order parameter $m$ is

$$4Nm\epsilon + NkT \ln [(1+m)/(1-m)] = 0.$$

Solve this to find the transition temperature $T_c$ and plot $m$ versus $T$ curve. Also plot $F$ versus $m$ curves for $T = 0$, $T_c/2$, $T_c$, $5/4\, T_c$ and interpret them. Discuss the order of the phase transition. [Hint : Expand near the transition, $4Nm\epsilon + 2NkTm = 0$, to get $T_c = -2\epsilon/k$, with $\epsilon < 0$ for the transition to occur.]

11.9 Obtain the free energy and the chemical potentials of each component of a two component solution. Use a lattice model for the liquid and work in the Bragg-Williams approximation.

11·10 Consider a plane square lattice having $N$ points. At each point place a dipole with its axis perpendicular to the plane. The dipole can have two opposite orientations. Thus to each lattice point $(k, l)$ we assign a variable $\sigma_{Kl}$ which can take two values $\pm 1$. If interactoin between adjoining dipoles is considered,

$$E(\sigma) = -\epsilon \sum_{k,l=1}^{L} (\sigma_{kl}\,\sigma_{k,l+1} + \sigma_{kl}\,\sigma_{k+1,l}),$$

where $L$ is the number of points on a lattice line, $N = L^2$. Show that

$$Z = (1-x^2)^{-N} S,$$

$$S = \left[\sum_{(\sigma)} \prod_{k,l=1}^{L} (1 + x\sigma_{kl}\sigma_{k,l+1})(1 + x\sigma_{kl}\sigma_{k+1,l})\right].$$

where $x = \tanh(\epsilon/kT)$.

11.11 Each term of the polynomial $S$ in Prob. 11·9 can be uniquely correlated with a set of bonds joining various pairs of adjoining lattice points. Try to estimate $S$.

11.12 The magnetic induction **B** vanishes (Meissner effect) inside a superconducting metal ($T < T_c$). Use this fact to show that the total magnetic moment $M_s$ of the superconducting sample of volume $V$ is $M_s = -(V/4\pi)\, H$, where $H$ is the applied field. Then use $dF_n = dF_s$, where $F_n$ is the free energy in the normal phase and $F_s$ in the superconducting phase to obtain

$$S_n - S_s = -\frac{V}{4\pi} H \frac{dH}{dT}.$$

Compare it with the Clausius-Clapeyron equation. Discuss this result for $dH/dT \leqslant 0$. As $T \to 0$, the third law of thermodynamics requires that $S_n - S_s \to 0$. Use this fact to discuss the shape of $H$ versus $T$

curve, and the role of latent heat. For most of the superconductors

$$H_c(T) = H_c(0)\left[1 - \left(\frac{T}{T_c}\right)^2\right],$$

where $H_c$ is the critical field at which the transition from the normal phase to the superconducting phase (or vice versa) occurs. At $T = T_c$, we have $H_c = 0$ and so $S_s = S_n$. Use this to show that (i) the $F_s$ and $F_n$ curves do not cross but merge at $T_c$, (ii) the two energies are the same, and (iii) there is no latent heat associated with the transition at $T = T_c$. Calculate and plot $C_s$ and $C_n$. Show that $C_s(T_c)/C_n(T_c) = 3$.

# 12
# CRITICAL PHENOMENA

## 12.1 INTRODUCTION

Part of our daily experience is the observation of phase changes resulting from change in the temperature of a substance—for example, the freezing of water at 273 K and the boiling of water at 373 K, both at a fixed pressure of $P = 1$ atm. Similar phase changes occur at different temperatures if we allow the pressure to vary. We can construct a phase diagram in the $T, P$ plane where boundaries between the various phases are delineated by the locus of the appropriate $T, P$ points. For a given point on the bounding curve, the substance can exist in either of two phases. For example, at $T = 373$ K and $P = 1$ atm, water can exist as a high-density liquid or as a low-density vapour. Latent heat added to liquid at constant pressure and temperature converts it into vapour. Following the liquid-vapour coexistence curve to higher temperatures, we encounter a remarkable new region where the density difference between the liquid and vapour goes to zero and water and steam become indistinguishable. Accompanying this decreased distinction between phases is a decreased surface tension and a growth of patches of each phase immersed in the other, drops of water in steam and bubbles of steam in water. The presence of drops and bubbles with sizes comparable to the wavelength of visible light causes the system to scatter light strongly, so that critical opalescence occurs and the system appears milky. The region where the liquid-vapour coexistence curve terminates is called the critical region, and critical phenomena describe characteristic behaviour observed in this region surrounding the critical point at $T_c$ and $P_c$. For water, the critical point occurs at $T = 647$ K and $P = 218$ atm.

The reason for the great interest in elucidating criticial phenomena lies in the fact that it is representative of a class of physical problems in which microscopic behaviour persists and cannot be averaged out to obtain macroscopic descriptions. Far from the critical point the distance over which density fluctuations are correlated is small compared with distances of macroscopic interest, and one can obtain a satisfactory macroscopic description by averaging over these small-scale fluctuations. In contrast, as one approaches the critical point, the distance over which density fluctuations are correlated increases without bound, and one needs to include a large number of coupled degrees of freedom in order to describe the system adequately. A key feature observed for critical phenomena is the emergence of non-analytic dependence of thermodynamic quantities on their variables. For example, as $T$ approaches the critical temperature $T_c$, the density difference across the coexistence curve $\rho_L - \rho_V$ varies as $(\rho_L - \rho_V) \sim (T_c - T)^\beta$ where $\beta$, a so-called *critical exponent*, is found to be of the order of 1/3. Similar dependence of many variables in a variety of systems exhibiting critical behaviour has been measured, and it is clear that non-integral critical exponents are a general characteristic of critical behaviour. The implication of the occurence of non-analytic behaviour is that standard methods of analysis which make use of perturbative techniques are not applicable, and problems involving the singular behaviour cannot be approximated with these techniques. What is more remarkable is that critical exponents for systems with vastly different microscopic structure may be the same, and that within a small number of classes, critical exponents display a remarkable regularity.

## 12.2 CRITICAL EXPONENTS

Critical behaviour has been identified in many systems. In addition to the water-steam system, liquid-gas systems generally display critical behaviour where the density difference between the liquid and gas phases tends to zero. Other systems displaying critical behaviour include ferromagnets, ferroelectrics, binary liquid mixtures, binary alloys, superfluids and superconductors. In many of these systems, one phase is ordered, while the second is disordered. It has become customary to label a parameter which vanishes at the critical point and above as the order parameter. The critical exponent associated with all order parameters is designated as $\beta$. For the liquid gas system, the order parameter is chosen as the density difference between the liquid and gas phases at a given point $T$, $P$ on the coexistence curve, which approaches zero according to

$$(\rho_L - \rho_V) \sim (T_c - T)^\beta, \quad T < T_c . \tag{12.1}$$

Above the critical temperature the order parameter is zero. For the ferromagnetic system, the order parameter is the homogeneous magnetization, while for the antiferromagnetic system it is the magnitude of the alternating magnetization. In either case, the source of magnetization is the spin-related magnetism of electrons in unfilled $d$ and $f$ shells of transition metals such as iron, cobalt and nickel. The combination of Coulomb repulsion between electrons and the Pauli exclusion principle, which keeps like spins separated, leads to a lower energy for parallel spins. Below the critical temperature, but in the absence of an external magnetic field, the thermodynamically stable state is the one with a significant number of spins aligned along a common direction, producing a net macroscopic vector magnetization Above the critical temperature, the entropy contribution dominates the free energy and the equilibrium state has no residual macroscopic magnetization in the absence of an external magnetic field. In the critical region about the critical point $T = T_c$ and $H = 0$, the macroscopic magnetization is either quite small or zero, depending on whether $T$ is below or above $T_c$. Even though the macroscopic magnetization vanishes above $T_c$, there are still regions of considerable extension, whose extent depends on how close $T$ is to $T_c$, in which a significant number of spins are lined up. The macroscopic magnetization vanishes because the microscopic magnetizations in different regions are not aligned along a common direction. The situation is not so radically different for $T$ below $T_c$ since the only significant difference is that the ordering tendency wins out—if just barely—over the disordering tendency, and a small but finite number of the correlated spin regions are aligned along a common direction, producing a macrosopic magnetization. The common thread connecting the behaviour of a liquid-vapour critical system and a ferromagnetic critical sytem is the dominance of fluctuations in the critical regions. In statistical mechanics, fluctuations are quantified by calculating the correlation function. For a stationary (non-time varying) system which possesses translational symmetry,the correlation function for density fluctuation can be expressed as

$$\Gamma_{\delta\rho\,\delta\rho}(x) = \langle \delta\rho\,(x)\ \delta\rho\ (0)\rangle \tag{12.2}$$

where the bracket $\langle\rangle$ indicates an ensemble average. For many systems, away from the critical point, the correlation function falls off exponentially with distance, and is given by

$$\Gamma(x) = \exp\,(-|\,x\,|/\xi), \tag{12.3}$$

where $\xi$ defines the correlation length and provides a measure of distances over which fiuctuations are strongly correlated. In the critical region, it is found that the long-distance behavior of $\Gamma(x)$ as a function of $|\,x\,|$ involves a critical exponent, $\eta$, according to

$$\Gamma(x) \sim |\,x\,|^{-(d-2+\eta)},\ T = T_c,\ \xi \gg |\,x\,| \gg a \tag{12.4}$$

where $d$ is the dimension of the space and $a$ is a measure of the microscopic structure of the system. The correlation length diverges as $T$ approaches $T_c$ with the critical exponent $\nu$ according to

$$\xi \sim |(T-T_c)\,|^{-\nu}. \tag{12.5}$$

The Fourier analysis of the correlation function allows one to represent the fluctuating density as a superposition of sinusoidal density waves whose amplitudes reflect the intensity of the fluctuation as a function of wavelength. These waves may be sensed by scattering waves of light, X-rays, or neutrons off of the system, where the diffraction grating provided by each wavelength produces constructive interference patterns which can be analyzed to obtain the amplitude of a particular Fourier mode. Using the inversion techniques, one can obtain both the correlation function and the correlation length.

Three additional critical exponents are found to be useful. The coefficient $\alpha$ is associated with the behaviour of specific heat in the vicinity of the critical temperature according to

$$C \sim |(T-T_c)|^{-\alpha}, \tag{12.6}$$

while the coefficient $\gamma$ is related to the critical behaviour of the generalized susceptibility $\chi$ according to

$$\chi \sim |(T-T_c)|^{-\gamma}. \tag{12.7}$$

For the liquid-gas system $C_V = -T(\partial^2 F/\partial T^2)$, and $K_T = (1/V)(\partial V/\partial P)_T$ are characterized by critical exponents $\alpha$ and $\gamma$, while for the ferromagnet $C_H$ and $\chi_M = (\delta M/\delta H)_T$ are the appropriate variables with these exponents.

Finally, the critical exponent $\delta$ occurs either in the relation between the external magnetic field and the magnetization, at the critical temperature according to

$$H \sim M^{\delta}, \quad T = T_c, \tag{12.8}$$

or to the relation between pressure and density at the critical temperature according to

$$(P-P_c) \sim |\rho-\rho_c|^{\delta}, \quad T = T_c. \tag{12.9}$$

Traditionally, one made a distinction between the exponents $\alpha$, $\gamma$ and $\nu$ for values of $T > T_c$ and those for $T < T_c$. Although it appears now that these exponents have the same value on either side of $T_c$. However, it is not correct to assume that the critical behaviour is symmetric about $T_c$, since the coefficients in front of the power-bearing term are different for the regions above and below $T_c$.

The six critical exponents defined here are connected by *scaling relations*. These are:

Rushbrooke scaling law: $\alpha + 2\beta + \gamma = 2$ (12.10)

Widom scaling law: $\gamma = \beta(\delta - 1)$

Fisher scaling law: $\gamma = (2-\eta)\nu$

and Josephson scaling law: $\nu d = 2-\alpha$.

These laws, which were originally derived as inequalities, are converted to equalities by the so-called scaling hypothesis. It appears that only two of the critical exponents are independent and the others can be determined using the scaling relations. Several inequalities involving the critical exponents can be derived using phenomenological equilibrium thermodynamics stability considerations. For example, the standard thermodynamic result for the difference between the specific heat at constant magnetization $C_M$, and the specific heat at constant magnetic field $C_H$, for a magnetic system is

$$C_H - C_M = T(\partial H/\partial M)_T \, (\partial M/\partial T)^2 . \tag{12.11}$$

Thermodynamic stability requirements dictate that $C_M$ be non-negative, leading to the inequality

$$C_H \geqslant T(\partial H/\partial M)_T \, (\partial M/\partial T)_H^2 = (T/\chi) \, (\partial M/\partial T)_H^2 . \tag{12.12}$$

For temperatures in the vicinity of the critical temperature, one can make use of the power law dependence of $C_H$, $\chi$ and $M$ on $(T_c - T)$, to obtain the inequality

$$(T_c - T)^{-\alpha'} \geqslant A(T_c - T)^{2\beta+\gamma'-2}, \tag{12.13}$$

where $A$ is a function independent of $(T_c - T)$ and $\alpha'$ and $\gamma'$ refer to the critical exponents for $C_H$ and $\chi$ for $T < T_c$, since the critical point is approached from below. For infinitesimal values of $(T_c - T)$, the inequality can only be maintained if

$$\alpha' + 2\beta + \gamma' \geqslant 2, \tag{12.14}$$

resulting in the Rushbrooke inequality. The Griffith inequality, $\alpha' + \beta(1 + \delta) \geqslant 2$, can be derived following similar thermodynamic arguments. The scaling hypothesis allows one to make a stronger statement and replace these inequalities with the corresponding equalities. Furthermore, the critical exponents can be shown to be the same regardless of whether one is above or below the critical point.

## 12.3 SCALING HYPOTHESIS

The *scaling hypothesis* is a conjecture which proposes that in the neighbourhood of a critical point, the singular parts of the appropriate thermodynamic functions should transform homogeneously under a change of scale. If $\lambda$ is an arbitrary scale change, then a generalized homogeneous function satisfies the following relation:

$$f(\lambda^{x_1} x_1, \lambda^{x_2} x_2, \ldots, \lambda^{x_n} x_n) = \lambda^y f(x_1, x_2, \ldots, x_n). \tag{12.15}$$

This scaling hypothesis leads to relationships among the critical exponents, and also leads to predictions about the equation of state.

The physical ideas which make the scaling hypothesis plausible are most easily visualized in the framework of correlation function for the order parameter, rather than in terms of macroscopic thermodynamic quantities, such as specific heat or magnetization. Scattering experiments yield direct information on the Fourier transfrom of the correlation function in the form of the square of the amplitudes of a Fourier mode with wavenumber $k$, where $\hbar k$ is the change in momentum on scattering of the probe beam of neutrons or X-rays. The singular behaviour exhibited by both macroscopic thermodynamic quantities and the correlation function in the vicinity of the critical point have their common origin in the critical fluctuations and the fluctuation-dissipation theorem relates the correlation function to the macrscopic thermodynamic response variables.

For illustrative purposes, consider a magnetic system in which the fine scale structural detail have been averaged out since they are not significant for the description of critical behaviour. The magnetization vector is considered as a vector magnetization field $M(x)$. In general, one defines at every point $x$ of a $d$-dimensional space a field variable $M(x)$ having $n$ components. If the system Hamiltonian in the absence of an external field is $H_0$, the application of a weak external magnetic field $H_e$ introduces the perturbation $-H_e \int dx M(x)$ and the Hamiltonian becomes $H = H_0 - H_e \int dx\, M(x)$. The ensemble average magnetization is given by

$$\langle M \rangle = \frac{1}{Z} \sum_{\text{states}} \int dx\, M(x) \exp(-\epsilon/k_B T), \tag{12.16}$$

where $Z$ is the partition function and $k_B$ the Boltzmann constant. The magnetic susceptibility $\chi = \partial\langle M\rangle/\partial H_e$ is easily obtained as

$$\chi = \frac{1}{k_B T} \int dx\, \{\langle (M(x) - \langle M(x)\rangle)\, (M(0) - \langle M(0)\rangle)\rangle\} = \frac{1}{k_B T} \int \Gamma_{MM}(x)\, dx. \tag{12.17}$$

Use has been made of translational invariance property required for passage to the thermodynamic limit of infinite system volume, leading to the independence of the correlation function on the choice of origin. Experimentally $\Gamma_{MM}(k)$, the Fourier transform of $\Gamma_{MM}(x)$, is observed in static scattering experiments. $\Gamma_{MM}(k)$ is peaked strongly about $k = 0$ and diverges at $T_c$ as

$$\Gamma_{MM}(k) \sim k^{2-\eta} \tag{12.18}$$

where $\eta$ is the same critical exponent as appears in (12.4) and the extra factor of $d$, the spatial dimension, follows from the dimension involved in Fourier transformation.

The fluctuation-dissipation theorem result can be modified by replacing $\Gamma_{MM}(x)$ by its Fourier representation $\Gamma_{MM}(x) = \int \Gamma_{MM}(k)\, e^{ikx} dk$, leading to the expression for the susceptibility

$$\begin{aligned}\chi &= \frac{1}{k_B T} \iint dx\, dk\, \Gamma_{MM}(k)\, e^{ikx} \\ &= \frac{1}{k_B T} \int \Gamma_{MM}(k)\, \delta(k)\, dk = \frac{\Gamma_{MM}(0)}{k_B T}.\end{aligned} \tag{12.19}$$

The width $\Delta k$ of the peak in $\Gamma_{MM}(k)$ around $k = 0$ is a measure of the correlation length $\xi \sim 1/\Delta k$, where $\xi$ is the average characteristic length of a region of correlated magnetization.

Applying the scaling hypothesis to the function $\Gamma_{MM}(k)$ results in

$$\Gamma_{MM}(\lambda^y k) = \lambda^w \Gamma_{MM}(k). \tag{12.20}$$

With no loss in generality, this can be expressed as

$$\Gamma_{MM}(\lambda k) = \lambda^z \Gamma_{MM}(k). \tag{12.21}$$

Equating the arbitrary scale parameter $\lambda$ to $\xi$, and noting that the critical behaviour of $\xi$ in terms of the reduced-temperature variable $\tau = |T-T_c|/T_c$ can be written as $\xi \sim \tau^{-\nu}$, we have

$$\Gamma_{MM}(\xi k) = \xi^z \Gamma_{MM}(k) = \tau^{-\nu z}\Gamma_{MM}(k). \tag{12.22}$$

The critical exponent $\gamma$ associated with $\chi$ allows us to relate $\gamma$ to $\nu z$ as follows:

$$\chi \simeq \frac{\Gamma_{MM}(0)}{k_B T} = \tau^{-\gamma} = \tau^{\nu z}\Gamma_{MM}(\xi k), \tag{12.23}$$

which in turn requires that

$$z = -\gamma/\nu. \tag{12.24}$$

Using the $k$-dependence of $\Gamma_{MM}(k) \sim k^{2-\eta}$, we find from (12.8)

$$\Gamma_{MM}(\xi k) = \xi^{-\gamma/\nu}\, k^{2-\eta}, \tag{12.25}$$

so that the exponents of $\xi$ and $k$ are equal, which requires that

$$\gamma/\nu = 2-\eta, \tag{12.26}$$

resulting in the Fisher scaling law.

The scaling relations for the critical exponents $\sigma$, $\beta$, $\delta$, $\gamma$ and $\nu$, can be obtained from the scaling of the Gibbs free energy $G$ expressed as a function of $H$, the external magnetic field, and $\tau = (T-T_c)/T_c$, the reduced temperature. The thermodynamic relations for obtaining $M$, $\chi$, and $C_H$ from $G$ are $M = -(\partial G/\partial H)_\tau$, $\chi = -(\partial^2 G/\partial H^2)_\tau$, and $C_H = (1/T_c)\,(\partial^2 G/\partial \tau^2)_H$. The requirement that $G$ be a generalized homogeneous function can be expressed by the condition

$$\lambda G(H, \tau) = G(\lambda^a H, \lambda^b \tau), \tag{12.27}$$

which, in turn, leads to the following scaling for $M$, $\chi$ and $C_H$:

$$\lambda M(H, \tau) = \lambda^a M(\lambda^a H, \lambda^b \tau), \tag{12.28a}$$

$$\lambda \chi(H, \tau) = \lambda^{2a} \chi(\lambda^a H, \lambda^b \tau), \tag{12.28b}$$

$$\lambda C_H(H, \tau) = \lambda^{2b} C_H(\lambda^a H, \lambda^b \tau). \tag{12.28c}$$

If, in (12.28a), the arbitrary scale parameter is chosen so that $\lambda^b = -(1/\tau)$, and $H$ is set equal to zero, we get

$$M(0, \tau) = (-\tau)^{(1-a)/b}\, M(0, 1) = (-\tau)^{\beta}\, M(0, 1),$$

which allows us to assert that

$$\beta = (1-a)/b. \tag{12.29}$$

However, by choosing $\lambda^a = 1/H$ in (12.28a), and setting $\tau$ to zero, one obtains

$$M(H, 0) = H^{(1-a)/a}\, M(1, 0) = H^{1/\delta}\, M(1, 0),$$

and consequently can assert that

$$\delta = a/(1-a). \tag{12.30}$$

If, in (12.28b), the scale parameter is chosen so that $\lambda^b = 1/\tau$, and $H = 0$, one obtains

$$\chi(0, \tau) = \tau^{(1-2a)/b}\, \chi(0, 1) = \tau^{-\gamma}\, \chi(0, 1),$$

and

$$\gamma = (2a-1)/b. \tag{12.31}$$

Finally, choosing $\lambda^b = 1/\tau$ and $H = 0$ in (12.28c) results in

$$C_H(0, \tau)^{(1-2b)/a}\, C_H(0, 1) = \tau^{-\alpha}\, C_H(0, 1),$$

from which we deduce that

$$\alpha = (2b-1)/b. \tag{12.32}$$

Combining (12.29), (12.30), and (12.31), results in the Widom equality for critical exponents:

$$\gamma = \beta(\delta - 1). \tag{12.33}$$

Combining (12.29), (12.31) and (12.32) results in the Rushbrooke equality for critical exponents:

$$\alpha + 2\beta + \gamma = 2. \tag{12.34}$$

The arguments given above can be applied, when appropriate, to both $T > T_c$ ahd $T < T_c$, and it can easily be shown that the critical exponents must be the same for $\tau > 0$ and $\tau < 0$.

Solving (12.29) and (12.30) for $a$ and $b$, we can write (12.28a) as

$$M(H, \tau) = \lambda^{a-1}\, M(\lambda^a H, \lambda^b \tau) = \lambda^{-1/1+\delta}\, M(\lambda^{\delta/1+\delta} H, \lambda^{1/\beta(1+\delta)}\, \tau). \tag{12.35}$$

Choosing the value of $\lambda$ so that $\lambda^{1/\beta(1+\delta)} = 1/|\tau|$, we obtain

$$M(H, \tau) = |\tau|^{\beta}\, M[(H/|\tau|^{\beta\delta}), 1], \tag{12.36}$$

which leads to the form of the equation of state:

$$M/|\tau|^{\beta} = f_{\pm}(H/|\tau|^{\beta\delta}), \tag{12.37}$$

where the function $f_{\pm}(H/|\tau|^{\beta\delta})$ is different for $\tau > 0$ and $\tau < 0$. If $M$ is scaled in terms of $H^{1/\delta}$ and $\tau$ is scaled in terms of $H^{1/\beta\delta}$, then $\tilde{M} = M/H^{1/\delta}$ can be plotted as a function of $\tilde{\tau} = \tau/H^{1/\beta\delta}$ and leads to a single function valid for all $\tau$. Experimental results obtained on many systems have been found to obey the equation of state predicted by these scaling requirements.

Unlike the three other scaling laws, the Josephson scaling law involves $d$, the dimensionality of the space. Since we are dealing with an infinite system, the $G$ function we have used is really the Gibbs function per unit volume and as such entails a dimensional factor of $1/L^d$, where $L$ is the length unit. In the critical region, the proper length scale is $\xi$, so that the density dimension of $G$ is given by $1/\xi^d$. Recalling the relation $C_H = (1/T_c)(\partial^2 G/\partial \tau^2)_H$ and making use of (12.5) and (12.6), we find

$$\tau^{-\alpha+2} \sim A/\xi^d \sim A/\tau^{-\nu d}, \tag{12.38}$$

and as a consequence, obtain the Josephson scaling relation

$$\nu d = 2-\alpha. \tag{12.39}$$

The scaling hypothesis can be clarified through the use of the ideas of scale transformation and dimensional analysis. The idea is that in the critical region, the significant length is the correlation length $\xi$, however, this length increases as $\tau^{-\nu}$ as we approach the critical point. We assume that the basic physics remains the same as $\xi$ changes, provided we transform the length scale accordingly. If we were to replace our standard measuring rod with one of twice the length, a distance interval would be halved in terms of the new unit of measurement. If $L'$ is the length measured after the scale is transformed by a factor $s$, we find $L' = L/s$, for an original length $L$. Generalizing this idea, we introduce scale dimension $\lambda_Q$, associated with the scale transformation of some quantity $Q$ according to

$$Q' = S^{\lambda_Q} Q. \tag{12.40}$$

For example, the volume element in $d$ dimensional space transforms as $(L')^d = (L/s)^d = S^{-d} L^d$ and has scale dimension of $-d$. The scale dimension of the Gibbs free energy density is $d$, since it contains the reciprocal of the $d$-dimensional volume element. From (12.18), the scale dimension of the correlation function $\Gamma_{MM}(X)$, $\lambda_G$ is $(d-2+\eta)$, from which it follows that $\lambda_M$, the scale dimension of the parameter is $(d-2+\eta)/2$. Since $M = -(\partial G/\partial H)_\tau$, the scale dimension of $H$, $\lambda_H$ is $(d+2-\eta)/2$. Similarly, dimensional analysis and the thermodynamic relation $\chi = -((\partial^2 G)/(\partial H^2))_\tau$ leads to the dimension $\lambda_\chi = \eta - 2$. Once the scale dimension of a quantity is known, the dependence of $\tau$ follows from the $\tau$ dependence of $\xi$. For $G$, which varies as $1/L^d$, we get the $\tau$ dependence of $\xi^{-d}$ as $\tau^{\nu d}$, when we adjust our scale according to $\xi$. Similarly, the order parameter $M$ varies as $\xi^{-(d-2+\eta)/2}$ leading to a $\tau$ dependence: $\tau^{\nu(d-2+\eta)/2}$. The critical exponent $\beta$ associated with the order parameter satisfies the relation $\beta = \nu(d-2+\eta)/2$. Continuing in this fashion, we can recover the scaling laws derived previously.

The predictions of the scaling hypothesis are in excellent agreement with experimental results, taking into consideration both the crudeness of the assumptions and the experimental difficulties associated with measurements in the critical region. Although the scaling hypothesis is not a proper theory, any proper theory must predict the scaling relations and give a satisfactory basis for understanding them.

## 12.4 THEORY OF CRITICAL PHENOMENA

Initial efforts to account for observed phase transitions and critical phenomena centred on developing phenomenological models which, when fitted with appropriate choices of parameters, explained classes of phase transitions over a range of physical parameter space. Perhaps the best known example of this kind of model is the Van der Waals equation of state which describes, when properly interpreted, the liquid-gas phase transition as well as the critical point, observed for systems of real gases. With the development of statistical mechanics emphasis shifted to describing phase transitions occurring in a many-body system, such as an imperfect gas, by calculating the partition function for the system when employing reasonable physical functions for the two-body intermolecular potential energy of the particular system involved. Although some degree of success has been achieved in describing the deviation from perfect gas behaviour in terms of the so-called virial expansions, no demonstrations of phase transitions have been forthcoming based on such calculations. The problem involves more than just calculational difficulties. From a general point of view, a satisfactory of critical phenomena must account for the observed non-analytic behaviour at $T = T_c$ and should predict the observed values of the critical exponents as well as the equation of state in the critical region. The origin of the already noted singular behaviour at the critical point is puzzling since the partition function for a finite system, calculated on the basis of physically reasonable Hamiltonians is a finite sum of analytic functions and thus itself analytic and cannot yield the observed non-analytic behaviour. Some insight into the problem was given by Lee and Yang who considered the grand canonical partition function, $\Omega$, for a finite system such as an imperfect gas with volume $V$, with $N < N_{\max}$ (since for a fixed volume $V$, the hard sphere character of the molecules limits the maximum number of molecules),

$$\Omega = \sum_{N}^{N_{\max}} \eta_a{}^N Z_N$$

where $\eta_a = e^{\mu/k_B T}$, is the activity, and $Z_N$ is the $N$ particle partition function. From elementary considerations it is clear that the appearance of singularities in thermodynamic quantities requires the presence of singularities in either $\Omega$ or $\ln \Omega$; however, since $\Omega$ is a polynomial in $\eta_a$ with positive coefficients, it has no singularities for any finite, real or complex values of $\eta_a$, and no zeros on the positive real $\eta_a$ semiaxis, the physical domain of $\eta_a$. For this reason, the only possible explanation for the occurrence of phase transitions associated with singular behaviour lies in the assumption that in passing to the thermodynamic limit ($N \to \infty$, $V \to \infty$, but $N/V = n$ remaining finite), the zeros of the grand canonical partition function coalesce to form sharp lines in the complex $\eta_a$ plane and the edge of these lines touch the real $\eta_a$ axis at $\eta_{a0}$, a point in parameter space at which the phase transition occurs. The character of the phase transition is related to the density of the zeros along the locus of zeros as the real axis is approached and second order phase transitions, associated with critical phenomena, have a vanishing density of zeros as the axis is approached with an isolated root on the axis. Although the Lee-Yang approach indicates how non-analytic behaviour can appear, it does not provide a practical frame work for investigating specific phase transition problems, since it is generally easier to evaluate the partition function than it is to trace the behaviour of its zeros in the complex plane in the thermodynamic limit.

Important theoretical results on the theory of phase transitions have been obtained by studying model systems with Hamiltonians permitting either exact evaluation, or reasonably accurate numerical calculation of the partition function and the resulting thermodynamics of the system. For example, a Hamiltonian describing a system of quantized spins arrayed on a $d$-dimensional lattice, with adjacent spins coupled by an exchange interaction forms the basis of the Heisenberg model. In general,

$$E = \sum_{i,j} \sum_{\alpha}^{n} \epsilon_{ij}^{\alpha} \sigma_i^{\alpha} \sigma_j^{\alpha}, \tag{12.41}$$

where $\epsilon$ specifies the strength of the exchange interaction, $i$ and $j$ label the interacting lattice sites and $\alpha$ labels the various components of the spin vector $\sigma$. The Ising model follows by assuming

$$\epsilon_{ij}^{x} + \epsilon_{ij}^{y} = 0, \; \epsilon_{ij}^{z} = \epsilon\delta_{j,\, i+1}, \tag{12.42}$$

while a model called $X$-$Y$ model assumes

$$\epsilon_{ij}^{z} = 0, \; \epsilon_{ij}^{x} = \epsilon_{ij}^{y} = \epsilon\delta_{j,\, i+1}. \tag{12.43}$$

Studies of critical phenomena based on both Heisenberg and Ising models have been seminal in the field, yielding results which support the non-integral values for critical exponents as suggested by experiments and exhibiting the insensitivity of critical exponent values to certain details of the model, supporting the ideas of universality, and the sensitivity of critical phenomena to the value of $n$, the number of components of the spin vector, which serves as the order parameter, and $d$ the dimension of the space covered by the lattice, in agreement with observation.

## 12.5 MEAN FIELD THEORY

The easily solved many-body problems of physics are those in which a coordinate transformation allows the uncoupling of the various degrees of freedom and the system reduces to $N$ independent one-body problems which are generally immediately soluble. The reduction of motion of a lattice of coupled particles to a system of uncoupled normal modes is a familiar example of the procedure. Often the coordinate transformation fails to completely uncouple the new degrees of freedom, but results in weak couplings whose effects may be incorporated by perturbative techniques.

Another approach to the reduction of the many-body problem to a soluble form involves treating each particle as moving in a force field which is the ensemble average of the fields seen by each of the individual particles. The $N$ body system is reduced to a system of $N$ non-interacting particles, each moving in the same force field, the so-called *mean field*. Landau developed a mean field theory which contained the elements necessary for the description of critical behaviour. Although this theory is correct only in some limited cases it serves as a general model for what is called classical behaviour and yeilds results which are equivalent to those obtained from a variety of models developed to describe specific physical system embodying critical behaviour.

The Landau theory was couched in the language of magnetic systems and assumes that for the description of critical phenomena details such as the magnitude of the spins, their geometric array and the range of their interaction are irrelevant. The model employs an $n$-component magnetization field variable $M(x)$ distributed continuously at points $x$ spanning a $d$-dimensional space. Since the energy of a state of the system is a function of $M(x)$, it is convenient to integrate over all configurations of $qM(x)$ rather than summing over the states of the system in evaluating the partition seen. Defining $W\{M(x)\}$ as the density of states corresponding to a given configuration $M(x)$, the canonical partition function can be written as

$$Z = \int dM(x)\, W\{M(x)\} \exp\left[-E\{M(x)\}/k_BT\right]. \tag{12.44}$$

Introducting the entropy $S = k \ln W\{M(x)\}$, the partition function can be expressed in the local free energy $F_L\{M(x)\}$ as

$$Z = \int dM(x) \exp\left(-\int F_L\{M(x)\}\, dx/k_BT\right). \tag{12.45}$$

The system free energy $F\{M(x)\} = E\{M(x)\} - T \ln W\{M(x)\}$ is given by the volume integral of the local free energy. Landau assumed that the principal contributions to $Z$ came from a narrow range of $M(x)$, the particular values of $M(x)$ being those which minimized $F_L$. The integral could then be replaced by

$$Z = \int dM(x)\, \delta[M(x) - M_0(x)] \exp\left(-\int F_L\{M(x)\}\, dx/k_BT\right), \tag{12.46}$$

where the field values $M_0(x)$, were those which minimized the function $F_L\{M(x)\}$. In the critical region $F_L$ was assumed to be an analytic function of $M(x)$, containing only even powers of $M$ since in the absence of an external field the system is unchanged when $M(x)$ is change to $-M(x)$. The Landau-Ginzburg form for $F_L$ is

$$F_L = \sum_{i=1}^{n} \left\{\frac{A}{2} M_i^2(x) + \frac{B}{4} M_i^4(x) + C \sum_{\alpha=1}^{d} \left(\frac{\partial M_i}{\partial x_\alpha}\right)^2\right\}. \tag{12.47}$$

Because the gradient term is positive definite and the system is specially homogeneous, the minimum of $F_L$ will occur for $M(x) = M$ independent of $x$, and is found by minimizing $\sum_{i=1}^{n} \left\{\frac{A}{2} M_i^2 + \frac{B}{4} M_i^4\right\}$; requiring $\frac{\partial F_L}{\partial M_i} = 0$ and $\frac{\partial^2 F_L}{\partial M_i^2} > 0$. These conditions lead to the equations

$$M_i(A + BM_i^2) = 0, \tag{12.48}$$

$$A + 3BM_i^2 > 0. \tag{12.49}$$

Equation (12.48) has two solutions $M_i = 0$ and $M_i = \sqrt{-A/B}$, the first root associated with the paramagnetic state, $\tau < 0$. Thermodynamic stability requirements dictate that $B > 0$, otherwise $M = -\infty$ would minimize $F_L$. From (12.49) it is clear that for the paramagnetic root $(M_i = 0)$ $A$ must be positive, while, for the ferromagnetic state $A$ must be negative. It follows that at the critical point $A = 0$. Landau assumed that $A$ and $B$ were analytic functions of temperatnre so that in the critical region, the leading terms in the expansion are

$$A = (\partial A/\partial\tau)_0\tau + \ldots,\ B = B(0) + \ldots. \tag{12.50}$$

It is interesting to note that the paramagnetic root $M = 0$, leads to a $\partial^2F/\partial M_l^2 < 0$, if $\tau < 0$, so the paramagnetic state is unstable for $T < T_c$. This is in contrast to the behaviour of phases for a first order phase transition where the phases are metastable on crossing the phase transition boundary line. Identifying $M_l$ as an order parameter, the critical exponent $\beta$ is found to be 1/2, since

$$M_l = \sqrt{-A/B} = \sqrt{(\partial A/\partial\tau_0)/B(0)}\,(-\tau)^{1/2}. \tag{12.51}$$

To complete the thermodynamic description of the system, the response to an external magnetic field $H$ must be included. From thermodynamics $H_l = \partial F/\partial M_l$ and

$$H_l = M_l[A + BM_l^2]. \tag{12.52}$$

The scalar magnetic susceptibility $\chi = (\partial M_l/\partial H_l)_{H\to 0}$ in the critical region can be calculated from $\partial H_l/\partial M_l = (A + BM_l^2) + 2BM_l^2 = A + 3BM_l^2$. For $\tau < 0$, $M_l = \sqrt{-A/B}$, $\chi^{-1} = -2A$ and while for $\tau > 0$, $M_l = 0$, $\chi^{-1} = A$. The behaviour of the susceptibility around the critical point is

$$\chi(\tau) = \begin{cases} (\partial A/\partial\tau)_0^{-1}.\ \tau^{-1} & \tau > 0, \\ [2(\partial A/\partial\tau)_0^{-1}(-\tau)^{-1} & \tau < 0. \end{cases} \tag{12.53}$$

So the critical exponent $\gamma$ has the value $-1$. From (12.52) the variation of magnetization at $\tau = 0$ as a function of external field is

$$H_l = BM_l^3, \tag{12.54}$$

so the value of the critical exponent $\delta = 3$.

For $\tau \neq 0$, (12.52) can be expressed as

$$H_l/M_l^3 = B(0) + (\partial A/\partial\tau)_0\tau/M^2. \tag{12.55}$$

Using the values critical exponents $\beta$ and $\delta$, gives

$$H/M^\delta = B(0) + (\partial A/\partial\tau)_0\tau/M^{1/\beta} = f(\tau/M^{1/\beta}), \tag{12.56}$$

the scaling form of the equation of state.

The specific heat, calculated from $C_H = (1/T_c)\,(\partial^2F/\partial\tau^2)_H$, vanishes above the critical point, while for $\tau < 0$ it is constant. Thus the critical exponent $\alpha = 0$ both above and below the critical point, however, there is a discontinuity in the value of the specific heat at the critical point. Finally, the critical behaviour of the correlation function can be obtained by examining the spatial magnetization response to a non-uniform external magnetic field, $H(x) = H_0\delta^3(x)$. We consider only a one component $M$ field and neglect the $M^4$ term in $F_L$, retain the gradeint term and add the term $-H(x)\,M(x)$. The condition for the minimum of

$$F = \int d^3 x \left\{\frac{A}{2} M^2(x) + C\,(\nabla M(x))^2 - H(x)\,M(x)\right\}, \quad (12.57)$$

follows from the Euler-Lagrange equation

$$-C\nabla^2 M(x) + AM(x) = H(x), \quad (12.58)$$

which for the delta function $H(x)$ field reduces to the equation for the Green's function for the Helmholz equation, with the well known solution

$$M(x) = De^{-\sqrt{A}\,|x|}/\,|\,x\,|\,, \quad (12.59)$$

which gives the correlation length $\xi = 1/\sqrt{A}$ and the critical exponent, $\nu = 1/2$.

## 12.6 THE RENORMALIZATION-GROUP THEORY OF CRITICAL PHENOMENA

The heart of the difficulty in treating problems dealing with critical phenomena lies in the fact that the problems involve a large number of degrees of freedom with no clear-cut procedure available for decoupling them in a systematic way. Stated another way: the increase without limit of the correlation length $\xi$, as $T$ approaches $T_c$, means that elementary parts of the system are coupled over distances of the order of $\xi$, enclosing volumes containing many particles and so requiring the taking into account of more than two body interactions. The renormalization group technique establishes a correspondence between systems having different correlation lengths. Through repeated use of the transformations generated by the renormalization group which change the initial system to an equivalent system having a smaller correlation length, the effective correlation length is reduced to a value where methods which consider only pair interactions again can be used allowing the problem to be solved by more or less standard methods. Once the problem is solved for $\xi \simeq 1$, the correspondences can be followed in reverse to yield the solution of the original problem. The renormalization-group technique provides a systematic way of reducing the number of degrees of freedom of the system, the reduction resulting from scale changes which in turn reduce the correlation length. Consider as an illustration a two-dimensional square Ising lattice of spins with coupling constant $K = \epsilon/k_B T$, and lattice spacing $a$. The first step consists in grouping the individual spins into cells containing blocks of spins and to consider the Ising lattice constructed from these block cells, treating each cell as an equivalent spin. The coupling constant $K'$ between the equivalent cells is redefined by a renormalization transformation so that the form of the partition function is unchanged. As a result of this coarse graining of the lattice, the correlation length is reduced, along with the number of degrees of freedom of the system. Such a renormalization transformation can be repeated, resulting in further reduction in scale and yielding in turn a set of recursion relations, relating the coupling constants at different scales. The key feauture for the occurrence of critical behaviour is the existence of fixed points associated with these recursion relations, that is, specific points which are unchanged under the renormalization transformation. When the renormalization transformation is viewed as a symmetry operation, the fixed points represent invariants under the symmetry operation of dilatation. Starting with the symmetry at the fixed points allows for a simple treatment in the region close to fixed $a$, the point where the symmetry is broken but still approximately valid. This takes the form of a scale dilatation according to some power law.

A geometric picture can be constructed in which the state of the system is represented by a point in multidimensional parameter space, with parameter such as temperature, magnitude of applied fields, and strength of interaction between spins. The locus of the system point as temperature alone is varied is called the physical line. This line crosses orthogonally a sequence of surfaces, on each of which the correlation length $\xi$ is constant, the value of $\xi$ increases as $T$ approaches $T_c$, and the surface at $T_c$ corresponding to $\xi = \infty$ is called the critical surface. Applying the renormalization-group transformation to any point on the physical line generates a trajectory which may first approach the critical surface, but after repeated applications veers away towards regions of low $\xi$. The system states along these trajectories generated by the renormalization group in general, have parameters differing from those on the physical line, from which the trajectory started. From their definition, fixed points, which are unchanged by the renormalization-group transformation, must lie on surface with either $\xi = 0$ or $\xi = \infty$, $\xi' = \xi/s = \xi$ the only sulutions for $\xi' = \xi/s = \xi$, for arbitrary $s$. In particular, the critical points are associated with fixed points in the surface $\xi = \infty$.

The fixed points which lie on the critical surface are located at either a local maximum, a local minimum, or a saddle point, and these in turn give rise to different types of critical behaviour. The fixed point which lies at a local minimum on the critical surface corresponds to an ordinary critical point while saddle points are associated with tricritical points. In the neighborhood of a fixed point, it is possible to linearize and diagonalize the renormalization transformation and use the eigenfunctions to construct scaling variables, a function of system parameters that only changes in scale under the application renormalization-group transformation. Associated with a scaling variable is an anomalous dimension $y$, so that under scale change $s$, the scaling variable is changed by a factor $s^y$. If the anomalous dimension for a given scaling variable is positive, a trajectory along a direction in parameter space associated with the eigenvector linked to that scaling variable will diverge from the fixed point under the application of the renormalization-group transformation and the corresponding scaling field is said to be relevant. If the anomalous dimension is negative, a trajectory along an axis defined by that scaling field will be driven toward the fixed point by the renormalization group, and the scaling variable is said to be irrelevant. The critical behaviour associated with each fixed point is determined by the local scaling variables and their anomalous dimensions.

The point at which the physical line intersects the critical surface is not necessarily a fixed point, however, the renormalization-group trajectory issuing from this point is restricted to remaining on the critical surface. Depending on the anomalous dimension of the appropriate scaling fields associated with fixed points in the neighborhood of this intersection, the trajectory will be drawn to or repelled away from the fixed point. From this simple picture, it is possible to visualize the origin of the concepts of scaling and universality.

## *PROBLEMS*

12.1 If $C_P \propto (T - T_c)^{-\alpha}$ then show that $\alpha < 1$.

[Hint: $\int C_P \, dT$ is always finite.]

12.2 If not $C_P$, but $\partial C_P/\partial T \to \infty$, then show that $-1 < \alpha < 0$.

12.3 For $(T - T_c) \to 0$, more accurately,

$$C_P = T_c \frac{dP_c}{dT} \left(\frac{\partial V}{\partial T}\right)_P + a,$$

$$\left(\frac{\partial V}{\partial T}\right)_P = -\left(\frac{\partial V}{\partial P}\right)_T \frac{dP_c}{dT} + \frac{b}{T_c} \frac{dT_c}{dP},$$

where $a$ and $b$ are constants. If $C_P \propto (T - T_c)^{-\alpha}$ tends to $\infty$ with $\alpha > 0$, then show that $\partial C_V/\partial T \propto |T - T_c|^{-(1-\alpha)}$.
[Hint: $C_V \simeq a - b - b^2/C_P$.]

# APPENDICES

## I. PROBABILITY

### Concept of Probability

Consider a molecule $A$ moving chaotically inside a box. A *random event* is defined to be a phenomenon that in an experimental set up for its observation either occurs or does not occur. For example, at any given instant of time, the finding of $A$ into the volume element $\Delta V$ selected inside the box is a random event. An experiment involving the observation of a random event is called a *trial*.

The *probability* of a random event is given by the ratio of the number of trials $n$ at which the given event occurred to the total number of trials $N$, provided that $N$ is large enough. If $P(A)$ is the probability that the event $A$ occurs, then

$$P(A) = \lim_{N \to \infty} (n/N).$$

### Mutually Exclusive Events

Two events are *mutually exclusive* if the happening of one excludes the possibility of happening of the other. For example, the event 1 that at a given instant the molecule $A$ will be in $\Delta V_1$ and the event 2 that at the same instant it will be in $\Delta V_2$ are mutually exclusive if the two volume elements do not intersect. The probability of occurring of one of two mutually exclusive events equals the *sum of the probabilities* of occurring of each of them.

### Independent Events

Two events are *independent* if the happening of one of them does not affect the probability of happening of the other one. The probability of the joint happening of independent events equals the *product of the probabilities* of each of them.

For example, consider a rarefied gas. Suppose on $n$ trials the molecule $A$ is found $n_1$ times in the volume $\Delta V_1$ and the molecule $B$ is found $n_2$ times in the volume $\Delta V_2$, that is,

$$P(A) = n_1/n, \qquad P(B) = n_2/n.$$

Among all the traials $n_1$ in which $A$ got in $\Delta V_1$ there would be some in which $B$ also gotinto $\Delta V_2$. Their number is $n_1\,(n_2/n)$. Therefore, the probability of the joint happening of the events $A$ and $B$ is

$$P(AB) = \frac{n_1(n_2/n)}{n} = \frac{n_1}{n}\cdot\frac{n_2}{n} = P(A).\,P(B).$$

**Mean Value**

If in a total of $N$ observations on a system we find that $n_1$ observations give a value $R_1$, $n_2$ a value $R_2$, etc., for a physical quantity of interest $R$, then he *mean value* $\bar{R}$ of $R$ is defined as

$$\bar{R} = \frac{n_1R_1 + n_2R_2 + \ldots}{n_1 + n_2 + \ldots} = \frac{\Sigma\, n_iR_i}{\Sigma n_i} = \frac{\Sigma\, n_iR_i}{N}$$

$$= \frac{n_1}{N}R_1 + \frac{n_2}{N}R_2 + \ldots$$

$$= P_1R_1 + P_2R_2 + \ldots = \sum_i P_iR_i. \tag{1}$$

**Permutations and Combinations**

(i) The number of possible permutations of $n$ distinct (distinguishable) objects is $n!$

Think of $n$ empty positions in a straight line. We have $n$ choices of an object to put into the first position, $n-1$ choices from the remaining objects to put into the second position, and so on. The total number of ways therefore $n\,(n-1)\,(n-2)\,\ldots 2 = n!$

(ii) We wish to arrange $n$ objects in $r$ groups, such that there are $n_1$ objects in the first group, $n_2$ in the second group, etc., so that $\sum_{i=1}^{r} n_i = n$. We are not concerned with the order of the objects within the various groups. For the first group the number of ways is $n\,(n-1)\,\ldots\,(n-n_1+1)/n_1!$, where the denominator gives the number of ways the objects could be arranged in the group. For the second group the $n_2$ objects can be selected out of the remaining $N-n_1$ objects in $(n-n_1)\,(n-n_1-1)\ldots(n-n_1-n_2+1)/n_2!$ ways, etc. Therefore, multiplying these numbers we get the total number of different permutations as

$$M = \frac{n!}{n_1!n_2!\ldots n_r!}. \tag{2}$$

(iii) The number of *permutations* of $n$ distincts objects taken $r$ at a time, denoted by ${}^nP_r$ is given by

$${}^nP_r = n\,(n-1)\ldots(n-r+1)$$

$$= n\,(n-1)\ldots(n-r+1)\times\frac{(n-r)\,(n-r-1)\ldots 2.1}{(n-r)\,(n-r-1)\ldots 2.1}$$

$$= \frac{n!}{(n-r)!}. \tag{3}$$

Note that ${}^nP_n = n!$.

(iv) If order is of no importance, then we have a *cambination* rather than a permutation. The basic formula is

Total number of combinations $\times$ Number of permutations in *each* combination = Total number of permutations.

If ${}^nC_r$ denotes the number of combinations of $n$ objects taken $r$ at a time, without regard to the order in which they are selected, then

$${}^nC_r . \, r! = {}^nP_r,$$

or

$${}^nC_r = \frac{1}{r!}\, {}^nP_r = \frac{n!}{r!\,(n-r)!}. \tag{4}$$

**Binomial Distribution**

Consider $N$ boxes each containing $R$ red balls and $B$ black balls. The probability of drawing a red ball from a box is $r = R/(R+B)$, and the probability of drawing a black ball is $b = B/(R+B) = 1-r$. The probability of taking $s$ red balls from a particular group of $s$ boxes and $N-s$ black balls from the group of $N-s$ other boxes is then $r^s b^{N-s}$. There are $N!/s!\,(N-s)!$ ways of selecting $s$ boxes out of $N$, without regard to the order in which they are selected. Therefore, the total probability $P(s, N)$ that $s$ of the balls will be red and $N-s$ black, when one ball is taken from each box is

$$P(s, N) = \frac{N!}{s!\,(N-s)!}\, r^s\, b^{N-s}. \tag{5}$$

This is called the *binomial distribution* because we can use the binomial expansion to prove $\sum_s P(s, N) = 1$,

$$\sum_{s=0}^{N} P(s, N) = \sum_{s=0}^{N} \frac{N!}{s!\,(N-s)!}\, r^s\, b^{N-s} = (r+b)^N = 1^N = 1. \tag{6}$$

## II. STIRLING APPROXIMATION

We have

$$\int_1^n \ln x\, dx = \Big[\, x \ln x - x \,\Big]_1^n = n \ln n - n + 1. \tag{1}$$

If we plot (Fig. II.1) the curve $y = \ln x$ and draw ordinates to it for $x = 2, 3, \ldots, n$, then the sum of the areas of strips is approximately

$$\ln 2 + \ln 3 + \ldots + \ln n = \ln (1 \times 2 \times 3 \times \ldots \times n) = \ln n!. \tag{2}$$

For large $n$, the error incurred in regarding the strips as rectangles can be neglected, and sum (2) of the areas of rectangles can be equated to the area (1) of the curve itself,

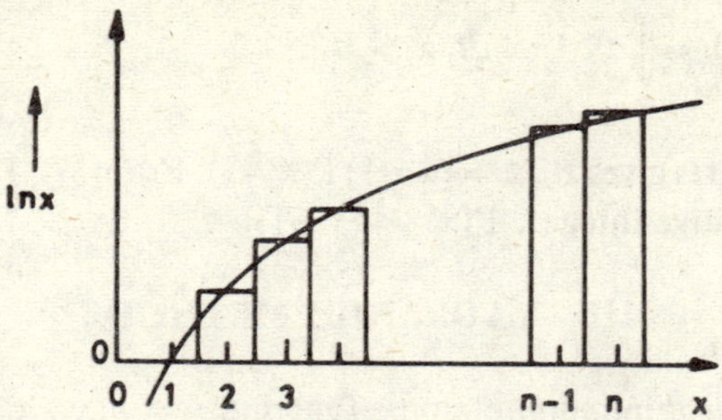


Fig. II.1. Plot of ln $x$ versus $x$.

$$\ln n! \simeq n \ln n - n, \quad \text{or} \quad n! \simeq (n/e)^n, \tag{3}$$

where 1 on the right hand side of (1) has been neglected in comparison to $n$. This is Stirling's approximation in its simple form. The approximation is fairly satisfactory. For example, for $N = 100$, we have $\ln N! = 363.7$, and $N \ln N - N = 360.5$.

If $n$ is a positive integer,

$$\int_0^\infty t^n e^{-t}\, dt = n(n-1)\ldots 2.1. = n!. \tag{4}$$

In the integrand $F = t^n e^{-t}$ for large $n$, $t^n$ is a rapidly increasing function of $t$ and $e^{-t}$ is a rapidly decreasing function of $t$. Hence $F$ exhibits a sharp maximum for some value of $t$.

Put $t = xn$, so that

$$\begin{aligned} F &= n^n x^n e^{-nx} \\ &= n^n e^{-n} \exp(n - nx + \ln x^n) \\ &= n^n e^{-n} \exp[-n(x - 1 - \ln x)] \\ &= n^n e^{-n} \exp\{-n[y - \ln(1+y)]\}, \end{aligned}$$

where in the last step $x = y + 1$. Then using $\ln(1+y) \simeq y - \frac{1}{2} y^2 + \ldots$,

$$F \simeq n^n e^{-n} \exp(-ny^2/2).$$

Thus $F$ has a sharp maximum for $y = 0$ and

$$\begin{aligned} n! &\simeq n^{n+1} e^{-n} \int_{-1}^{+\infty} \exp(-ny^2/2)\, dy = n^{n+1} e^{-n} \int_{-\infty}^{+\infty} \exp(-ny^2/2)\, dy \\ &= (2\pi n)^{1/2} (n/e)^n. \end{aligned} \tag{5}$$

$$\begin{aligned} \ln n! &= n \ln n - n + \tfrac{1}{2} \ln(2\pi n) \\ &\simeq n \ln n - n. \end{aligned} \tag{6}$$

For $n = 100$, $\frac{1}{2} \ln(2\pi n) = 3.2$. Thus, (5) is a better approximation than (3).

The gamma function is given by

$$\Gamma(x) = \int_0^\infty t^{x-1} e^{-t}\, dt, \; x > 0 \tag{7}$$

Integration by parts gives $\Gamma(x) = (x-1)\,\Gamma(x-1)$. From (7), $\Gamma(1) = 1$, $\Gamma(\frac{1}{2}) = \pi^{1/2}$. If $x$ is a positive integer, $\Gamma(x) = (x-1)!$.

## III. LAGRANGE METHOD

The maximum, or minimum, of some function $f = f(x_1, x_2, \ldots, x_n)$ is given by the ordinary conditions of extremum

$$\delta f = \sum_{i=1}^{n} \frac{\partial f}{\partial x_i}\, \delta x_i = 0, \tag{1}$$

where the $\delta x_i$ are the variations of the $x$'s. Since $x_i$ are arbitrary, each coefficient of the $\delta x_i$ must vanish,

$$\frac{\partial f}{\partial x_i} = 0, \quad \text{(for all } i\text{)}. \tag{2}$$

If we have constraints on the $x$'s, such as

$$g(x_1, x_2, \ldots, x_n) = 0, \tag{3}$$

or, a condition

$$\delta g = \sum_{i=1}^{n} \frac{\partial g}{\partial x_i}\, \delta x_i = 0, \tag{4}$$

then the $\delta x_i$ are no longer all independent being interrelated by (3). For example, if we knew $(n-1)$ of them, we can evaluate the $n$th from (4).

In the method of Lagrange, we multiply (4) by an undetermined parameter $\lambda$ and add the result to (1),

$$\sum_{i=1}^{n} \left( \frac{\partial f}{\partial x_i} + \lambda \frac{\partial g}{\partial x_i} \right) \delta x_i = 0. \tag{5}$$

Here only $(n-1)$ of the $\delta x_i$ are independent, for example, $\delta x_1 \ldots \delta x_{n-1}$. But we still have the value of $\lambda$ at our disposal. We can choose it so as to eliminate the coefficient of $\delta x_n$,

$$\frac{\partial f}{\partial x_n} + \lambda \frac{\partial g}{\partial x_n} = 0. \tag{6}$$

This method of elimination can be used for any of the terms. We therefore conclude that

$$\frac{\partial f}{\partial x_i} + \lambda \frac{\partial g}{\partial x_i} = 0, \quad \text{for all } \; i = 1, \ldots, n. \tag{7}$$

In other words, after $\lambda$ has been introduced, (5) can be regarded as if all the $\delta x_i$ are mutually independent.

## IV. MATHEMATICAL RESULTS

(A) A frequently occurring sum is that of a geometric series

$$S = c + cf + cf^2 + \dots + cf^n. \tag{1}$$

Multiply it by $f$,

$$fS = cf + cf^2 + \dots + cf^n + cf^{n+1}, \tag{2}$$

so that

$$S = c\,\frac{1-f^{n+1}}{1-f}. \tag{3}$$

If $|f| < 1$ and $n \to \infty$, the series converges, because $f^{n+1} \to 0$. Thus

$$S = \frac{c}{1-f}. \tag{4}$$

(B) We wish to evaluate $I_n(a) = \int_0^{+\infty} x^n \exp(-ax^2)\,dx,\ a > 0.$

$$I_0(a) = \int_0^{\infty} \exp(-ax^2)\,dx = \frac{1}{a^{1/2}}\int_0^{\infty} \exp(-t^2)\,dt = \frac{1}{2a^{1/2}}\int_{-\infty}^{\infty} \exp(-t^2)\,dt. \tag{5}$$

We can write

$$\left(\int_{-\infty}^{+\infty} \exp(-t^2)\,dt\right)^2 = \int_{-\infty}^{\infty} \exp(-x^2)\,dx \int_{-\infty}^{\infty} \exp(-y^2)\,dy$$

$$= \int_{-\infty}^{\infty}\int_{-\infty}^{\infty} \exp[-(x^2 + y^2)]\,dxdy. \tag{6}$$

Changing to polar coordinates $(r, \theta)$ over the $(x, y)$ plane, $dxdy = rdrd\theta$, the double integral (6) becomes

$$2\pi\int_0^{\infty} \exp(-r^2)\,rdr = 2\pi\int_0^{\infty} (-\tfrac{1}{2})\,d\,[\exp(-r^2)]$$

$$= -\pi\Big[\exp(-r^2)\Big]_0^{\infty} = -\pi\,(0-1) = \pi. \tag{7}$$

Thus

$$I_0(a) = 2^{-1}(\pi/a)^{1/2}. \tag{8}$$

$$I_1(a) = \int_0^{\infty} x\exp(-ax^2)\,dx = \frac{1}{2a}\int_0^{\infty} e^{-t}\,dt = \frac{1}{2a}. \tag{9}$$

Differentiation of $I_n(a)$ with respect to $a$ yields

$$\frac{d\,I_n(a)}{da} = \int_0^\infty x^n \exp(-ax^2)(-x^2)\,dx = -I_{n+2}(a). \tag{10}$$

Application of this recurrence relation repeatedly to $I_0(a)$ and $I_1(a)$ gives

$$I_n(a) = \begin{cases} \dfrac{1.3\ldots(n-1)}{(2a)^{n/2}}\,\dfrac{1}{2}\left(\dfrac{\pi}{a}\right)^{1/2}, & n = 2, 4, 6, \ldots, \quad (11) \\[2ex] \dfrac{2.4\ldots(n-1)}{(2a)^{(n+1)/2}}, & n = 3, 5, 7, \ldots. \quad (12) \end{cases}$$

Thus

$$\begin{aligned} I_0(a) &= 2^{-1}(\pi/a)^{1/2}, \; I_1(a) = (2a)^{-1}, \\ I_2(a) &= (4a)^{-1}(\pi/a)^{1/2}, \; I_3(a) = (2a^2)^{-1}, \end{aligned} \tag{13}$$

and so on.

Note that

$$\int_0^\infty x^{1/2}\exp(-ax)\,dx = 2\int_0^\infty t^2\exp(-at^2)\,dt = (2a)^{-1}(\pi/a)^{1/2}. \tag{14}$$

(C) We can evaluate

$$I = \int_0^\infty \frac{x^3\,dx}{(e^x-1)} \tag{15}$$

in an elementary way as follows. Since $e^{-x} \leqslant 1$ in the range of integration,

$$\frac{1}{e^x-1} = \frac{e^{-x}}{1-e^{-x}} = \sum_{n=1}^\infty e^{-nx}, \tag{16}$$

$$\begin{aligned} I &= \sum_{n=1}^\infty \int_0^\infty x^3 e^{-nx}\,dx = \sum_{n=1}^\infty \frac{1}{n^4}\int_0^\infty e^{-y}y^3\,dy \\ &= \sum_{n=1}^\infty \frac{3!}{n^4} = 6\sum_{n=1}^\infty \frac{1}{n^4} = 6\times 1.082, \end{aligned} \tag{17}$$

where we have used (II.4), and numerically evaluated the rapidly convergent series.

To sum the series (17) in closed form, consider the Fourier series

$$f(x) = \sum_{n=1}^\infty a_n \sin nx \tag{18}$$

for the function

$$f(x) = \begin{cases} \pi/4, & 0 < x < \pi \\ -\pi/4, & -\pi < x < 0. \end{cases} \tag{19}$$

Using

$$\int_{-\pi}^{\pi} \sin mx \sin nx \, dx = \begin{cases} \pi, & m = n, \\ 0, & m \neq n, \end{cases} \tag{20}$$

we get

$$a_n = \begin{cases} 0\ , & n = 2, 4, 6, \ldots \\ 1/n, & n = 1, 3, 5, \ldots, \end{cases} \tag{21}$$

$$\frac{\pi}{4} = \sum_{n=1, 3, 5, \ldots} \frac{\sin nx}{n}, \ 0 < x < \pi. \tag{22}$$

Integrating (22) from 0 to $x$, in succession, we find

$$\frac{\pi}{4} x = \sum_{n=1, 3, 5, \ldots} \frac{1}{n^2} (1-\cos nx), \tag{23}$$

$$\frac{\pi}{8} x^2 = \sum_{n=1, 3, 5, \ldots} \left(\frac{x}{n^2} - \frac{\sin nx}{n^3}\right), \tag{24}$$

$$\frac{\pi}{24} x^3 = \sum_{n=1, 3, 5, \ldots} \left[\frac{x^2}{2n^2} - \frac{1}{n^4} (1-\cos nx)\right]. \tag{25}$$

For $x = \pi/2$, (23) and (25) give

$$\frac{\pi^2}{8} = \sum_{n=1, 3, 5, \ldots} \frac{1}{n^2}, \tag{26}$$

$$\frac{\pi^4}{192} = \frac{\pi^2}{8} \sum_{n=1, 3, 5, \ldots} \frac{1}{n^2} - \sum_{n=1, 3, 5, \ldots} \frac{1}{n^4}$$

$$= \frac{\pi^4}{64} - \sum_{n=1, 3, 5, \ldots} \frac{1}{n^4},$$

or

$$\frac{\pi^4}{96} = \sum_{n=1, 3, 5, \ldots} \frac{1}{n^4}. \tag{27}$$

Using

$$\sum_{n=2, 4, 6, \ldots} \frac{1}{n^4} = \frac{1}{16} \sum_{n=1}^{\infty} \frac{1}{n^4}, \tag{28}$$

we finally get

$$\sum_{n=1}^{\infty} \frac{1}{n^4} = \left(\sum_{n=1, 3, 5, \ldots} + \sum_{n=2, 4, 6, \ldots}\right)\frac{1}{n^4} = \frac{\pi^4}{96} + \frac{1}{16} \sum_{n=1}^{\infty} \frac{1}{n^4}$$

or

$$\sum_{n=1}^{\infty} \frac{1}{n^4} = \frac{\pi^4}{90}. \tag{29}$$

Thus

$$I = \int_0^{\infty} \frac{x^3 dx}{e^x - 1} = \frac{\pi^4}{15}, \tag{30}$$

$$I' = \int_0^{\infty} \frac{x^4 dx}{(e^x-1)^2} = -\int_0^{\infty} x^4 d\left(\frac{1}{e^x-1}\right)$$

$$= -\left[\frac{x^4}{e^x-1}\right]_0^\infty + 4\int_0^\infty \frac{x^3dx}{e^x-1} = 4I = \frac{4}{15}\pi^4. \tag{31}$$

## V. VOLUME OF A HYPERSPHERE

Consider the integral

$$I = \int_{-\infty}^{\infty} \exp\left[-(x_1^2 + x_2^2 + \ldots x_n^2)\right] dx_1 dx_2 \ldots dx_n$$

$$= \left[\int_{-\infty}^{\infty} \exp(-x^2)\, dx\right]^n = \pi^{n/2}. \tag{1}$$

For the $n$-dimensional sphere of volume $V_n$, the surface area can be written as $r^{n-1}S_n$. Therefore, we also have

$$I = \int_0^\infty \exp(-r^2)\, r^{n-1} S_n\, dr$$

$$= \frac{1}{2} S_n \int_0^\infty \exp(-t)\, t^{(n-2)/2}\, dt = \frac{1}{2} S_n \Gamma\left(\frac{n}{2}\right), \tag{2}$$

where we have used (II.4).

From (1) and (2)

$$S_n = \frac{2\pi^{n/2}}{\Gamma\left(\frac{n}{2}\right)}, \tag{3}$$

$$V_n = \int_0^R S_n R^{n-1} dR = \frac{\pi^{n/2}}{\Gamma\left(\frac{n}{2}+1\right)} R^n \tag{4}$$

For a positive integer $n$, $(n-1)! = \Gamma(n)$.

## VI. USEFUL FUNCTIONS

Consider

$$F_s(\alpha) = \frac{1}{\Gamma(s)}\int_0^\infty \frac{x^{s-1}}{e^{x+\alpha}-1} = \sum_{n=1}^{\infty} \frac{1}{n^s}\, e^{-n\alpha}, \tag{1}$$

For $\alpha = 0$,

$$F_s(0) = \sum_{n=1}^{\infty} \frac{1}{n^s} \equiv \zeta(s), \tag{2}$$

where $\zeta(s)$ is called the *Riemann-zeta function.*

If $s$ is positive and not an integer,

$$F_s(\alpha) = \alpha^{s-1}\,\Gamma(1-s) + \sum_{n=0}^{\infty} (-\alpha)^n \frac{\zeta(s-n)}{n!}. \tag{3}$$

Useful functions are

$$F_{1/2}(\alpha) = 1.773\ \alpha^{-1/2} - 1.460 + 0.208\alpha - 0.0128\ \alpha^2 - \ldots, \tag{4}$$

$$F_{3/2}(\alpha) = -3.545\ \alpha^{1/2} + 2.612 + 1.460\ \alpha - 0.104\ \alpha^2 + \ldots, \tag{5}$$

$$F_{5/2}(\alpha) = 2.363\ \alpha^{3/2} + 1.342 - 2.612\ \alpha - 0.730\ \alpha^2 + 0.0347\ \alpha^3 - \ldots \tag{6}$$

We can invert* the series

$$x \equiv F_{3/2}(\alpha) = e^{-\alpha} + 2^{-3/2}\,e^{-2\alpha} + 3^{-3/2}\,e^{-3\alpha} + \ldots \tag{7}$$

to get

$$e^{-\alpha} = x - 2^{-3/2}\,x^2 + (2^{-2} - 3^{-3/2})\,x^3 - \ldots \tag{8}$$

$$\alpha = -\ln x + 2^{-3/2}\,x - 3\,(2^{-4} - 3^{-5/2})\,x^2 + \ldots \tag{9}$$

We can also calculate $y/x \equiv F_{5/2}(\alpha)/F_{3/2}(\alpha)$ by finding $y = y(x)$. If we try the power series

$$y = \sum_k a_k x^k,$$

the coefficients can be obtained by Taylor's theorem

$$a_k = \frac{1}{k!}\left(\frac{d^k y}{dx^k}\right)_{k=0},$$

where $d^k y/dx^k$ can be derived from $x(\alpha)$ and $y(\alpha)$ by using

$$\frac{d^k y}{dx^k} = \frac{d}{d\alpha}\left(\frac{d^{k-1}y}{dx^{k-1}}\right)\Big/\left(\frac{dx}{d\alpha}\right).$$

This yields an expansion of $y = F_{5/2}(\alpha)$ in powers of $x = F_{3/2}(\alpha)$,

$$\begin{aligned} y &= x - 2^{-5/2}\,x^2 - 2(3^{-5/2} - 2^{-4})\,x^3 - \ldots \\ &= x - 0.17678\,x^2 - 0.00330\,x^3 - \ldots \end{aligned} \tag{10}$$

A similar calculation gives

$$\begin{aligned} x &= y + 2^{-5/2}\,y^2 + 2(3^{-5/2} - 2^{-5})\,y^3 + \ldots \\ &= y + 0.17678\,y^2 + 0.06580\,y^3 + \ldots \end{aligned} \tag{11}$$

and putting this in (10)

---

*If $y = x\,(1 + a_1 x + a_2 x^2 + a_3 x^3 + \ldots)$, then inverting it

$$\begin{aligned} x &= y\,(1 + b_1 y + b_2 y^2 + b_3 y^3 + \ldots) \\ &= (x + a_1 x^2 + a_2 x^3 + \ldots) + b_1 (x + a_1 x^2 + b_2 x^3 + \ldots)^2 + \ldots \\ &= x + (a_1 + b_1) x^2 + (a_2 + 2a_1 b_1 + b_2) x^3 + \ldots \end{aligned}$$

Therefore

$$b_1 = -a_1,\ b_2 = -a_2 - 2a_1 b_1,\ b_3 = -a_3 - 2b_1 a_2 - b_1 a_1^2 - 3a_1 b_2, \text{ etc.}$$

$$\frac{y}{x} = 1 - 0.17678\,y - 0.03455\,y^2 - 0.01291\,y^3 - \ldots \tag{12}$$

From (2)

$$\zeta(3/2) = 1 + \frac{1}{2^{3/2}} + \frac{1}{3^{3/2}} + \ldots = 2.612, \tag{13}$$

$$\zeta(5/2) = 1 + \frac{1}{2^{5/2}} + \frac{1}{3^{5/2}} + \ldots = 1.342. \tag{14}$$

**SI (Systeme International) Base Units**

Length: meter (m); Time: second (s); Mass: kilogram (kg)

Temperature: Kelvin (K); Amount: mole (mol); Current: ampere (A)

Luminous intensity: candela (cd).

**Useful SI Derived Units**

| | | | |
|---|---|---|---|
| Force | newton | (N) | 1 kg m/s$^2$ |
| Energy | joule | (J) | 1 kg m$^2$/s$^2$ = 1N m |
| Pressure | pascal | (Pa) | 1 kg/m s$^2$ = 1N/m$^2$ |
| Power | watt | (W) | 1 kg m$^2$/s$^3$ = 1 J/s |
| Charge | coulomb | (C) | 1 A s |
| Potential difference | volt | (V) | 1 kg m$^2$/A s$^3$ = 1 AΩ |
| Resistance | ohm | (Ω) | 1 kg m$^2$/A$^2$ s$^3$ = 1 V/A |
| Capacitance | farad | (F) | 1 A$^2$ s$^4$/kg m$^2$ = 1 C/V |
| Magnetic flux | weber | (Wb) | 1 kg m$^2$/A s$^2$ = 1 V s |
| Inductance | henry | (H) | 1 kg m$^2$/s$^2$ A$^2$ = Wb/A |

**Some Prefixes for SI Units**

$10^{-9}$ nano (n); $10^{-6}$ micro (μ); $10^{-3}$ milli (m)

$10^{-2}$ centi (c); $10^{3}$ kilo (k); $10^{6}$ mega (M)

$10^{9}$ giga (G)

**Values in CGS and SI Units**

| | | *Value* | *CGS* | *SI* |
|---|---|---|---|---|
| Velocity of light | $c$ | 2.9979 | $10^{10}$ cm s$^{-1}$ | $10^{8}$ m s$^{-1}$ |
| Charge | $e$ | 1.6022 | — | $10^{-19}$ C |
| | | 4.8033 | $10^{-10}$ esu | — |
| Planck constant | $h$ | 6.6262 | $10^{-27}$ erg s | $10^{-34}$ J s |
| Atomic mass unit | amu | 1.6606 | $10^{-24}$ g | $10^{-27}$ kg |
| Electron mass | $m$ | 9.1095 | $10^{-28}$ g | $10^{-31}$ kg |
| Bohr magneton | $\mu_B$ | 9.2741 | $10^{-21}$ erg G$^{-1}$ | $10^{-24}$ J T$^{-1}$ |

| | | | | |
|---|---|---|---|---|
| 1 electron volt | eV | 1.6022 | $10^{-12}$ erg | $10^{-19}$ J |
| | eV/$hc$ | 8.0655 | $10^{3}$ cm$^{-1}$ | $10^{5}$ m$^{-1}$ |
| | eV/$k$ | $1.1605\times10^{4}$ K | — | — |
| Boltzmann constant | $k$ | 1.3807 | $10^{-16}$ erg K$^{-1}$ | $10^{-23}$ JK$^{-1}$ |
| Permittivity (free space) | $\epsilon_0$ | — | 1 | $10^{7}/4\pi c^{2}$ |
| Permeability (free space) | $\mu_0$ | — | 1 | $4\pi\times10^{-7}$ |
| Molar gas constant | $R$ | 8.3144 | $10^{7}$ erg mol$^{-1}$ K$^{-1}$ | J mol$^{-1}$ K$^{-1}$ |
| Molar volume ideal gas at 273.15 K, 1 atm | | 22.4138 | $10^{3}$ cm$^{3}$ mol$^{-1}$ | $10^{-3}$ m$^{3}$ mol$^{-1}$ |

**Useful Conversion Factors**

| | |
|---|---|
| 1 A | $10^{-8}$ cm |
| 1 N | $10^{5}$ dyn |
| 1 J | $10^{7}$ erg = $10^{7}$ dyn-cm = 0.2390 cal |
| 1 cal | 4.184 J |
| 1 eV | $1.602\times10^{-12}$ erg = $1.602\times10^{-19}$ J=$1.1605\times10^{4}$ K = 23.06 kcal mol$^{-1}$ |
| 1 kWh | 3600 J |
| 1 BTU | 1055 J |
| 1 atm | 760 mm Hg = $1.013\times10^{5}$ N m$^{-2}$ = $1.103\times10^{6}$ dyn cm$^{-2}$ = 1.013 bar |

$\nu = 10^{10}$ cycles s$^{-1}$

associated energy $h\nu = 4.14\times10^{-5}$ eV

associated temperature $h\nu/k = 0.48$ K.

**Physical Constants in Useful Forms**

| | |
|---|---|
| Loschmidt number | $2.68719\times10^{19}$ molecules cm$^{-3}$ |
| Avogadro number ($N_a$) | $6.022\times10^{23}$ mol$^{-1}$ |
| Boltzmann constant ($k$) | $1.381\times10^{-23}$ J K$^{-1}$= $8.62\times10^{-5}$ eV K$^{-1}$ = 1 eV/11604.5 K |
| Gas constant ($R$) | 8.314 J mol$^{-1}$ K$^{-1}$ = 1.99 cal mol$^{-1}$ K$^{-1}$ |
| Molar volume at NTP (or STP) | $22.4\times10^{3}$ cm$^{3}$ mol$^{-1}$ |
| Planck constant ($h$) | $6.626\times10^{-34}$ J s |
| ($\hbar = h/2\pi$) | $6.582\times10^{-22}$ MeV s = $1.055\times10^{-27}$ erg s |
| Electron mass | $9.109534\times10^{-28}$ g = 0.511 MeV |
| Proton mass | 938.28 MeV |

**Numerical Constants**

| | |
|---|---|
| $\pi = 3.14159$ | $\pi^{1/2} = 1.772$ |
| $e = 2.71828$ | $2^{1/2} = 1.414$ |
| ln 2 = 0.6931472 | $3^{1/2} = 1.732$ |
| ln 10 = 2.3025851 | $10^{1/2} = 3.162$ |

**Bibliography**


Allis W. P. and Herlin M.A., *Thermodynamics and Statistical Mechanics*, McGraw-Hill, New York, 1952.

Andrews F.C., *Equilibrium Statistical Mechanics*, John Wiley, New York, 1963.

Atkins K.R., *Liquid Helium*, Cambridge U.P., Cambridge, 1959.

Becker R., *Theory of Heat*, 2nd end., Springer, Berlin, 1967.

Chisholm J.S.R. and de Borde A.H., *An Introduction to Statistical Mechanics*, Pergamon, New York, 1958.

Desloge E.A., *Statistical Physics*, Holt, Rinehart and Winston, New York, 1966.

Donnan F.G. and Arthur H., editors, *Commentry on the Scientific Writings of J.W. Gibbs*, Yale U.P. London, 1936.

Eyring H., Henderson D. and Eyring E.M., *Statistical Mechanics and Dynamics*, John Wiley, New York, 1963.

Feynman R.P., *Statistical Mechanics*, Benjamin, Reading, Massachusetts, 1972.

Hill T.L., *Statistical Mechanics*, McGraw-Hill, New York, 1956.

Huang K., *Statistical Mechanics*, John Wiley, New York, 1963.

Kittel C., *Elementary Statistical Physics*, John Wiley, New York, 1968.

Kittel C. and Kroemer H., *Thermal Physics*, Freeman, San Francisco. 1980.

Kubo R., *Statistical Mechanics*, North-Holland, Amsterdam, 1965.

Landau L.D. and Lifshitz E.M., *Statistical Physics*, Pergamon, London, 1959.

London, F., *Superfluids*, Vol. II, John Wiley, 1954.

Mandl F., *Satistical Physics*, John Wiley, London, 1971.

Morse P.M., *Thermal Physics*, Benjamin, New York, 1964.

Reif F. *Fundamentals of Statistical and Thermal Physics*, McGraw-Hill, New York, 1965.

Schrödinger E., *Statistical Thermodynamics*, Cambridge UP, Cambridge, 1946.

Sommerfeld A., *Thermodynamics and Statistical Mechanics*, Academic Press, New York, 1956.

Ter Haar D., *Elements of Statistical Mechanics*, Rinehart, New York, 1954.

Tolman R.C., *The Principles of Statistical Mechanics*, Oxford UP, Oxford, 1938.

# INDEX